Mechatronics

Mechatronics

and the design of intelligent machines and systems

David Bradley

Professor of Mechatronic Systems, School of Science and Engineering
University of Abertay Dundee

Derek Seward

Senior Lecturer, Engineering Department
Lancaster University

David Dawson

Senior Lecturer, Engineering Department
Lancaster University

Stuart Burge

BHW Partnership

Published in 2000 by
CRC Press
Taylor & Francis Group
6000 Broken Sound Parkway NW, Suite 300
Boca Raton, FL 33487-2742

International Standard Book Number-10: 0-7487-5443-1 (Softcover)
International Standard Book Number-13: 978-0-7487-5443-4 (Softcover)

Library of Congress Cataloging-in-Publication Data

Catalog record is available from the Library of Congress

Taylor & Francis Group
is the Academic Division of Informa plc.

Visit the Taylor & Francis Web site at
http://www.taylorandfrancis.com

and the CRC Press Web site at
http://www.crcpress.com

Contents

Preface

The past few years have seen mechatronics have an increasing impact on engineering and engineering education as a defining approach to the design, development and operation of an increasingly wide range of engineering systems. In addition, mechatronics is now recognised as involving not only the technical aspects of its core disciplines – mechanical engineering, electrical and electronic engineering and software – but also aspects of organisation, training and management. Indeed, in its approach to engineering product development, mechatronics has much in common with concurrent engineering strategies.

The WHAT of mechatronics
It can be argued that it is the development of the microprocessor and the subsequent growth in microelectronics technologies and the support that it provides for the 'transfer of complexity' from the mechanical to the electronic and software domains that is at the heart of mechatronics. The resulting systems are more complex with higher levels of performance and greater reliability than their predecessors, all achieved at significantly reduced real costs.

The WHY of mechatronics
In an increasingly competitive and global market, companies need to have the ability to increase the competitiveness of their products through the use of technology and must be able to respond rapidly and effectively to changes in the market place. Mechatronic strategies have been shown to support and enable the development of new products and markets such as the compact disc player as well as through enhancing existing products while responding to the introduction of new product lines by a competitor.

The influence of mechatronics as a driver of the product development process is perhaps most strongly witnessed in the automotive industry where vehicle systems have become increasingly more mechatronic in nature with features such as engine management systems, traction control and ABS now commonplace. Future developments include drive-by-wire, collision avoidance systems, lane tracking and navigational control.

However, and whatever the level of technology, the motivation for the adoption by a company of a mechatronic approach to product development and manufacturing must be one of providing the company with a strategic and commercial advantage either through the development of new and novel products, through the enhancement of existing products, by gaining access to new markets or some combination of these factors.

The HOW of mechatronics

The achievement of a successful mechatronics design environment essentially depends on the ability of the design team to communicate, collaborate and integrate. Indeed, a major role of the mechatronics engineer is often that of acting to bridge the communications gaps that can exist between more specialist colleagues in order to ensure that the objectives of collaboration and integration are achieved. This is important during the design phases of product development and particularly so in relation to requirements definition where errors in interpretation of customer requirements can result in significant cost penalties.

In writing this book, we have drawn upon some 16 years of experience in teaching mechatronics on the Masters courses at Lancaster University and at the University of Abertay Dundee as well as on research programmes covering areas as diverse as construction robotics, systems for the disabled, manufacturing technologies, modular robotics and especially in engineering design. This has also involved extensive involvement with a wide variety of industries from the small to multi-nationals with projects ranging from the design of advanced instrumentation and systems for the disabled to domestic appliances, smart homes, manufacturing automation, safety systems and construction robotics.

A single book covering all aspects of mechatronics would not fit on any conventional bookshelf! Instead, we have structured the text to reflect what we believe are the current significant areas of development in mechatronics and to provide the reader with guidance as to problems and developing techniques in these areas. Thus the book begins in Chapter 1 with a review of the evolution of the mechatronics technologies, of design tools and methodologies and the link to concurrent engineering, Chapter 2 introduces the concepts of machine intelligence and its role in modern mechatronic systems. Chapters 3 and 4 then concentrate on the mechatronic design process and particularly the procedures for requirements interpretation. Chapters 5 and 6 further explore the concepts of artificial intelligence and the applications of neural networks and fuzzy logic in the control and operation of mechatronic systems while Chapters 7 and 8 introduce the reader to software concepts and the human-machine interface. Chapter 9 deals with a particularly important, but often relatively neglected, area of system design and development, that of safety, while Chapter 10 looks at manufacturing technology. Chapter 11 takes a look at the future of mechatronics in a number of areas of technology while Chapter 12 provides a series of short case studies covering various aspects of current mechatronic systems.

A particular theme running throughout the book is the use of the autonomous and robotic excavator, LUCIE, as a running case study to illustrate points made in the various chapters. The LUCIE project began in 1987 at Lancaster and has at some point or other in its development involved all of the authors. It has evolved over the years to a point where it is capable of locating itself on site and of digging a trench, including the removal of certain classes of obstacles from the trench line, without operator intervention. Thus, Chapter 2 contains an introduction to the problems of autonomous trenching while Chapter 4 shows how some of the techniques discussed therein could be applied. In Chapters 5 and 7 the artificial intelligence and software aspects are presented while Chapter 8 considers the operator interface design, with safety aspects being discussed in Chapter 9.

Much of the material in the book is based on lecture notes and other material prepared for the undergraduate, masters and industrial courses with which we have been associated as well as on research programmes extending over many years. Our thanks therefore go to all those students; undergraduate, masters and doctoral, whom we have taught and worked with for their input in support of the development of much of the material presented. In terms of research, particular mention must be given, but in no particular order, to Rob Bracewell, Patrick Langdon, Les Dungworth, Peter Green, Stephen Quayle, Ajaz Ahmed, Simon Butterworth, Costas Giannopoulos, Jon Goda, Peter Griffith, Richard Walters, Johnny da Silva, Thomas Olbrich, Allan Gardam, Jim Mann, Mark Goodwin, Simon Brownsell, Gareth Williams, Sa'ad Mansoor, Jason Scott, Frank Margrave and Linda Chua.

Thanks must also be given to those companies who have provided material and information for the book and especially to the various people who provided help and advice on its content and structure. Particular mention must be given to Bill Scarfe and Gordon Humphries for their direct and indirect input over many years which helped to ensure that we retained an industrial perspective in our approach to mechatronics.

Finally, but perhaps most importantly, thanks go to Professor Michael French for his help and advice over a period of over 25 years. Michael French founded the Engineering Department at Lancaster University in the late 1960s and was responsible for establishing a design-based, and ultimately mechatronic, culture among both staff and students which culminated in the establishment in 1990 of the Lancaster Engineering Design Centre investigating the development and provision of computer based tools to support the design of mechatronic products. Without his work in creating the environment within which we, the authors, have all at some time worked this book would never have been written.

David Bradley
Derek Seward
David Dawson
Stuart Burge

Of machines and mechatronics 1

1.1 INTRODUCTION

In 215 BC the Roman general Marcellus led an army to Sicily for the purpose of capturing the city of Syracuse taking with him a range of military technology that included siege towers and a catapult so big that it required eight ships lashed together to carry it. On arriving outside Syracuse, Marcellus discovered that its defences had been augmented by a range of machines devised by the Greek scientist and engineer Archimedes and that assaults were met by a barrage of missiles from giant catapults and by boulders dropped from cranes that swung out over the city walls. Most terrifying were the giant 'claws' that grasped and shook the Roman ships as they tried to enter the harbour. In the words of Pliny:

> 'The ships, drawn by engines within and whirled about, were dashed against steep rocks that stood jutting out under the walls, with great destruction of the soldiers that were aboard them. A ship was frequently lifted up to a great height in the air – a dreadful thing to behold – and was rolled to and fro, and kept swinging, until the mariners were all thrown out, when at length it was dashed against the rocks, or was dropped.'

As to the role of Archimedes, Livy commented:

> 'An operation launched with such strength might well have proved successful, had it not been for the presence in Syracuse at that time of one individual – Archimedes, unrivalled in his knowledge of astronomy, was even more remarkable as the inventor and constructor of types of artillery and military devices of various kinds, by the aid of which he was able with one finger, as it were, to frustrate the most laborious operations of the enemy'.

Archimedes was born in Syracuse around 287 BC. The son of the astronomer Phidias, he studied at the 'Museum' or university in Alexandria that had been founded by King Ptolemy II based on the teachings of Strato the philosopher, the head of the Lyceum in Athens from 287 BC to 269 BC. Following his studies, Archimedes later returned to Syracuse where he acted as adviser to its ruler, Hieron II.

In addition to the war engines referred to above and the discovery of the principle of floatation that carries his name, Archimedes was responsible for a series of developments in mechanics and was reputed to have single handedly launched the Royal ship *Syracusa* when fully loaded using a system of pulleys and levers of his own devising. When Syracuse finally fell, Archimedes was killed by Roman legionaries and was buried with full honours by Marcellus who had inscribed on his tomb a design depicting the ratio between the volumes of a sphere and a cylinder as Archimedes himself had requested be done.

Though the lever, pulley, wedge and windlass were known and in use prior to the end of the 4th century BC and the screw was an invention of the 3rd century BC, traditionally by Archimedes himself, Archimedes was the first of the great scientist/engineers whose works have passed down through history and was one of the first to appreciate and understand the working relationships between men and machines and the ways in which machines could be used to support all areas of human function and activity.

Of others of this early period in the development of the relationship between machines and men, four in particular are worthy of note. The first of these is Ctesibius of Alexandria. A contemporary of Archimedes, Ctesibius lived and worked in Alexandria in the middle of the 3rd century BC and though his writings have been lost he is credited with the invention of a range of mechanisms including a fire engine (Fig. 1.1), a pneumatic 'gun', a water organ and a clock. Slightly later than Ctesibius comes Philo of Byzantium whose work 'On Artillery Construction' dates from around 200 BC and details features such as the use of twisted skeins of material as an energy store as well as experimentation that was undertaken to determine the most effective combination of parameters for artillery. Ctesibius and Philo were followed in the later part of the 1st century BC by Marcus Vitruvius Pollio who in his writings details a range of mechanical systems including developments in artillery, water and wind power and a description of an odometer for measuring distance.

After Archimedes, perhaps the greatest of these early engineers of whom a record has survived is Heron (or Hero) of Alexandria who in the 1st century AD produced descriptions of a series of inventions and mechanisms such as a constant head water clock (Fig. 1.2), a system for automatically opening temple doors (Fig. 1.3), a coin operated dispenser for holy water, a miniature theatre complete with simple automata as the players and many others. Perhaps Heron's best known invention is the aeolopile, a simple steam turbine. Shown in Fig. 1.4 this consists of a large, sealed cauldron of water connected by pipes to a sphere with two opposed nozzles. As the water in the cauldron was heated, steam was transferred to the sphere where it exited via the nozzles, causing the sphere to rotate.

By the end of the 1st century AD in Europe the study of mechanics was well developed and applied, particularly by the Romans, to systems including treadmill operated cranes such as that shown in Fig. 1.5 and the large scale use of water power for milling. By the 5th century, the overshot wheel had appeared while tidal mills were in operation along the Atlantic coast of France and floating mills were reported in operation on the Tiber in the 6th century. Similar developments with water power were also taking place in China though, unlike the Romans, the Chinese used their watermills to drive bellows and trip hammers as well as to mill corn.

Over the next 1500 years however technological progress was to to be relatively slow despite works such as the 'Book of Ingenious Devices' written by the three Ban Musa brothers of Baghdad in the 9th century. This work describes a range of devices including an oil lamp with a self adjusting wick and automata such as the pond surrounded by moving statues of warriors reportedly built for the Caliph at the start of the 10th century. This interest in automata in the Arab world was continued by others including al-Jazari who

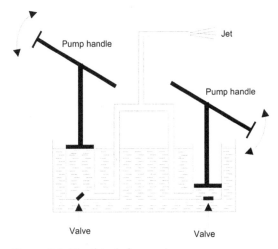

Figure 1.1 Ctesibius's fire engine.

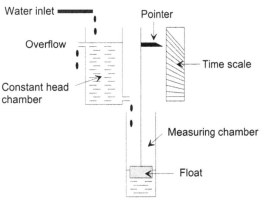

Figure 1.2 Heron's constant head water clock with variable scale for changes in the length of day (from sunrise to sunset).

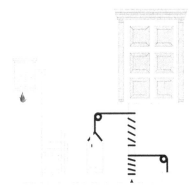

Figure 1.3 Heron's mechanism for opening temple doors.

Figure 1.4 Heron's aeolipile.

in 1206 described clocks that marked the passage of time by using figures to strike drums, blow trumpets or drop stones onto cymbals. Developments in China also included automata and Su Sung's 30 foot high 'Cosmic Engine' built in Hunan in 1090 was powered by a water-wheel and marked the time by figures which moved in and out of doors.

The next great development of scientific thought in Europe took place with the advent of the Renaissance and the work of individuals such as Copernicus, Galileo, Brahe and Leonardo da Vinci. da Vinci in particular turned his attention to many aspects of engineering and technology and has left a legacy of drawings and sketches encompassing subjects as diverse as catapults and crossbows, bridges, the manufacture of cannon, ballistics, lock systems, excavating machines, a water powered saw, gears and gearing, a screw cutting machine, pulleys, ball and roller bearings, power transmission systems, screw jacks, drills, file cutting machines and many others. These individuals were followed in their turn by others such as Pascal, Leibnitz and Newton, each expanding and defining the scientific and mathematical framework on which the industrial revolution of the 18th century was to be based. This

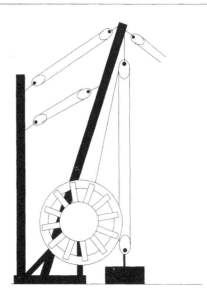

Figure 1.5 Treadmill powered crane.

development and the subsequent growth of technology to the present day is illustrated by Table 1.1 which sets out some of the major technological developments over this period.

Unlike most revolutions, the onset of the industrial revolution was not marked by a single explosive event, or even a series of such events. In Britain it gathered momentum throughout the first half of the 18th century before its main effects began to be felt from around 1760, after which change occurred at an increasing rate. That this was the case may be illustrated by the fact that before this date it was normal to take work to people for it to be done by them in their own homes and yet by 1820 it was usual to bring workers into factories to carry out work under supervision.

The driving force behind much of this change was the availability of a new source of power in the form of the steam engine which, when combined with developments in the technology of weaving, made the concentration of labour into factories the most economic means of production. The demand for greater precision in the manufacture of components for steam engines in turn brought about developments in the manufacturing processes themselves leading to the development and introduction of precision lathes and boring machines. The concentration of manufacturing into factories was also a mechanism for the standardisation of parts and components. Thus, in 1808, Marc Isambard Brunel and Samuel Bentham were able to open a factory in Portsmouth to manufacture pulley blocks for the Royal Navy which replaced the 110 skilled blockmakers previously required by 10 unskilled men to supervise the machines.

A further significant development towards the automation of production in the early part of the 19th century was the introduction by Jacquard of a system of punched cards to control the operation of an automated loom, enabling the pattern to be pre-programmed and changed simply by replacing one set of cards by another. The idea of using punched cards to control the operation of a machine was taken up by Charles Babbage who used the idea as the basis of his 'Analytical Engine'. Though this proposed steam powered computer was never built it nevertheless served to introduce many of the concepts that are

Table 1.1 Developments in technology

1452	Birth of Leonardo da Vinci
c1540	A mandolin playing female figure attributed to Giannello Torriano, watch-maker to Emporor Charles V
1543	Copernicus' work on planetary orbits, 'De Revolutionibus Orbium Celestium' is published
c1550	Tartaglia develops the science of ballistics
1610	Galileo uses a telescope to make astronomical observations of the moons of Jupiter
c1642	Blaise Pascale produces his automatic calculator 'La Pascaline'
1614	John Napier publishes his work 'A Description of the Marvellous Rule of Logarithms'
1687	Newton develops his theory of gravitation in 'Philosophiae Naturalis Principia Mathematica'
1694	Leibnitz completes his 'stepped reckoner'
1699	Thomas Savery demonstrates his steam engine to the Royal Society
1700	Development of precision lathes by clockmakers
1709	Abraham Darby establishes his iron foundry at Coalbrookdale
1712	Thomas Newcomen installs his first engine at Dudley, UK. This had a cylinder of 19 inches (48.26 cm) internal diameter and a stroke of around 6 feet (1.82 metres) and at each stroke raised 10 gallons (45.5 litres) of water 153 feet (46.6 metres)
1725	Basile Bouchon devised a silk loom which used holes punched onto a role of paper to control the production of a pattern
1733	The flying shuttle loom invented by John Kay
1738	Jacques de Vaucanson created a flute player capable of playing a dozen tunes
c1740	The introduction of the 'Leyden Jar' as a means of storing charge
c1750	Screw driven tool carriage for lathes introduced by Antoine Thiout
1759	John Harrison completes his marine chronometer No 4 which ultimately led to his being awarded the prize of £20 000 offered by the Board of Longitude, London
1764	James Hargreaves introduces the 'Spinning Jenny' to spin multiple threads
1765	An improved steam engine using a separate condenser is devised by James Watt
1769	Richard Arkwright introduces his water frame spinning machine
1770	Screw cutting lathe developed by Jesse Ramsden
1774	Pierre and Henri-Louis Jaquet-Droz create an automata capable of writing and drawing
1775	John Wilkinson establishes a boring mill in Denbighshire used to manufacture cylinders for Boulton and Watt engines are installed at Bloomfield Colliery in Staffordshire and New Willey in Shropshire, UK
1779	The first iron bridge constructed across the River Severn at Coalbrookdale
	Samuel Crompton introduces the 'mule' combining features of both the Spinning Jenny and Arkwright's water frame
1782	Watt introduces the double-acting steam engine
1784	Oliver Evans introduces a waterpowered automated grain mill outside Philadelphia, USA
1783	The cotton gin is invented by Eli Whitney to comb the seeds from cotton increasing production more than 10 fold
1794	Eli Witney receives the patent for the cotton gin
c1800	Maudslay produces a bench micrometer with an accuracy of 0.0001 inch (0.00254 mm)
19th Century	A variety of 'difference engines' are produced to calculate tables for astronomy, navigation and tides
1800	Maudslay introduces his large screw cutting lathe
	Alessandro Volta describes the 'Voltaic Pile' in a communication to the Royal Society, London
1801	Eli Witney demonstrates principles of mass production to President-elect Thomas Jefferson by assembling muskets from a random set of standard parts
c1804	Joseph Marie Jacquard devised the Jacquard loom which used punched cards to control up to 1200 needles
1808	Marc Isambard Brunel and Samuel Bentham open a plant in Portsmouth to manufacture pulley blocks for the Royal Navy. Driven by a 30 horse power (23 kW) steam engine the machines produced 130 000 blocks a year and replaced the 110 skilled blockmakers previously required by 10 unskilled men

1818	Thomas Blanchard introduces a copying lathe for the manufacture of gun stocks
1819	Hans Christian Oersted observes the deflection of a compass needle by wire carrying an electric current. The relationship is later quantified by Ampere
1821	Michael Faraday demonstrates that a current carrying conductor in a magnetic field experiences a force
1826	Thomas Telford completes his suspension bridge across the Menai Straights, Wales
1829	Robert Stephenson's Rocket wins the Rainhill trials for the Liverpool to Manchester railway
1831	Michael Faraday and Joseph Henry demonstrate electromagnetic induction
1831	Charles Babbage produces the plans for his 'Analytical Engine'
	Hippolyte Pixii produces a generator based on Faraday's discoveries
1836	The development of the 'Daniel Cell', the first reliable battery
1837	Samuel Morse obtains a patent for his telegraphy system
	Charles Wheatstone and William Cooke install an experimental electronic telegraph system between Euston and Camden Town on the London and North Western Railway
	The *Great Western* floated out at Bristol, UK
1838	Cooke and Wheatstone's electric telegraph installed between Paddington and West Drayton on the Great Western Railway
1839	James Nasmyth develops the steam hammer
1840	A remotely controlled dividing engine produced by Henry Gambey is used to graduate a 2 metre mural circle for the Paris Observatory, remains in use until 1920
1843	Morse telegraph is installed between Washington and Baltimore, USA
c1850	Introduction of the turret lathe in America
1853	Production of revolvers by Samuel Colt used 1400 machine tools
1856	Patent granted to Henry Bessemer for the 'Bessemer Converter' to produce steel from pig iron
	Friedrich von Siemens patents the regenerative furnace
	Ernst Werner von Siemens patents a practical electric generator
1857	The *Great Eastern* is launched on the Thames
1862	Brown and Sharpe introduce the first true universal milling machine
c1865	Introduction of automatic screw cutting lathes in America
1866	The first successful trans-Atlantic cable is laid by the *Great Eastern*
1876	Alexander Graham Bell demonstrates the telephone
1877	Thomas Alva Edison demonstrates the phonograph
1879	Practical electric lighting demonstrated by Edison and Swan
1880	Edison opens a power generating station at Holborn Viaduct in London
1881	Frederick W. Taylor begins studies on the organisation of manufacturing operations at Midvale Steel Company
	Frank B. Gilbreth and Lillian M. Gilbreth use 'motion picture' technology to analyse worker's movements and work areas
1882	The Edison Electric Illuminating Company power generating station at Pearl Street in New York begins operation. By mid-1883 it is supplying 431 customers
1885	The first petrol driven automobile built by Karl Benz
1887	First patent for a gear hobbing machine taken out in America
1889	Sebastian Ziani de Ferranti opens a power station at Deptford, London, operating at 10 volts
1890	Herman Hollerith wins the competition held by the US Census with a system which used punched cards to record and sort data
1895	Wireless telegraphy invented by Guglielmo Marconi
1897	Charles Parsons demonstrates the steam turbine ship *Turbinia* at Queen Victoria's Diamond Jubilee Review at Spithead
1901	Marconi transmits a radio message across the Atlantic
1903	Orville and Wilbur Wright make the first sustained powered flight
1904	The vacuum tube is developed by Fleming
1906	Lee De Forrest adds a control grid to Fleming's valve to produce the triode valve
1908	Ford Model T introduced. In 1909 it costs $950 falling to $360 by 1916 and $290 in 1920
	General Motors founded
1909	Bakelite announced by its inventor, Leo Hendrik Baekeland

	Louis Bleriot succeeds in flying across the English Channel
c1910	Mechanical analogue computers are developed for naval gunnery control
1913	Henry Ford introduces the production line for the manufacture of the Model T
1919	First trans-Atlantic flights
1920	Karel Capek writes the play 'Rossum's Universal Robots'. The Czech work 'robota' means servitude or forced labour
1920–1930	Widespread introduction of live radio in the United States and Europe
1930	Vannevar Bush develops a 'differential analyser', the first electrical analogue computer
	A patent for a jet engine is taken out by Frank Whittle
1935	Robert Watson-Watt begins the series of experiments which were to result in the development of radar
	The first jet propelled aircraft is flown in Germany
1936	The BBC launches the first public televison service
1939	The development of the cavity magnetron by Randall and Boot enabling centrimetric radar
1940	Isaac Asimov published the first of his robot stories in 'Super Science Stories'. Asimov went on to develop the 'Three Laws of Robotics'
	Nylon and Terylene, the first synthetic fibres produced commercially
1944	Colossus, the first electronic digital computer, enters service at Bletchley Park for cryptography
1945	The first atomic bombs are exploded
1946	The Electronic Numerical Integrator and Calculator (ENIAC) enters service. Designed by J. Presper Eckhert and John W. Mauchly ENIAC weighed around 30 tonnes, occupied a room 9 metres by 15 metres and used some 18 000 vacuum tubes, 70 000 resistors, 10 000 capacitors and 6000 switches
1947	The Bell X-1 flown by Charles Yeager breaks the sound barrier
1948	Baby, the first true stored programme computer, developed at Manchester University in the UK
	The transistor is developed at Bell Laboratories by John Bardeen, Walter H. Brattain and William B. Shockley
	George C. Devol develops a controller which used information recorded on magnetic tape to operate a machine
1951	Work on teleoperated manipulators for the nuclear industry for handling radioactive materials
	The Ferranti Mark 1 becomes the world's first commercially available computer
1952	A prototype numerically controlled machine is demonstrated at Massachusetts Institute of Technology
	Colour television broadcast using the NTSC system begins in the United States
1956	FORTRAN, the first high level computer language, appears
1957	Sputnik 1, the first artificial satellite, is launched
1958	Texas Instruments introduce the first commercial integrated circuit
1959	Planet Corporation introduces the first commercial robot based on limit switches and cams
1960	The first Unimate hydraulically driven robot is introduced. This is based on Devol's 'programmed article transfer' system and used numerical control
	T. H. Maiman produces the first laser
1961	A Unimate robot is installed by Ford to service a die casting machine
	The part programming language APT (Automatically Programmed Tooling) is released
	Yuri Gagarin becomes the first man in space
1962	Telstar I, the first commercial communication satellite, is launched
1963	The American Machine Company introduces the Versatran robot
1966	In Norway, Tralfa design and install a paint spraying robot
	Colour television broadcasts using the PAL system begin in Europe
1968	Burroughs produces the first computers to use integrated circuits
	Stanford Research Institute introduce a mobile robot. Named 'Shaky' this robot was equipped with a range of sensors including a camera and a rangefinder enabling it to find its way around a room. Communication with the main computer was by radio
1969	The US Department of Defense introduces Arpanet, the forerunner of Internet
	Apollo 11 lands on the moon

1971	A small electrically powered robot arm is developed at Stanford University
1972	The first colour VCR is introduced
	The first 8-bit microprocessors appear on the market
1973	Large Scale Integration with 10 000 components on a 1 sq-cm chip
	Stanford Research Institute demonstrates the WAVE programming language. This was followed in 1974 by AL. The two languages were merged into the VAL language by Unimation
1974	ASEA introduce the all electric drive IRb6 robot
	Kawasaki install arc-welding robots for the production of motorcycle frames
	'The Tomorrow Tool', better known as the T3 robot, is introduced by Cincinnati Milicron
	UNIX computer system described in the Communications of the Association for Computing Machinery
1975	A Sigma robot from Olivetti is used in an assembly operation, an early example of this type of operation
1976	The first Cray-1 super computer delivered to Los Alamos Laboratory
	Viking spacecraft land on Mars
	Whitfield Diffie, Martin Hellman and Ralph Merkle define the basis of public/private key encription
1977	Apple Computer introduces the Apple II, the first personal computer incorporating a keyboard and power supply to generate colour graphics
	CompuServe goes on line
	Ronald Rivest, Adi Shamir and Leonard Adelman define the RSA algorithm for public/private key encription
1978	The PUMA (Programmable Universal Machine for Assembly) is introduced by Unimation
	A T3 robot is used for drilling and routing operations as part of the US Air Force Integrated Computer Aided Manufacturing programme
	16-bit microprocessors introduced
1979	The SCARA (Selective Compliance Arm for Robotic Asembly) arm is developed at Yamanshi University in Japan
1980	Intel introduce the first 32-bit microprocessor
	Telenet introduces a commercial computer networking service
1981	IBM introduce a personal computer with an industrial standard disc operating system (DOS)
	Carnegie-Mellon University demonstrates a direct drive robot using electric motors located at the joints
	Sony introduce the Walkman personal stereo system
1982	IBM introduces the RS-1 assembly robot
	Hewlett-Packard introduce the first 32-bit personal computer
	Institute for New Generation Computer Techniques established in Japan to investigate 'fifth-generation' computer systems
1983	Motorola receive the licence for the first cellular telephone system in the USA
	Cray Research introduce the Cray X-MP supercomputer
1984	Philips and Sony develop the compact disc read-only memory (CD-ROM)
	Apple Computers introduce the Macintosh computer system
	Introduction of CMOS technology for integrated circuits (ICs)
	Sumitomo demonstrates an organ playing robot developed at Waseda University in Japan
1985	The equivalent of 300 000 telephone calls are transmitted using an optical fibre
1987	Integrated Services Digital Network extends the scope of its field trials
	The number of subscribers to cellular systems reaches 1000 000
1988	Intel introduce the 80386 processor
	SUN Microsystems introduce the SUN 4 based on the use of a Reduced Instruction Set Computer (RISC) processor
	IBM introduce the PS/2 personal computer together with the OS/2 operating system
1990	The World Wide Web is set up on the Internet by Tim Berners-Lee at the European Particle Physics Laboratory in Switzerland
	Intel introduce the i860 chip
	Hewlett-Packard introduce workstations based on the Precision Architecture-Reduced

	Instruction Set Computer (PA-RISC) processor
1991	Thinking Machines introduce the massively parallel CM-5 computer
1993	Sun introduce the SPARCstation 10 capable of being configured around multiple CPUs
	Subscribers to cellular systems reach 10 000 000
1993	Intel introduce the Pentium processor
	Mosaic, a free software program, is introduced to search the World Wide Web
	Apple introduce the Newton using a 'write-on' screen instead on a keyboard
1994	Netscape creates its Navigator software for browsing the Internet
	Digital Satellite System provides up to 175 TV channels
	IBM and Apple introduce the Power Mac
	DEC introduces the Alpha processor operating at 275 MHz
1995	Microsoft launch Windows 95
1997	Intel launch the Pentium II processor
	Pathfinder mission lands the Sojourner vehicle on Mars
1998	Microsoft introduce Windows 98
	450MHz Pentium II processors available
	Digital television introduced in UK
	Work begins on the construction of the first permanent manned Earth orbital station
	Apple introduce the iMAC computer
1999	Bluetooth 1.0 specification for wireless communication of data and voice released
	High performance flat screen desktop displays available
	700 MHz plus processors available
	Sony produce a robot 'puppy' which responds to commands and can 'play' with a ball
	Dyson introduces a robot vacuum cleaner
2000	The 'Millennium Bug' fails to appear

found in modern computers including the ideas of a program containing both instructions and data and a processing unit in which the data is manipulated according to the instructions contained in the program. Babbage's work was followed by the development of a number of hand operated 'difference engines' used for the calculation of tide tables, planetary motion and insurance rates. Then, in 1890, Herman Hollerith won a contract from the US Census Office to produce an electrically operated sorting machine working with punched cards to collate census data, thus founding IBM.

Though levels of industrial automation were to increase steadily throughout the 19th century, perhaps the most significant developments took place in another area of technology, that of electricity. Although static electricity and its effects had been known for centuries, it was only with the development by Volta of the 'voltaic pile' at the end of the 18th century that a ready source of electricity became available. Driven by the experimental work of Faraday, Henry and others, the first half of the 19th century saw dramatic developments in this new science such that by the time of the Great Exhibition in London in 1851, telegraph systems were common and experiments with arc lighting were underway. The final years of the 19th century saw electricity establish itself as a major technology with the establishment by Edison, Siemens, Ferranti and others of generating stations in London, New York and elsewhere and the development by Edison and Swan of practical electric lighting systems.

Developments in technology continued to gather pace in the later part of the 19th century and the early part of the 20th century, a period which saw the introduction of the motor car, aircraft, radio, mass production and, perhaps most significantly of all, the vacuum tube valve; this later enabling the development by Vanevar Bush in 1930 of the first electronic analogue computer.

By 1930 the many of the technologies that were to drive the second half of the 20th century were in place but were still essentially separate, existing within their own individual context and with relatively little interaction. The decade leading up to the start of World War II did however see a number of further major developments in electronics including the early experiments in radar, the first public television service and the development of the magnetron. Also, in 1940 Isaac Asimov published the first of his robot stories, leading eventually to the introduction of the 'Three Laws of Robotics' which for the first time defined a behavioural framework for intelligent, human-like machines.

Electronics made a major contribution to the conduct of World War II in the form of both radio and radar and led to the growth of 'electronic warfare' as each side tried to intercept or jam their opponents communications networks. A major, secret, battle was also fought to break and read the codes and cyphers used by both sides as for instance the messages generated using the Enigma machine and its variants and later by the Gehieimschreiber or 'secret-writing machine', code named 'Fish' by the Allies. These efforts led in turn to the introduction at Bletchley Park in the UK in articular of a series of special code-breaking 'computers'.

The first of these were the 'bombes' whose task was to analyse intercepted messages to determine, by what was essentially a process of trial-and-error, the wheel settings for the Enigma machine. The bombes were 'programmed' by means of 'menus' generated by the operating staff using 'cribs' based on other intercepted messages to provide an initial guess at the current settings.

The first attempt a deciphering messages generated using the Fish machine, which used a random binary sequence as the basis of its encryption process, was a machine known at Bletchley Park as 'Heath Robinson', after a well-known cartoonist of the time. This used some 80 valves and suffered from a range of operational problems leading to its replacement in 1943 by the machine known as Colossus. This was the worlds first electronic computing system and in its original version used some 1500 valves, a number which was to rise to 2500 in later versions. Programming was interactive by means of switches and a plugboard and by the end of the war operation of the number of Colossi then in service involved some 20 cryptanalysts, 20 engineers and over 250 operators. Included their number was the mathematician Alan Turing who pre-war had determined the basic structure of a programmable, stored program computer.

The first two decades following the end of World War II saw electronics gradually establish itself as perhaps the dominant technology, impacting on all aspects of industry and home life. The development of the transistor in 1948 by Shockley, Bardeen and Brattain coincided with the entry into service of the Electronic Numerical Integrator and Calculator (ENIAC). Intended for ballistic calculations, ENIAC occupied a room 9 metres by 15 metres and used some 18 000 vacuum tubes, 70 000 resistors, 10 000 capacitors and 6000 switches. ENIAC was followed two years later by the development at Manchester University in the UK of the first true stored program computer. Developed by a team led by F.C. Williams and including Turing, their work led directly to the introduction in 1951 of the world's first commercial computer, the Ferranti Mk I.

At the same time that computers were finding their feet, developments in manufacturing automation were beginning which were to lead ultimately to the integration of computers with machine tools and the subsequent growth of mechatronics. By the mid-1950s, the first numerically controlled (NC) machines were appearing in industry and by the end of the decade the first viable commercial robots were under production by Unimation, the company formed by George C. Devol and Joseph Engleberger. The first major commercial application was in 1961 by Ford to service a die casting machine.

Also by this time the first integrated circuits, combining multiple transistors and other components onto a single chip, had started to appear and the first satellites had been launched, including in 1961 the first commercial communications satellite, Telstar I. Development of electronics technologies continued throughout the 1960s. However, although the first computer based on the use of integrated circuits had been introduced by Burroughs in 1968, the discrete transistor still formed the basis of most electronic systems at the end of the decade.

The development in the early 1970s of the large scale integration (LSI), enabling by 1973 the incorporation of 10 000 components onto a 1 square centimetre chip, led directly to the introduction of the 8-bit microprocessor. By 1980, 'home computers' such as the Apple II were available and 16-bit microprocessors were being produced. These were followed in 1980 by the introduction by Intel of the first 32-bit microprocessor while in 1981 IBM introduced the first personal computer (PC). Over the same period minicomputers such as the PDP-8 and PDP-11 produced by the Digital Equipment Corporation were increasingly being used for control purposes in industry leading in turn to the introduction of computer numerically controlled (CNC) machine tools. Dedicated microprocessors were also being used to control the operation of an increasing range and number of systems, both at home and in industry.

Throughout the period of the 1980s and 1990s, the technologies of electronics and computing resulted in more and more processing power becoming available in the form of advanced processors such as the Pentium of Fig. 1.6. The result is that there are now few items which are not designed using computers or which do not deploy computers, microprocessors or microcontrollers in some way in support of their operation. This integration of the technologies of electronics and software with mechanical engineering is referred to as mechatronics and offers the design engineer the opportunity not only to improve the performance of existing systems such as cars and aircraft but enables the introduction of new products and systems such as the compact-disc (CD) player which would not be achievable by conventional means.

Figure 1.6 Pentium processor chip. (*Courtesy of Intel.*)

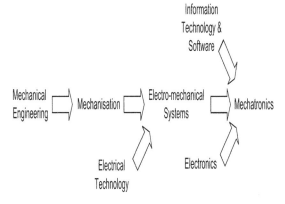

Figure 1.7 The evolution of mechatronics.

1.2 MECHATRONICS AND TECHNOLOGY

Mechatronics provides both a title and a focus for the design and development of a wide range of engineering systems, both products and processes, in which the technologies of electronics, software engineering and information systems are integrated with mechanical engineering. The resulting combination of what have until relatively recently been considered as separate, and often competing, disciplines and the associated transfer of functionality between solution domains has been responsible for the appearance of systems in which the previous, essentially mechanical, solutions have been replaced by ones based on the integration of electronics and software with the mechanical functions.

In considering the technical development of mechatronics suggested by Fig. 1.7 it is important to note that the general structure and concept of mechatronics and mechatronic systems as represented by Fig. 1.8, in which the system is separated into an energetic and an information domain together with a world interface, can exist at a variety of levels. Consider for instance the manufacturing system of Fig. 1.9(a). This can be redrawn in the form of the context diagram of Fig. 1.9(b) from which it is seen that the nature and characteristics of the mechatronic system remain the same whatever the viewpoint adopted, only the scale changes. Thus, for instance, the world of a robot joint, which may itself constitute a mechatronic system in its own right, is the robot itself, while the world of the robot is the manufacturing cell of which it forms a part. The world of the manufacturing cell is then the factory.

As well as integration, a feature of many mechatronic systems is their transparency of operation when in use. Thus, the driver of a car with an advanced engine management system has no direct perception of its operation other than in terms of the improvement in performance that results. The effect of this transparency of operation and the associated devolution to the system level of responsibility for its functional behaviour is that the operator or user of a

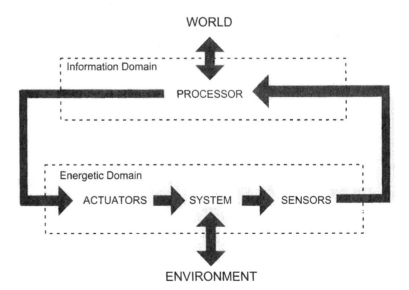

Figure 1.8 A generalised mechatronic system.

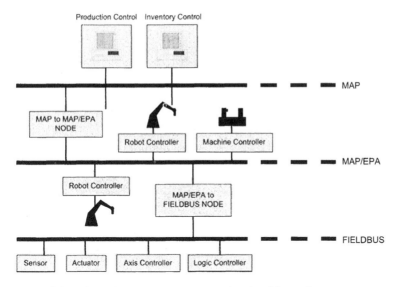

Figure 1.9a Manufacturing systems communication hierarchy.

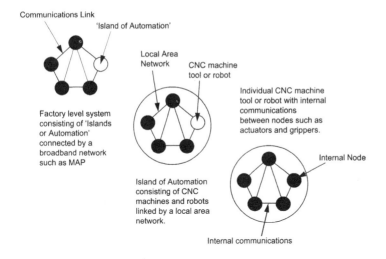

Figure 1.9b Manufacturing system context diagram.

mechatronic system is free to concentrate on the higher level functions associated with its use.

This can be illustrated by reference to the functional diagram of an automatic, autofocus reflex camera with interchangeable lenses shown in Fig. 1.10 in which each of the major 'hard' elements of the system; body, lens and flashgun, constitutes a mechatronic system in its own right. From the point of view of the user, once the appropriate operating mode has been selected the process of determining the required combination of aperture and shutter speed and of maintaining focus becomes the responsibility of the camera. The user is therefore freed to concentrate on achieving the desired composition. Changing part

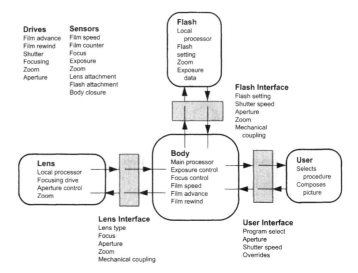

Figure 1.10 Camera system.

of the system, for instance by replacing a lens with one of a different focal length, is automatically recognised by the system and the internal set up modified accordingly.

The major characteristics of an intelligent mechatronic system can therefore be summarised as:

- generally complex systems exhibiting high levels of integration;
- increased functionality with respect to conventional systems;
- transfer of function from the mechanical to the electronics and software domains;
- system assumes responsibility for process allowing operator to concentrate on procedures;
- distributed processing with devolved intelligence;
- multi-sensor environment;
- system operation generally transparent to user;
- multi-program environment with user selection.

In order to further understand the development and importance of mechatronics, the fundamental technologies involved in the development of both the concept and the practice need to be considered.

1.2.1 Basic mechatronics technologies

Microprocessors and integrated circuits

The development of intelligent and mechatronic systems has been supported by increases in the performance, at reducing real costs, of electronics and processing power which has in many cases led to the development of dedicated controller architectures based on the use of microprocessors, microcontrollers and application specific integrated circuits (ASICs). Unfortunately, the improvements in the performance and cost of processing power have not necessarily been paralleled by similar developments in the cost of sensors and actuators, as suggested by Fig. 1.11.

For mechatronic applications, there is a fundamental need for interface electronics functioning in relation to the sensors and actuators. Both of these elements require an analogue circuit capability and in the case of actuators there will be a power requirement at the output stage. In addition, local processing, transparent to the remainder of the system, may well be incorporated to shape individual device characteristics.

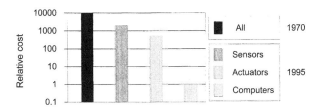

Figure 1.11 Changes in the relative costs of sensors, actuators and computing power, 1970 to 1995.

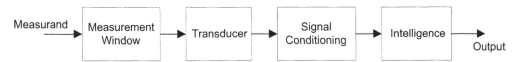

Figure 1.12 The basic structure of a smart sensor.

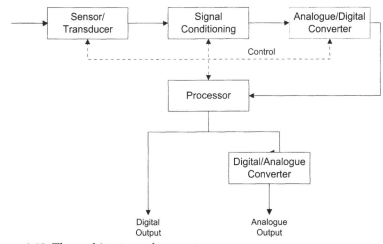

Figure 1.13 The architecture of a smart sensor.

In electronics, advances in computer aided design and simulation have increased access to on-chip solutions across a wide range of circuit types and applications. These application specific integrated circuits (ASICs) are playing an increasingly important part in the development of mechatronic systems, resulting in a growing need to integrate the design of ASICs and other electronic elements more fully into the overall system design process.

Sensors and sensing

In an intelligent mechatronic system, sensors are used to provide information about both system and world conditions and as engineering systems of all types become more mechatronic in nature, the demands on sensors and the associated signal processing will continue to increase. The expansion of silicon technology and the growth of silicon micromachining is leading to the development of low cost, high performance sensors which will, in turn, make possible new developments in intelligent systems which will place an increased emphasis on the incorporation at the sensor level of local signal processing and the ability to communicate with other sensors and sub-systems. A typical configuration for such a 'smart sensor' is shown in Fig. 1.12 while Fig. 1.13 suggests one possible architecture for such a sensor.

In safety critical areas such as flight control systems or car airbags it is important to ensure system reliability and performance integrity. In terms of the sensors and sensing systems this means the incorporation at the chip level of built-in-self-test (BIST) structures together with increased support for sensor, data and system validation processes.

Actuators

While the most common actuators for mechatronic applications are based on conventional and established technologies such as electric motors, including servomotors and stepper motors, fluid power and mechanical drives other forms of actuator are becoming increasingly available. These include:

Ultrasonic or piezoelectric motors
These typically use an array of piezoelectric elements to create a travelling wave which interacts with a friction plate to produce motion. Applications include camera focusing drives and micropositioning systems.

Micromotors
Created by the micromachining of silicon, micromotors with dimensions of the order of a few hundred microns have been produced with possible applications in medical instruments, office and domestic equipment and cameras.

Piezoelectric actuators
Piezoelectric actuators are capable of providing high forces over small distances with fast response times. Though the range of movement is generally too small for most applications this can be amplified by means of levers and hydraulics. Applications include micropositioning systems, inkjet printers and Braille output devices.

Magnetostrictive actuators
Magnetostrictive actuators utilise the dimensional change that occurs in certain materials when subject to a magnetic field. Performance and applications are similar to piezoelectric actuators.

Shape memory alloys
A shape memory alloy will move between a current state and a 'remembered'

state in response to temperature variations. Applications include positioning systems and transfer devices.

Information structures

Figure 1.14 shows in simplified form a possible information structure for a hierarchy of mechatronic systems. Within this structure, the information contained can be considered in terms of that which is associated with a particular level within the system and is therefore concerned with the process or processes carried out at that level together with a vertical flow of procedural data between levels.

In terms of Fig. 1.14, process data is concerned with the level of the system at which it is generated and may be either active, in which case it is directly associated with the control of the process being executed, or passive in which case it plays no part in normal operation, for instance an alarm. Examples of process data at the plant level of Fig. 1.14 would include the co-ordination of the operation of the machine tools and robots making up a manufacturing cell while the operation of the individual machine tools and robots is defined at the servo-control level.

In the context of Fig. 1.14, procedural data flows vertically between levels and can range from being highly-conditioned to ill-conditioned in form. Highly-conditioned procedural data is bounded and contains all the data necessary to define the process or processes to be executed at any particular level. For example, in a manufacturing cell, the task definition for the cell provides the procedural data necessary to establish the processes to be carried out by an individual machine tool or robot. On the other hand, ill-conditioned procedural data contains either or both of redundancies or omissions and requires interpretation by the receiving agent. This is particularly the case involving communications between humans where reliance is often placed on the understanding and knowledge of the parties involved to interpret the transmitted data correctly. The ability to deal with incomplete data using a heuristic approach based on a combination of knowledge, experience and understanding is inherent to much of human intelligence and similar structures will be required at the human–machine boundary if an effective human–machine collaboration based on an understanding and sharing of their individual capacities and capabilities is to be achieved.

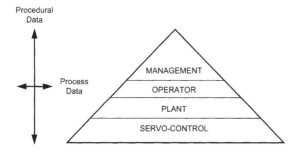

Figure 1.14 The relationship between procedural and process data in a mechatronic context.

Communications

Communications, and particularly digital communications, form an essential component of mechatronics as the means by which information of all forms is transferred between the individual system elements and sub-systems. Communications systems would typically be based on the International Standards Association (ISO) Open Systems Interconnection (OSI) model of Fig. 1.15.

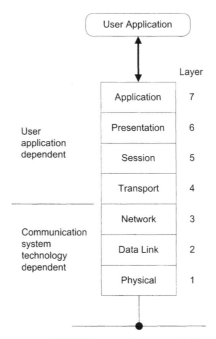

Figure 1.15 OSI/ISO seven layer model.

Control

Table 1.2 shows that the behaviour of intelligent mechatronic systems must be considered at several levels, and an appropriate control strategy is therefore required for each level. Additionally, the control strategy adopted must take account of the fact that each of the levels in the control hierarchy operates on a different time scale and to a different spatial resolution with respect to the other levels. This temporal and spatial separation is necessary to ensure safe and consistent operation. Further, it has been postulated that at each higher level:

Table 1.2 Levels of control and operation for intelligent machines

Strategic level	User specified production goals
Tactical level	Production goals are analysed to establish specified goal
Task level	Define tasks to be completed in relation to the specified goal
Action level	Decompose individual tasks into an appropriate sequence of actions
Trajectory level	Define the motion path required from the current position
Servo level	Decompose actions into the drive commands for individual joints

- control bandwidth decreases by about an order of magnitude;
- perceptual resolution decreases by about an order of magnitude;
- planning horizons increase by about an order of magnitude;
- planning models decrease in resolution and expand in range by about an order of magnitude.

For instance, a time constant of about 300 ms may well be appropriate for a heavy manipulator and this response time can be achieved by a servomotor with a sampling interval of about 30 ms and a command update interval of about 3 ms. Considerations of this kind lead to relationships such as those in Table 1.3 for the temporal and spatial resolution required at each level. Similar relationships can be defined for other types of system.

Table 1.3 Temporal and spatial resolution requirements for different levels of control

Control level	Event detection interval(s)	Spatial resolution (mm)
Task	30	3000
Action	3	300
Trajectory	0.3	30
Servo	0.03	3

Also in the area of control, the development of robust control strategies at the actuator level are being combined with high level, real-time artificial intelligence (AI) based structures, perhaps involving neural networks and fuzzy logic, to enable systems to assume an increasing degree of responsibility for their own actions. The operation of such goal oriented, task driven systems may also be integrated with operator control through the provision of appropriate interfaces to reduce the load on, or to provide additional functionality to, the operator.

Software

The provision of properly structured software is essential to the operation of an intelligent mechatronic system as it is in the software that flexibility and capability of the system generally resides. For instance, in the case of the 'software cam' of Fig. 1.16 the response characteristic is readily modified by changing within the software the parameters defining the desired motion profile.

In many applications, the cost of the software development forms a major, if not *the* major, cost component, requiring careful attention to and control of the software definition, planning and development processes. Well engineered software should:

- be written and documented so as to support, at minimum cost, development and maintenance over its life;
- have reliability built in to its design and coding;
- be as efficient as possible without adding complexity and hence increasing maintenance costs;
- support appropriate user interfaces.

In a mechatronic system, the software specification and design must be integrated with that of the electronic and mechanical elements, ensuring the appropriate 'transfer of functionality' between the individual domains.

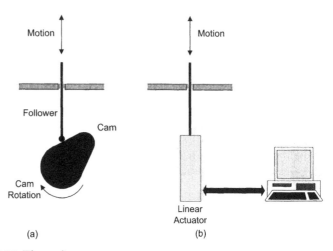

Figure 1.16 The software cam.

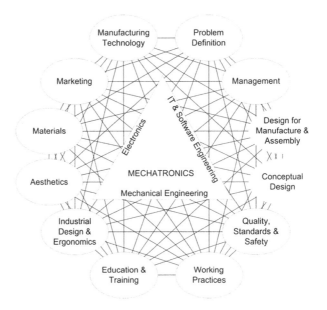

Figure 1.17 Mechatronic relationships.

1.3 ENGINEERING DESIGN AND MECHATRONICS

'.... design is taken to mean the process of conception, invention, visualisation, calculation, marshalling, refinement and specifying of details which determines the form of an engineering product.' M.J. French, *Conceptual Design for Engineers*, The Design Council

'Total design is the systematic activity necessary, from the identification of the market/user need to the selling of the successful product to satisfy the need – an activity that encompasses product, process, people and organisation.'

S. Pugh, *Total Design*, Addison-Wesley (Reprinted by permission of Addison-Wesley Longman)

To make effective use of the opportunities presented by a mechatronics approach to product design and development presents a challenge both to engineering designers and the design process itself. Though the technological basis of mechatronics lies in the integration of mechanical engineering, electrical and electronic engineering with software engineering and information technology, the successful implementation of a mechatronic approach to product design and development requires the bringing together of a much wider range of factors as implied by Fig. 1.17. This in turn requires a company level commitment in order to put the necessary elements into place.

The importance of design to the success of a product is illustrated by Fig. 1.18 and Tables 1.4 and 1.5 in which the balance between the commitment of expenditure during the design phase is compared with the actual spend, primarily during the pre-production and production phases, at which later point the costs of changes to the design increase rapidly.

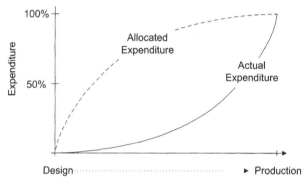

Figure 1.18 The relationship between allocated and actual expenditure from design to production.

Engineering design may be considered as a goal-oriented problem-solving exercise which has as its objective the conversion of a need, often expressed in the first instance as an abstract concept in terms of general functionality, into a product. As suggested by French, the design process can be represented by Fig. 1.19 by a conceptual phase during which possible solutions are considered and an embodiment phase in which the chosen solution is implemented. The associated problem solving process typically follows the path shown in Fig. 1.20 during which the designer may be supported in the development and evaluation of ideas and concepts by a variety of tools and techniques, some of which are also identified in Fig. 1.20.

Other forms of design process model have been suggested such as those by Meister, Ballay, Hubka and Rodenacker set out in Fig. 1.21. In addition, Hansen has attempted to position design in relation to politics, science, art and production as in Fig. 1.22.

However, the very generality of mechatronics across a wide range of engineering systems and environments raises questions in relation to the mechatronics design process, particularly:

• Do there exist certain principles or features of a mechatronic product

Table 1.4 Product development costs

Product development phase	Percentage of total cost (cumulative)	
	Incurred	Committed
Conceptual Design	3–5	40–60
Design Embodiment	5–8	60–80
Testing	8–10	80–90
Process Planning	10–15	90–95
Production	15–100	95–100

Table 1.5 The cost of design changes

Time change is made	Relative cost
During design	1
During testing	10
During process planning	100
During pilot production	1 000
During final production	10 000

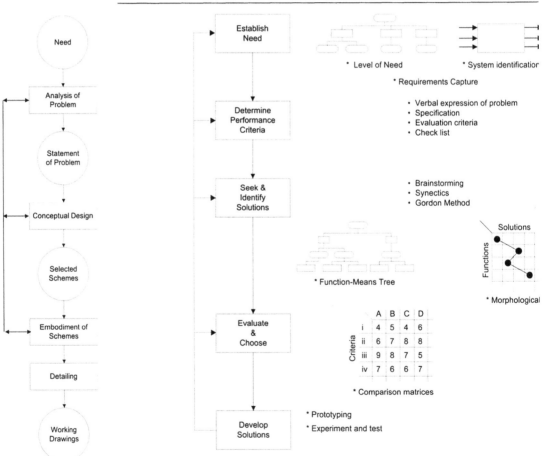

Figure 1.19 French's design model. **Figure 1.20** Problem solving in design.

which are valid across a range of different markets, products and environments?

● Are such features of themselves the means of gaining a competitive advantage from a mechatronic approach to product design and development?

Meisler's Conceptual Model	Ballay's Sequential Model	Hubka's General Procedural Model	Rodenaker's Structural Model
1. Formulation of the design process.	1. Criteria Formulation Collect and analyses information to establish design criteria.	1. Elaboration of assigned processes.	1. Clarify task. 2. Establish function structure.
2. Generation of alternative design solutions.	2. Information Transformation Translate information into written briefs, drawings, etc. in a format that can be used for problem solving.	2. Conceptual design. Establish: (a) Functional structure. (b) Design concepts.	3. Choose physical processes. 4. Determine embodiment. 5. Check logical, physical and constructional relationships.
3. Analysis and evaluation of alternatives.	3. Concept Generation Decisions affecting function, fit, aesthetics, etc.	3. Laying out. Establish: (a) Preliminary layout. (b) Dimensional layout.	6. Eliminate disturbing factors and errors. 7. Finalise overall design. 8. Review chosen design.
4. Selection of preferred alternative.	4. Detail Refinement Engineering refinement of previously defined concepts and decisions. Merging of concepts with aesthetic criteria. 5. Release Package Produce and assemble final document for release.	4. Detailing and elaboration.	

Figure 1.21 Design process models.

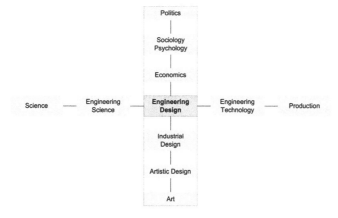

Figure 1.22 Positioning engineering design.

It must also be remembered that the motivation of a move by a company towards mechatronics is that of obtaining a strategic and commercial advantage, either by enhancing the performance or manufacture of an existing product, by gaining access to and developing new markets or by some combination of these. If therefore the answers to these questions is 'yes', then it is likely that the adoption of a mechatronic strategy will be of benefit to a company.

1.3.1 Design models and mechatronics

From the foregoing discussion, it is apparent that the design of a mechatronic system may well require the use of a number of overlapping descriptive and other 'models' in order to provide an effective basis for the design process. In order to be effective, the role of any individual model in describing a particular feature or features of the system must be properly understood together with its relationship with other model formats. Once this understanding has been obtained an appropriate model may then be chosen by the designer to emphasise desired features in the design linking this to the strategy to be used in the design process.

Models of various types play a major and important role in the design process. At one level, they provide a means of communication of matters such as function and purpose, something which is of particular importance when considering the multi-disciplinary nature of mechatronics, while at another they enable system properties and performance to be evaluated. Faced with the demands of industry to reduce development time scales for products of increasing technological complexity while meeting customer demands in areas such as quality and reliability, design models are likely to play an increasingly important role. Indeed, it is possible that the effectiveness of concurrent engineering may well ultimately depend upon the ability to develop effective means of modelling, or of 'rapid prototyping', in a particular design environment.

Mechatronics, by virtue of the need to integrate a range of technologies, and hence to define the interaction between those technologies, presents particular problems for the development of effective design models. In considering the requirements for the modelling of a mechatronic system it is therefore useful to begin by examining the characteristics of the primary areas of engineering and technology involved in the mechatronics design process and their associated solution domains. Taking a much simplified approach to these solution domains it can be argued that:

- The design of the mechanical elements of the system is primarily concerned with spatial relationships and interactions.
- The design of the electronic component of the system is primarily concerned with signal processing and the interconnection of discrete components and devices.
- The design and engineering of the associated software is primarily concerned with the development of data and information processing algorithms.

A particular problem highlighted by the above is the difficulty of ensuring appropriate communication between the individual specialists involved in the design of any mechatronic system since, as illustrated by Fig. 1.23, each will tend to think of the problem in terms of their own area of specialism. The role of the mechatronic engineer is often therefore to act as the link between the specialists and to provide the necessary channel of communication through which ideas and thoughts can be transferred.

Model formats

Figure 1.23
Communications problems.

There are available a number of different model formats that may be used in support of the design of a mechatronic system. These are described in the following sections.

Information and dataflow models
These enable the definition of the flow of information or data throughout the system at a variety of levels and include features such as the flow and storage of data and the definition of processes. Techniques from software engineering include those based on the use of Yourdon/De Marco and structured analysis. On a different scale, controlled requirements expression (CORE) was developed as a means of managing large systems by allowing the complete system to viewed from the aspect of each of the individual sub-systems and of defining the resulting connectivity.

Power and energy flow models
Bond graphs enable the flow of energy between individual multiport networks to be defined. Connections between individual system components are defined in terms of an 'effort' and a 'flow' component which are constrained by the nature of the individual system. At a high level, the bond graph model permits the description of the sub-systems in terms of their terminal or port characteristics in a form which is reducible for the purposes of simulation.

State and transition diagrams
The majority of systems exhibit a variety of states related to their mode of operation. The conditions pertaining to the individual states, the relationships between them and the transitions between individual states are important considerations in the definition of system behaviour.

Sequence diagrams
Sequence diagrams can be used to define the ordering of operations and the associated controlling logic.

Morphological charts
A morphological chart provides a means by which available solution options can be presented in a structured format with each row of the chart representing a particular group of options. A complete system or sub-system can then be constructed by selecting one option from each row of the chart and providing the appropriate interfaces.

Function-means tree
The function-means tree supports a top-down strategy for problem decomposition. At each level of the tree possible solutions, the means, are provided for the desired parent function. Individual solutions then lie at the farthest extremities of the tree from the original need. As it is possible for individual means, and indeed functions, to appear at several different points and at several different levels in the tree the problem of the associated 'combinatorial explosion' is very real.

Hierarchy diagrams
These defines the individual operations and activities associated with the operation of the overall system and indicates how they relate to each other on superior and subordinate levels.

Timing diagrams
In many instances, particularly in relation to logic and sequencing, the

timing of operations and transitions is a major element in defining the operation of the system and highlighting time constraints on the performance of tasks and activities. Tools such as Petri nets can be used to investigate the relationship between time dependent operations.

1.3.2 Top-down versus bottom-up design

In a top-down approach to design the designer starts at the highest level of need and gradually decomposes the problem down into its realisable elements, each of which can then be dealt with individually. This approach is typical of software design where the problem is decomposed into modules each with identified communication and interface requirements which can then be constructed individually before being assembled to form the complete system.

Using a bottom-up approach, a low level element of the system would initially be identified and defined and the remainder of the system constructed from this basis. Bottom-up design strategies are often found within electronics where there is a need to establish the operation and performance of a circuit element before the rest of the design can proceed.

In practice, most design follows a mixture of top-down and bottom-up strategies with the designer alternating between the two as the design progresses, possibly influenced by the need to evaluate a particular concept or approach before the design as a whole can proceed. Indeed, it is not unusual for designers to settle on some key feature within the overall design and then to adopt a top-down/bottom-up approach working outwards from their chosen key feature.

1.3.3 Design support tools

The nature and properties of individual areas of technology are to a significant degree reflected in the tools that have been developed to support the design process. Thus far however, relatively little has been done to integrate these individual design tools. The area where perhaps the most effort has been given to developing methods and methodologies for the representation of complex systems is that of software engineering, possibly reflecting the abstract nature of the software itself. Here, a variety of tools are available for describing the software structures and aiding and managing design and development.

In electronics, and in particular in VLSI design, computer based tools are now well established as a means of describing system structures and behaviour in the early stages of the design process while ECAD packages incorporate extensive simulation facilities.

In mechanical engineering, there now exist a wide range of simulation packages capable of providing a structured mathematical representation of the physical system and of the interaction between system elements. In addition, advanced CAD and other packages support the preparation of highly detailed three-dimensional (3D) representations of the physical systems and allow for the ready propagation of changes.

In describing a mechatronic system it is likely that some combination or mixture of simulation and modelling together with tools such as Yourdon, structured analysis, Petri nets, CORE diagrams, state transition diagrams, function-means trees and bond graphs will be required to provide the necessary level of detail in support of the allocation and distribution of functions between the individual technologies to be deployed.

The Schemebuilder® package developed in the Engineering Design Centre in the Engineering Department of Lancaster University, and described in more detail in Chapter 3, represents one such attempt to bring together a number of such systems to generate a design support environment which will enable the design engineer to rapidly explore and evaluate a wide range of 'schemes' or alternative solutions to a defined need.

1.3.4 Software design

The software lifecycle of Fig. 1.24 emphasises the fact that software when in use is subject to a process of continuous upgrading to take account of changes in the available processors or to correct errors found in earlier versions. The development phase of the software lifecycle can be expanded into requirements analysis, design, implementation and test stages.

Requirements analysis

This phase of the software development process is primarily associated with

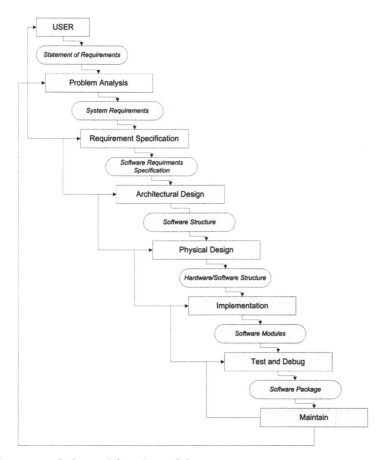

Figure 1.24 Software lifecycle model.

establishing the performance requirements of the software and of the system on which it is to run. The emphasis is therefore on the software engineer to work with the users to investigate and evaluate the context in which the software is to operate before moving to draw up the statement of requirements which will govern the next stages of the development process. The focus at this stage is therefore on 'what' should be done and not the 'how' of achieving it.

Design

During this phase of the development process the software is broken down into its constituent components or modules and the links between modules defined. The use of a modular structure enables the independent development and testing of individual modules prior to assembly into the complete systems and supports future updating on a modular basis.

Implementation and test

Within a modular design, the implementation of individual modules can be made the responsibility of a number of different programmers or groups. Indeed, in certain safety critical applications the same module may well be developed independently by a number of different groups, perhaps even using different languages, before integrating them as part of a redundant system.

Testing of software is a complex process usually involving a number of stages. For instance, at the alpha test stage a number of sites may be selected to evaluate and report on a version of the software while at the beta test stage those same and other sites may be used to test the pre-production version of the software. Despite extensive testing, it has been established that most software contains a significant number of errors, perhaps of the order of 2–5 errors for every 100 lines of code. Many of these errors will not influence the program during its normal operation while others may well cause the program to malfunction when in use. The goal of software engineering is to try to ensure that the presence of software errors does not affect the overall operation of the software, particularly when associated with safety critical environments.

Software design tools

Software engineering led the way in developing tools to support the software design and development process, many of which have themselves now been implemented in the form of computer aided software engineering (CASE) tools. In their more advanced form these are based around a number of approaches to system representation including dataflow diagrams, entity-relationship diagrams, data dictionaries and object orientation and support not only the analysis and design phases but also include code generators which convert the CASE model into appropriate high level code.

1.3.5 The role of simulation in design

Simulation involves the construction of a model of a system in order to better understand the behaviour of that system under a range of conditions and to evaluate operational strategies. Simulation is therefore a major element of the design process and usually involves a combination of one or more of the following forms or representations:

- physical model;
- mathematical model;
- computer graphics model.

A simulation based on the use of a physical model, such as a wind tunnel model, can be used to evaluate and demonstrate specific features of the system. The construction of physical models is generally relatively expensive, time-consuming, relatively difficult to modify and may be impractical in many instances. Where physical models are used they have the advantage of being easily understood and are effective in the identification of major features.

Simulations based on the use of a mathematical model represent the behaviour and performance of the system by means of its defining equations. Although the calculations provided by such a model can be very accurate, it is difficult to fully represent the large number of complex interrelationships associated with a mechatronic system and particularly the interaction between the various elements and sub-systems that make up the overall system. For this reason, there has been a move towards 'hardware-in-the-loop' simulations in which mathematical models are integrated with system hardware as suggested by Fig. 1.25.

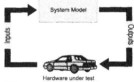

Figure 1.25 Hardware-in-the-loop simulation.

Computer graphics models support both analysis and the visualisation of the results and allow the presentation of the results of the simulation in a variety of forms, as for instance in the case of Fig. 1.26 which shows simulation of a robot workcell using the WORKSPACE package. The availability of cheap, fast computer power means that many of these simulations are capable of being run in real-time, enabling the system designer to 'see' the operation of the system as designed. The techniques of virtual reality (VR) have further extended the capability and use of such models by enabling the generation of enhanced visualisations for applications such as the design of aircraft cockpits and permitted the direct interaction of the user with the model environment.

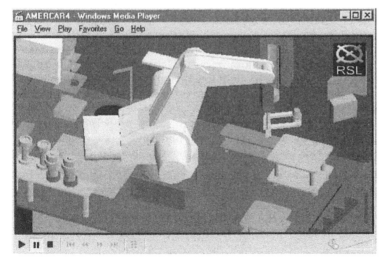

Figure 1.26 Simulation of a robot workcell.
(*courtesy of Robot Simulations Ltd*)

They may also allow the user to select their point of view or to 'walk through' the system while it is in simulated operation. Thus a designer might choose to view the operation of an engine from the point of view of its fuel system or an individual cylinder as well as from an external viewpoint.

Simulation programs are generally written in either a high-level general purpose language or a higher-level special-purpose simulation language. In addition, there are a number of specialised packages, such as the WORK-SPACE package referred to earlier, which are dedicated to the simulation of particular types of system. Packages such as MATLAB and MATHCAD support the rapid definition of mathematical relationships while others such as SIMULINK support the construction of simulations. More recently, the introduction of object-oriented packages such as SABRE and DYMOLA enable the user to define and select from a range of system objects in order to construct the model. In electronics, the introduction of tools such as SPICE, CADENCE and VHDL supports the simulation and design of complex integrated circuits (ASICs) while in software engineering the use of computer aided software engineering (CASE) tools supports the design and definition of the software element of a system.

1.4 MECHATRONICS AND CONCURRENT ENGINEERING

While mechatronics is centred around its core technologies, the achievement of an effective mechatronic approach to product design and development requires the bringing together of many other factors ranging from education and training through manufacturing technology to sales and marketing. Thus to be fully effective, the approach to the design of a mechatronics product must essentially be one which also adopts the precepts of concurrent or simultaneous engineering.

1.4.1 The development of concurrent engineering

In the early 1980s concern about perceived and potential deficiencies in the progress of defence related product development resulted in the Defense Advanced Research Projects Agency (DARPA) in the United States setting up a panel from industry, academia and government to study the situation. Among the concepts studied by this group was the Japanese 'Tiger Team' approach in which all stakeholders are brought together at the beginning of a project to try to ensure that the resulting solution meets the participant's individual needs as well as the global needs of the team.

As a result of these and other studies the panel proposed the adoption of a new approach to product development under the name of concurrent engineering (CE) which included the following definition:

> 'Concurrent engineering (CE) is a systematic approach to the integrated, concurrent design of products and their related processes, including manufacture and support. This approach is intended to cause the developers, from the outset, to consider all elements of the product life cycle from concept through disposal, including quality, cost, schedule and user requirements.' *Institute for Defense Analysis Report R-338*

In order to understand the impact of concurrent engineering on product design and development, it is necessary to place the DARPA programme into its historical context. As was suggested earlier, the growth of factory based production that occurred in the course of the Industrial Revolution saw a shift from craft production and the skilled artisan who was responsible for defining, designing, manufacturing and supporting product to the satisfaction of their customers to mass production. Factories were based around the use of increasingly complex and highly capital intensive buildings and machinery using increasingly sophisticated production processes which resulted in the establishment of specialist departments which gradually increased in independence. Thus design and manufacturing engineers became separated and artificial 'walls' or 'barriers' grew up between them which further reinforced the 'traditional' or 'sequential' product development process of Fig. 1.27. In the worst cases, each department concentrated totally on its own needs and requirements and had almost no perspective of the needs of the product, or indeed in some cases of the company as a whole! The result was a lengthening of the product development cycle and associated loss of competitiveness by the company.

In the early 1950s, a number of organisations, perhaps most famously the Lockheed 'Skunk Works' and General Electric, who began to use 'empowerment teams' to carry out design reviews to improve product development performance. However, the use of interdisciplinary 'task forces' to solve product development crises only began to take hold in the early 1970s, marking the beginning of the concurrent engineering concept. As described in Table 1.6, these initiatives were followed by several studies on competitive engineering

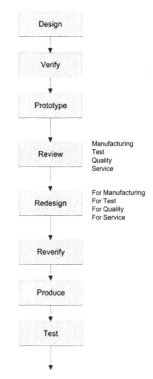

Figure 1.27 Sequential design.

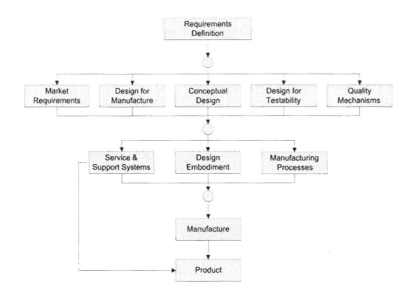

Figure 1.28 Concurrent engineering process.

which effectively set the scene for the DARPA Initiative in Concurrent Engineering (DICE).

Table 1.6 The development of the concepts of concurrent engineering

1978	Xerox, Hewlett Packard and Ford examine their design practices in relation to foreign competitors models
1979	Ford creates the supplier institute to work with its supplier organisations
1981	Xerox carries out benchmarking studies of design methods and practices
1984	Initial CALS study carried out by IDA Ford becomes the US supplier of Taguchi methods
1985	US Navy 'Best Practice' document evaluated Study of CALS policy office forms by Draper Labs First CALS report published
1986	DARPA workshop reports results IEEE R&M and CAE workshop programme FI&M 2000 study published
1988	IDA report R-338 on concurrent engineering CALS report 002 on 'Concurrent Engineering for Mechanical Systems' Report on Japanese manufacturing technology issued by Draper Labs
1990	CALS reports 003, 004 and 005

This five year programme had as its aim the spread of concurrent engineering throughout the US military and industrial base and supported upwards of a dozen industries, software houses and universities in the development of enabling technologies for concurrent engineering. The overall goal was the provision of an architecture for concurrent engineering which would enable people to communicate with each other and to access, share and store information and data without concern for geographic separation, organisational structure, product complexity and types of tool being used. As such, DICE was a forerunner for work on what is now referred to as computer support for collaborative working (CSCW) which has as its focus the creation of a shared working environment for individuals on separate sites.

1.4.2 The concurrent engineering process

Unlike the traditional, sequential approach to product development of Fig. 1.27, the concurrent engineering process of Fig. 1.28 weights aspects such as design for manufacture, testability, quality, serviceability and so forth equally with the engineering design of performance aspects. It therefore brings forward and highlights factors such as changes to the manufacturing, test and support procedures required to accommodate a new or novel design as well as helping to identify features that will increase competitiveness.

Concurrent engineering aims to integrate expertise from all disciplines, both technical and non-technical, during the product design phase with trade-offs regarding manufacturability, testability and serviceability being made in real time. Though these areas may be relatively small in terms of overall project cost, decisions made often have a high leverage and the right decisions made at the right time therefore have a significant impact on overall life cycle costs as was illustrated by Fig. 1.18 and Tables 1.4 and 1.5.

Traditionally, new products generate the most revenue early on in their life cycle, particularly if the product offers new features not present in its competitor's products. As the product matures and competitors enter the market, profit erosion will begin to occur as the competition for available customers increases. It is therefore important that products are designed and produced on time and that production rates are rapidly ramped up to mature levels. Any delays in the release of the product to the market will translate to lost sales that will not be recovered over the life of the product. A further problem in the introduction of new products is that of making sure that the release plan fits in with the overall production schedule in order to ensure that the transfer of customer orders from the old to the new product does not result in a fall in demand and consequent cash flow problems.

The demand for reduced time to market for new products is likely to continue and these products will be required reach the market fitting customer expectations, with high perceived quality and at a competitive price. Within these shortened time scales there is no room for the correction of design errors or for the re-engineering of the product for lower cost or higher quality. Customer loyalty will also suffer if the product is released too early still carrying defects which result in changes or recalls.

The focus of concurrent engineering is therefore on 'getting it right first time' rather than on the 'redo until right' basis associated with the sequential approach. Though in the concurrent engineering model, the conceptual design process may be lengthened, the elimination of design iterations reduces overall product development costs and reduces the overall time to service. Moreover, such products are more likely to be 'fit for purpose' as all factors will have been considered from the earliest stages of the design process which must therefore concentrate in the first instance on:

'Doing the right thing'

before setting out to:

'Do things right'.

Concurrent engineering therefore offers a number of benefits as follows:

- reduced product development time scale. This is important in terms of maximising the revenue from the product before competitors enter the market;
- reduced development costs as a result of the reduction in the number of design iterations and changes;
- improved product quality by building quality into the design from the outset;
- reduced manufacturing costs. By employing the techniques of design for manufacture and assembly (DFMA) products are likely to require fewer parts and be capable of assembly by automated means reducing labour costs. Also, as less defects are produced, there is will be also be a reduction in repair and scrap costs;
- reduced testing requirements by building in the test requirements as part of the design process;
- reduced servicing costs resulting from higher quality products;
- improvements in competitiveness as a result of trouble free product launches and the consequent creation of a reputation for high quality products;
- improved profit margins. Studies have suggested that companies can at

least double their profit margins by ensuring that designs can be manufactured, tested and serviced in the shortest time and at the lowest cost.

1.4.3 Design support tools in concurrent engineering

As set out in Fig. 1.29 there are a number of tools available to support the concurrent engineering process of which the most familiar are quality function deployment (QFD), robust design, failure mode effects and criticality analysis (FMECA), design for manufacture and assembly (DFMA) and design for testability, reliability and service.

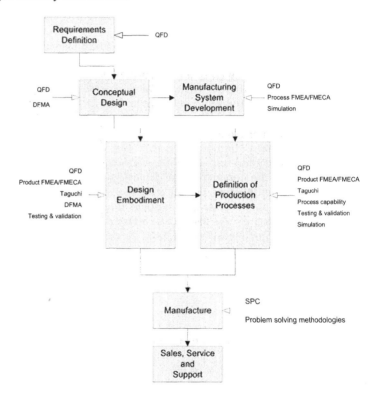

Figure 1.29 Design support tools in concurrent engineering.

Quality Function Deployment

Quality Function Deployment (QFD) is aimed at obtaining a customer input into the design process as part of the specification process and to focus and co-ordinate activities within an organisation. In operation, QFD uses a series of matrix charts in which the customer requirements (inputs) are listed on the left hand side of the matrix and the product features and functions (outputs) listed along the top. Relationships between the inputs and outputs are then classified using the matrix as strong, medium, weak or none taking into account organisational issues such as:

- the ranking of attributes in line with customer preferences;
- the mapping of the customer perception of a company's products against their view of its competitors;

- the mapping of actual measures of performance for a company's product against its competitors;
- the setting and defining of targets for a company's product base;
- the establishment of a measure of the difficulty and/or cost of meeting and achieving targets;
- the mapping of the interactions between product features and functions.

Full QFD involves four sequential phases. In the first phase, the customer requirements are converted into quality characteristics for the product after which a quality planning matrix is used in the second phase to define the characteristics of critical parts. In the third phase, the critical processes are identified while in the fourth and last phase the production planning is evaluated.

Robust Design

Developed by Dr G. Taguchi, Robust Design does not imply conservatism or ruggedness but rather seeks to reduce the sensitivity of a design to variations by a careful selection of design parameters and values. In order to achieve its goal, robust design combines elements from techniques such as brainstorming, design of experiments and analysis of variance together with two specific elements defined by Dr Taguchi; the quality loss function and signal-to-noise ratio.

Robust Design argues that the 'robustness' of a product is more a function of its design than of on-line quality control of the production process and therefore aims to 'build in' quality during the design process. This is expressed by the quality loss function which states that the loss of quality increases as the square of the deviation from the target value and that the associated costs are therefore those associated with warranty, repair and replacement costs together with loss of customer satisfaction and goodwill. The signal-to-noise ratio is used as a measure of robustness with the signal representing the desired performance and the noise then being the deviations from this target performance and the aim is therefore to select values which maximise the 'signal' in relation to the anticipated 'noise'.

In practice, there are so many design parameters and values and their interactions are so subtle and complex that their analysis presents significant difficulties. In order to reduce the problem Taguchi introduced the concept of the orthogonal array with several parameters being varied at a time according to a pattern which resembles an array of orthogonal vectors to reduce the total number of experiments required. It must be noted that the resulting set of experiments does not yield specific design values but serve to indicate to the designer how important each parameter is in establishing robustness.

Taguchi's approach would typically be applied at three levels:

- development of the product concept and system design;
- the determination of product parameters influencing the performance of the product;
- the setting of tolerances.

Failure Mode Effect and Criticality Analysis and Failure Mode and Effect Analysis

Failure Mode Effect and Criticality Analysis (FMECA) and Failure Mode and Effect Analysis (FMEA) are both aimed at providing the design team with a means of anticipating failures and hence of designing them out of the system. Using the techniques, all possible types of failure are first identified and then considered in terms of their effect on the performance of the product. Failures are then ranked in relation to the prospective effect on performance with catastrophic failures being given a high score and those that have little or no effect on performance obtaining a low score. Similarly, scores may be given during the assessment of the detection of failures and their probability of occurrence. Failures are then addressed in the order of their ranking and design changes introduced as necessary to reduce the chances of failure.

Design for Manufacture and Assembly

Automatic assembly systems are not as flexible or adaptable as their human counterparts and consideration must be given during the design process to the assembly methods to be used. Design for Manufacture and Assembly (DFMA) is essentially a means of looking at a product and evaluating it in terms of different means of assembly and, in particular, with regard to the number of parts and operations involved. The basic requirements of DFMA can be summarised as:

- use the minimum number of parts;
- use a modular approach;
- minimise the variation between parts;
- where variations are unavoidable, assemble the unique parts last;
- design using multifunctional parts;
- avoid separate fasteners;
- minimise handling;
- minimise the need for adjustments;
- avoid the use of flexible materials where possible.

When using DFMA, care must be taken that problems, and hence costs, are not simply moved to another part of the system with, for instance, a reduced part count resulting in a need for a more complex and costly mould. DFMA must therefore be used as part of a holistic approach to design and manufacture.

Design for testability, reliability and service

Testability, reliability and service performance are major contributors to the acceptance or otherwise of a product.

Testability is concerned with the ability of the components and sub-assemblies making up the finished product to be tested both individually and after assembly. This is particularly the case with safety critical elements of the system and complex integrated circuits where 'on-chip' structures have been developed to support testing during manufacture.

Reliability in products may be considered an evolutionary process requiring the setting of aggressive reliability targets or goals and making these the priority of the design team. This implies that reliability measures such as mean time between failures (MTBF), mean time to repair (MTTR) and warranty costs are clearly defined early on in the design process. Reductions in design complexity, product environment and manufacturing variability will all impact on reliability.

Other factors in achieving a reliable design include:

- selection of reliable components, processes and technologies. This implies that a full record is kept of components used in earlier designs and of their failure rates. Where new components and processes are used then their failure modes should be established and the design modified if necessary;

- reduction in part count, both to reduce assembly time and to increase reliability. Each part and connection is a possible source of failure and thus the fewer parts are used, the fewer opportunities there are for failure;
- analyse and review new designs continuously throughout the design life cycle and critically review each failure keeping a record of the causes of such failures;
- institute reliability reviews within the design process to eliminate poor design practices and review progress towards reliability goals.

Design for serviceability implies that:

- The inclusion of built-in test features reducing the need for complex and expensive diagnostic and test equipment.
- There is ease of access without interference from other, unrelated components. Thus, test points must be readily accessible, that there is provision for the use of extender cards, that sub-assemblies are pluggable or mounted on slides and that there is provision for strain relief on cables and sub-assemblies.
- The need for special tools and skills is reduced or eliminated to support ease of repair in the field.
- Product upgrading to incorporate new or enhanced features is facilitated. This can, in many instances, be a powerful marketing tool.

1.4.4 Design organisations in concurrent engineering

There are three main models for the organisation of product design and development:

- functional organisation;
- product-centred organisation;
- matrix organisation.

Functional organisation

This has the structure shown in Fig. 1.30 with members grouped together by their work, knowledge and specialism with the interest of each group represented to higher levels of management by the group leader. The work of the different groups is co-ordinated and integrated by means of the project guidelines and specification and by meetings between senior staff. This form of organisation tends towards overspecialisation and supports the development of narrow, specialist viewpoints making conflict resolution difficult.

Project-centred organisation

With this format, shown in Fig. 1.31, a relatively self-contained group is formed around a particular product development process by the secondment of individuals from specialist functional groupings on a temporary basis. An organisation of this form supports a greater focus on and attention to the individual project while informal co-ordination, which requires less organisational effort, serves to achieve and support interaction.

Figure 1.30 Functional organisation.

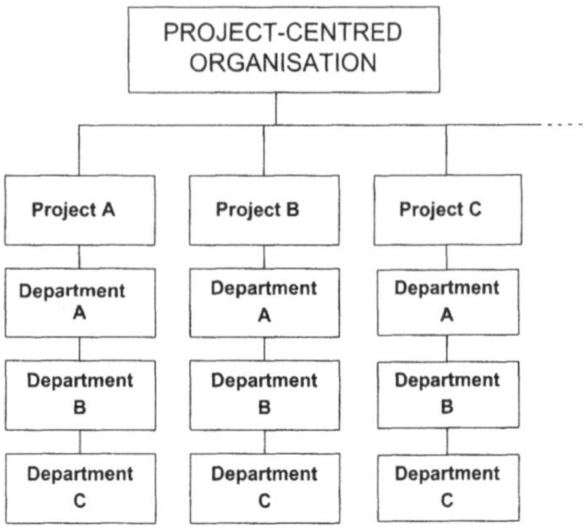

Figure 1.31 Project centred organisation.

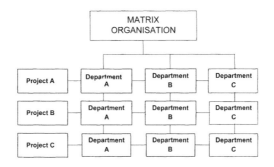

Figure 1.32 Matrix organisation.

Product-centred organisations can be used in support of product innovation by assigning increased responsibility to the team members. It does however tend to result in the duplication of resources and is not suited to centralised control.

Matrix organisation

The matrix organisation of Fig. 1.32 combines lateral co-ordination with a more conventional vertical chain command structure to avoid duplication of resources. Within the matrix organisation, the project team is made up of individuals from the individual specialist groups under the control of a project leader or manager responsible for co-ordinating the project. These individuals still however retain their membership of their specialist groupings, allowing their experience and knowledge to be made more widely available.

A matrix organisation is often suitable for the co-ordination of effort on large and complex projects involving several project lines. It is also suited to projects which have a limited time span requiring different people for each phase of the project. Team members must also report to two superiors, their function manager and the project leader, thus collaboration and conflict resolution are essential factors for a successful implementation.

1.4.5 The concurrent engineering team

A concurrent engineering team must be given ownership of and responsibility for the product and its development and should be empowered to take decisions about the way in which the product develops. There is therefore more to the construction of the team than simply gathering together specialists as to do so would place a heavy co-ordinating burden on the project manager. Instead, it is better to structure the team to provide both breadth and depth with overlaps between team members to ensure that the team is sufficiently flexible and adaptable to exploit developments as they occur.

Other factors relating to the operation of the concurrent engineering team are:

- People learn and develop specific skills in the course of a project which can then be transferred elsewhere. For instance, to the development of future products.
- It is important that the team retains 'ownership' of the project by moving with the product from design to the management of the production process.
- There should be a distribution and sharing of skills and responsibility within the team.

Belbin has suggested a number of guidelines for the construction of a team, including:

- Team members perform in a functional role by drawing on their professional expertise as the situation demands and they also have a team role to perform. Their team role describes a pattern of behaviour characteristic of the way in which one member interacts with others in facilitating the progress of the team.
- The team needs an optimum balance in both functional roles and team roles. The ideal blend will depend on the goals and the tasks the team faces.
- The effectiveness of the team will be promoted by the extent to which members correctly recognise and adjust themselves to the relative strengths within the team both in expertise and the ability to engage in specific team roles.
- Personal qualities fit members for some team roles while limiting the likelihood that they will succeed in others.
- A team can deploy its technical resources to best advantage only when it has the requisite range of team roles to ensure efficient teamwork.

(After R.M. Belbin, 1981, *Management Teams: Why They Succeed or Fail*, Heinemann)

1.5 SUMMARY

The increasingly mechatronic nature of many products, systems and processes when taken together with the demands for reductions in the time to market for new products is placing increasing demands on the engineering design process. In order to meet these demands, the mechatronic engineer must in many instances act as the link between specialists and provide support not only for the definition of the technologies to be used but also for the organisational structures employed.

In this context, the objectives of concurrent engineering have a close association with the requirements of mechatronics and consideration should therefore be given to their introduction and adoption as part of a mechatronic design process.

2 Mechatronics and intelligent systems

'Knowledge itself is power.' Francis Bacon (1561–1626), *Religious Meditations*

'If a little knowledge is dangerous, where is the man who has so much as to be out of danger.' Aldous Huxley (1894–1963), *Elementary Introduction to Physiology*

'Science is organised knowledge.' Herbert Spencer (1820-1903), *Education*

2.1 WHAT IS INTELLIGENCE?

A typical dictionary definition of intelligence takes the form:

intelligence, n. **1 a** the intellect; the understanding. **b** (of a person or animal) quickness of understanding; wisdom. **2 a** the collection of information, esp. of military or political value. **b** people employed in this. **c** information so collected. **d** *archaic* information in general; news. **3** an intelligent or rational being. **intelligential** *adj.* [Middle English via Old French from Latin *intelligentia* (as INTELLIGENT)]. *Concise Oxford Dictionary*, 9th edn, 1995

while a glance at a thesaurus produces the following information:

Synonym	*Antonym*
Knowledge	Ignorance
Information	Stupidity
News	Dullness
Tidings	Misunderstanding

From the above definition and the associated synonyms and antonyms it can be seen that the meaning of intelligence and hence of intelligent behaviour must of necessity contain subjective elements. Thus, certain patterns of behaviour can be interpreted as being characteristic of, or exhibiting, intelligence. This is particularly the case when considering the response of people or animals to external stimuli and their ability as a consequence to determine a particular course of action, often incorporating the setting of goals and the making of both strategic and tactical decisions in relation to the achievement of those goals.

This subjective element means that intelligence as such has to date proven difficult, if not impossible, to quantify in any absolute or measurable form. The concept of the 'intelligence test' as a means of measuring human intelligence relies upon the recording of responses to problems primarily in the video-spatial, alpha-numeric and linguistic domains and the expression of those responses as an 'intelligence quotient' or IQ which attempts to express intelligence in relation to a perceived norm.

Such tests rely on a concept of 'rightness' which itself is difficult to quantify. Consider the question taken from an IQ test and presented in Fig. 2.1 in which it is required to identify the 'odd one out' from among the group of numbered balls. The solution or 'right' answer as given was the ball numbered 15 as this was the only ball on which the digits did not add up to 7. However, when this question was presented to individuals with a mathematical, scientific or engineering background the answer given was more usually 43 as this is the only prime number in the group. On the basis of perceived 'rightness', this latter group would therefore have be recorded as being less intelligent than those who chose 15 as their answer! Indeed, researchers have reported many examples of socio-cultural biases in relation to IQ measurement and testing.

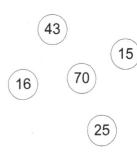

Figure 2.1 An intelligence test question.

How then might machines or systems be considered to be intelligent and what form would, or should, any expression of machine intelligence take? For instance, should an advanced chess playing computer be considered to be intelligent when it has no perception or concept of the game of chess or of the context or environment in which it carries out its analysis, but simply manipulates a set of rules against a defined set of objectives. Perhaps such systems would be more worthy of being called 'electronic idiot savants' than intelligent!

Nevertheless, it is clear that the ability to 'play chess' is highly regarded as a measure of human intelligence. This highlights a recurring problem in attempting to evaluate claims of having created an intelligent computer or machine. Sceptics point out that the machine has been programmed by a human and hence is not in itself intelligent but simply reflects the intelligence of its programmer. However, such individuals tend to forget that humans have been taught or 'programmed' to perform many of their so called intelligent functions. Furthermore, recent techniques such as fuzzy logic, 'biologically inspired' computing using artificial neural networks and genetic algorithms show the capacity of machines to independently evolve their own solutions to problems, often in ways which are not fully understood by humans. These techniques will be considered in detail later but it should be noted that some artificial intelligence (AI) researchers regard them as lying outside the field of 'classical AI'.

The concepts of artificial intelligence, digital computers, and perhaps increasingly of techniques such as those referred to in the previous paragraph, lie at the heart of what might be considered as an intelligent machine or system. If however the capabilities of humans and computers are considered as in Table 2.1, a further dichotomy arises in that those elements which often are used to characterises human intelligence are precisely those areas in which current computers are weakest. This suggests that machine intelligence will probably always be different from human intelligence but this, in itself, is not necessarily a reason for considering it to be inferior.

What Table 2.1 shows is that human intelligence has many aspects to it and that in many key areas machines have as yet made little impact. Once the idea of intelligent behaviour is seen as multi-faceted it becomes meaningful to make statements such as:

'A computer is more intelligent than humans as far as remembering telephone numbers is concerned but less intelligent at creating a novel.'

Intelligent behaviour in a complex, unstructured world requires proficiency in many of the aspects listed in Table 2.1 and, as such, could be said to 'emerge' from their complex interaction. Thus, if intelligent behaviour is relative and emergent it is perhaps meaningless to attempt to define a specific level of performance or activity at which behaviour suddenly becomes intelligent.

Table 2.1 Aspects of human and machine intelligence

Aspects of intelligence	Humans	Computers
Calculation	*	*****
Analysis	*	****
Reasoning and logic	**	****
Memory	**	****
Co-ordination	**	***
Control	**	***
Communication	***	**
Sensing	***	**
Organisation	***	**
Learning	***	*
Recognition	****	*
Abstraction and generalisation	****	?
Understanding	****	?
Judgement	****	?
Creativity	****	?
Intuition and insight	*****	?
Instinct	*****	?
Emotion	*****	?

Artificial intelligence has its origins in the early 1960s when work by Samuel, Newell, Simon and others began to investigate areas such as games playing and theorem solving in which the computer was able to reach a solution through the examination of possible solution paths using a variety of search algorithms. This early concept has been developed over the years using search algorithms of increasing sophistication to encompass more general problem solving and areas such as perception and context. Included in this development is the idea of the 'expert system' in which the computer assumes the role of the human expert in analysing and interpreting the data presented. In each case, and for each system, the underlying assumption is that the process of providing an intelligent response can be structured around a set of rules and associated probabilities which the computer can then use as the basis for its decision making process.

However, and despite the fact that there are a number of successful expert systems in operation, they are all attempting to reproduce, in some form or other, various aspects of human intelligence and, as a consequence, some measure of comparability with human intelligence is sought. In achieving this comparative assessment of intelligence, no better test has yet been developed than that proposed in 1950 by Alan Turing and known as the 'Turing test' which sets out the following framework for assessing machine intelligence.

'A human interrogator has two dumb terminals, one of which is connected to a computer and the other to a human operator as in Fig. 2.2. The interrogator can send and receive messages via these terminals. If after any length of time the interrogator cannot tell which of their two terminals is connected to the computer and which to the human operator, then the computer may be said to be intelligent.'

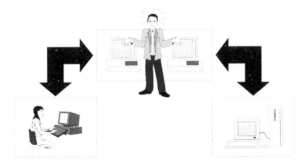

Figure 2.2 The Turing test.

Thus the Turing test requires that the computer fools the interrogator into believing it to be a human operator. This means that the computer must respond slowly and include (deliberate) spelling and numeric errors into its responses.

Though robots such as R2D2 and C3P0 and androids such as Isaac Asimov's Daneel Olivaw and Data in 'Star Trek: The Next Generation' are a major element of science fiction, the goal of achieving such 'human like' intelligence may in many instances be a misleading aim. In this context, it is perhaps worth noting that one of the major elements of Data's character, in the television series at least, was the search for an understanding of human emotions such as fear and humour, an absence of which would almost certainly result over an extended period in a failure to satisfy the conditions of the Turing test!

Perhaps therefore it is necessary to consider other aspects, definitions and concepts of intelligence which emphasise the capabilities of computers and do not rely on the ability to emulate aspects of human intelligence. As a consequence, to seek a single, all encompassing definition of machine intelligence is perhaps impractical and there is a need to consider the case for degrees of machine intelligence related to function. Consider for instance the categories of machine set out in Table 2.2, some of which exhibit characteristics of machine intelligence, suggesting that there is a need to consider machine intelligence within a defined contextual framework.

The following are two examples of intelligent mechatronic systems, one adaptive and the other goal seeking.

Table 2.2 Types of machine intelligence

Description	Example	Characteristic
Single Function	Drill	Capable of only one operation
Automatic	Pick and place mechanism	Carries out a single fixed sequence of operations
Programmable	CNC machine tool	Carries out a sequence of operations under programme control Includes branching
Adaptive	Fly-by-wire aircraft	Autonomously adapts to changing conditions using sensory feedback
Goal Seeking	Self guided vehicle	Adapts to changes in environment in relation to a defined goal
Advanced Autonomous	Data	Capable of strategic decision making and planning

1. *Automatic autofocus camera*. Historically, camera users were required to monitor lighting conditions by means of an independent light-meter and then transfer the readings to the camera in the form of specific settings of shutter speed and aperture. Consider typical automatic, autofocus cameras such as those shown in Fig. 2.3. As well as giving a basic 'point and shoot' capability the more advanced cameras will also allow the user to select from a number of operating modes including shutter priority, aperture priority and mixed or program mode.

Figure 2.3 Automatic, autofocus cameras.
(Courtesy of Minolta UK)

In addition, the incorporation of an autofocus capability means that the user no longer has to continually adjust the lens position and setting. The result is that the user, once the operating mode has been entered into the camera, can concentrate on matters such as picture composition and the exercise of their artistic capabilities.

Such a camera is readily seen to fit the class of adaptive systems of Table 2.2 in that, once the user has selected the operating mode, the responsibility for all functions such as selection of shutter speed and aperture, focusing and flash setting associated with the taking of the picture is assumed by the camera in response to data generated by its internal sensors.

2. *Intelligent paper roll manipulator*. The intelligent paper roll manipulator (iPRM) was developed by VTT in Finland to automatically stack rolls of paper in environments such as warehouses and ship's holds. In operation, the environment is first scanned to establish the position of the rolls as delivered on a trailer to the warehouse or ship's hold. Once the position of the rolls has been established, the iPRM establishes a plan to enable it to optimally stack the rolls, taking into account factors such as the number of rolls already stacked and the space available. Stacking is aided by sensors on the gripper which detect the presence of the roll as well as factors such as any slippage of the gripped roll and the position relative to already stacked rolls.

It may be considered by some observers that approaches such as those adopted in the automatic camera and the paper roll manipulator result in a deskilling of certain tasks and a subordination of the operator to the machine.

However, in many cases a more reasonable argument may be that the introduction of machine level intelligence opens up opportunities which previously did not exist. Indeed, many thousands of people who could not or would not use light-meters now successfully take photographs under a wide range of lighting conditions while the experts still have the opportunity to exercise their skills, albeit in more subtle and less obvious ways.

Thus, though the concept of machine intelligence suggested by the preceding arguments is perhaps broader in context and scope than many others, there still remain a large number of grey areas, particularly with regard to the way in which humans and machines interact. This influences not only the design of the operator interface but the way in which responsibilities for function are partitioned between humans and machine and the challenge is therefore to achieve the most effective use of the capabilities of each.

Consider for instance an intelligent wheelchair incorporating collision avoidance and warning routines. Such a wheelchair may at first glance seem to be essentially an autonomous guided vehicle which, once the goal is defined, will generate its own route to that goal. In practice, a wheelchair operating under operator command must in normal operation be subordinate to the wishes of the user and therefore be capable of capturing, in some form or other, the user's intent, a consideration which extends through to other forms of human–machine interaction, particularly where remote operation is concerned.

2.1.1 Intelligence and knowledge

Knowledge provides the framework of understanding within which human beings operate and is gained as part of the general learning process to which humans are subject. In any situation, a human will search their knowledge base to identify those conditions which most closely correspond to the current set of circumstances and will then extrapolate from the chosen set of conditions to determine their course of action. The ability of an individual to select a course of action leading to a successful conclusion may well be taken as an indicator of intelligence. Humans however accumulate a variety of different types and forms of knowledge with, in practice, a significant interchange taking place between these forms depending upon current circumstances. One possible definition of types of knowledge might be:

- catalogue knowledge: parametric data, components;
- physical knowledge: $E = mc^2$
- contextual knowledge: IF 'A' THEN 'B';
- abstract knowledge: 'To be or not to be … ';
- assumed knowledge: The earth is flat!
- observed knowledge: Birds eat worms.

A more formal definition of forms of knowledge would be:

- metaknowledge or knowledge about the structure of knowledge;
- knowledge;
- information;
- data;
- noise.

In either case:

Useful knowledge = Applicable knowledge

= Knowledge which can be directly applied to the current situation.

In the case of an advanced, autonomous machine, task based knowledge may be considered, initially at least, at two levels, namely as strategic knowledge and as tactical knowledge. At the strategic level, the knowledge system is concerned with the definition and setting of goals and objectives and with the planning, scheduling and management of the machine in relation to these. At the tactical level, the knowledge system is responsible for achieving the goals and objectives set at the strategic level and for the identification and execution of the required task structures, including the necessary control functions.

The strategic knowledge base for the machine must therefore contain all the necessary information and knowledge required for setting the goals and objectives as well as the ability to communicate with other high-level systems. The tactical knowledge base on the other hand contains information as to the performance parameters of the machine, and may be consulted by the strategic level knowledge system in the setting of goals, as well as knowledge about lower-level systems such as the sensors and actuators under command.

2.2 INTELLIGENT MACHINES

The first machines evolved from manual tools through the addition of an external source of energy which increased the effective force or effort that they could provide. As machines developed in the manner suggested by Table 2.3, additional layers of control were provided, initially based on mechanical electro-mechanical and pneumatic systems and latterly in a programmable form using first relay logic and then computers. Although many current machines such as computer numerically controlled (CNC) machine tools, machining centres and robots are highly capable and complex, they do not necessarily exhibit any degree of intelligence as they still rely upon an outside agency in the form of the operator or programmer to determine their actions and are only suited to deterministic or structured environments which are in turn often constructed and constrained to meet the requirements of the machine. As for instance in the case of a robot workcell which must be structured around the operating envelope of the robot being used.

Table 2.3 A hierarchy of machines

Tool	A device intended to supplement the capability of the human hand
Machine tool	A tool using an external energy source to increase its capability
Automated machine	A machine capable of independently carrying out a series of defined tasks
Programmable machine	A machine which can independently programmed to carry out a series of defined tasks
Intelligent machine	A machine which can sense and respond to its environment and which can make changes to its behaviour as a result

An intelligent machine is therefore considered to be one which will assume a degree of responsibility for its own actions and is as a consequence able to modify its operation in some way in response to external stimuli in the form of sensory data. Such machines will not require to be presented with a complete model of their environment but will be able take account of and respond to changes in that environment and will therefore be capable of operating in non-deterministic or unstructured environment on the basis of a general knowledge of their goal and of the type of environment in which they are operating.

Intelligent machines must therefore have the ability to receive, collate, store and process a wide variety of different types of data about both the machine and its environment in order to operate. This data may be provided by a range of different types of sensors chosen to reflect the nature of the information required by the machine in order to function in relation to its goal. In addition to its sensory input the intelligent machine will also receive through its world interface, command and information inputs which act to define the current goals of the machine and its operating parameters. The world interface also provides the communications link to the human operator as well as to other systems. The machine then interacts directly with its environment through the medium of its actuators and effectors.

An example of this interaction is seen in the Sojourner vehicle used in the Mars Pathfinder mission and shown in Fig. 2.4. Depending on the relative positions of Earth and Mars the transmission delays may be of the order of 4–20 minutes in each direction which means that direct control of Sojourner by an operator on Earth is impossible. Instead, the operator uses images generated by stereo cameras on the lander to plot a route by defining a series of waypoints with Sojourner then navigating itself from waypoint to waypoint using its onboard systems.

Figure 2.4 Sojourner vehicle on Mars.
(*Courtesy of NASA/JPL*)

2.3 ARTIFICIAL INTELLIGENCE AND EXPERT SYSTEMS

Artificial Intelligence has its origins in the work of Newell, Simon and others in the 1960s which culminated in the GPS or 'general problem solver' program. The GPS worked by defining a 'task environment' within which objects could be manipulated in accordance with a set of defined operators or rules and proved to be successful in handling a restricted class of situations with a relatively small number of states and reasonable well-defined rules. Central to the operation of GPS was the concept that problem solving could be based upon searching the possible solution space in order to reach the goal and that such searches could be directed by a set of heuristic rules in order to improve their efficiency.

Also at around this same time the concept of 'list processing' was introduced into AI as a means of handling diverse and complex data structures, leading in turn to the development by Feigenbaum of the concept of an 'expert system' which attempts to reproduce the function of a human expert over a limited and definable field of knowledge. One of the first of these early systems was DEN-DRAL, developed in the late 1960s to interpret data from mass-spectrographs in order to infer the structure of molecules. This was followed in the early 1970s by MYCIN which used a series of probabilistic rules to diagnose blood infections and which proved to be capable of making decisions on the basis of incomplete data. These early systems have since been followed by systems of increasing performance and complexity for applications in medicine, science and technology and finance.

2.3.1 Expert systems

Expert system programs differ from conventional computer programs in that they:

- are data driven and not procedure driven;
- manipulate both data and concepts;
- use heuristic structures to reach their goal;
- can in general work with incorrect or incomplete data to reach a conclusion.

The basic elements of an expert system are shown in Fig. 2.5 and consist of:

- the knowledge base which holds as a set of rules a representation of the knowledge that is required by the expert system in order to deal with a problem in its domain;
- the working memory consisting of a collection of the data required by the rules;
- the inference engine within which both knowledge and data is manipulated and is the means by which the expert system reaches its conclusion. Forms of inference are listed in Table 2.4;

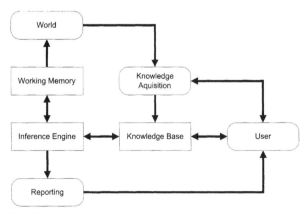

Figure 2.5 The elements of an expert system.

Table 2.4 Forms of inference

Induction	Reasoning from the specific case to the general
Intuition	Answer appears without any apparent connection to known facts. Perhaps based on subconscious reasoning and recognition and abstraction from patterns. A very human characteristic
Heuristics	'Rules of Thumb' determined by experience
Generate and test	Trial and error
Abduction	Reasoning from a true conclusion to the premises that may have resulted in the conclusion being reached
Autoepistemic	Knowledge about self
Non-monotonic	Revision of previous knowledge as a result of the availability of new data
Analogy	Generating a conclusion by reference to another, similar situation

- the knowledge acquisition system by which knowledge is entered into the system avoiding the need to explicitly code the knowledge;
- the reporting facility which interacts with the knowledge base and inference engine and is the means by which the expert system communicates the reasoning used in reaching a conclusion to the user.

2.3.2 Developing an expert system

The construction of an expert system is a complex task and the designer of such a system must resolve a number of problems.

Knowledge acquisition

The aim of an expert system is to reproduce the performance of a human expert in a particular field of knowledge given the same information and working from the same starting point. Unfortunately, different experts in the same field may well work in different ways in order to reach their conclusions, often using their individual knowledge and experience to arrive at a solution in what often may appear to an observer to be an intuitive approach to problem solving. The role of the knowledge engineer is therefore to extract information from the human experts as to how they approach a problem and then to translate that information into a form and format in which it can be encoded into the computer based expert system.

In order to achieve this, the knowledge engineer has available a variety of tools ranging from structured interviews with the experts to direct observation and recording of those experts when carrying out particular tasks. Working from this base, the knowledge engineer will then construct a set of rules and procedures which will form the basis of the expert system and will then work with the experts to refine these rules in order to achieve the desired level of performance.

A particular problem in knowledge acquisition is the ability of the chosen experts to articulate their expertise and to respond to the queries of the knowledge engineer about their progress to a final solution. This is particularly the case where the knowledge engineer is trying to capture physical skills and expertise such as an ability to manually control complex, three-dimensional motions such as are associated with the operation of remote manipulators.

Choice of domain

The correct choice and definition of the domain or field of expertise for an expert system is crucial to its chances of success or failure. If the domain boundaries are set too wide with too many interactions then the less likely it is that the expert system will be able to reach a conclusion. Similarly, if there is insufficient expertise associated with the chosen domain or experts disagree then it also becomes less likely that it is a suitable application for an expert system.

Uncertainty

Human experts are very good at reasoning on incomplete or sparse data, using their own experience to fill in gaps and to make reasoned assessments. Human experts also tend to be aware of their own limitations and of when to seek additional help or advice in order to solve a particular problem

Knowledge base

The knowledge base is typically structured around a set of rules which when used by the inference engine will enable it to reach a conclusion. Consider a knowledge base to be used by a garage to assist in diagnosing why a car will not start. Typically, the rules would be of the form:

- IF the engine turns over but does not start THEN check the petrol gauge.
- IF the engine turns over but does not start AND there is petrol in the tank THEN check the battery connections.
- IF the battery connections are alright THEN check the plug leads.

In some cases, probabilities may be added to the rule. Thus an expert system used to diagnose a patient's illness may contain rules of the form:

- IF the antibody count is positive THEN the probability that the patient has a fever is 0.8.
- IF the patient's skin has a yellowish cast THEN the probability that the antibody count is positive is 0.65.

Inference engine

The role of the inference engine is to manipulate knowledge and data using the information held in the knowledge base in order to reach a conclusion. In order to do this, the inference engine uses an inference rule base containing structures of the form:

- Begin by examining that part of the system which is most closely associated with the observed effect.
- Set up and carry out tests which will have the maximum effect on reducing the solution space.
- If the first selection does not account for the observed effects then carry out further tests.

The linkage between the knowledge base and the inference engine is therefore crucial to the effective operation of any expert system.

Search routines

Forward chaining
Using forward chaining an expert system would work from data defining the initial conditions to reach a conclusion. Forward chaining is used particularly where the number of possible conclusions is large in relation to the amount of observable data.

Backward chaining
Backward chaining is used where a small number of possible conclusions is supported by a much larger volume of observable data. The expert system will therefore start from a particular premise or conclusion and will then test this conclusion against the observed facts.

Depth-first searches
In a depth-first search the expert system will test each branch of the solution tree to its fullest extent before retreating up the tree to the point at which the next untested branch occurs and will then move down this branch to its end. A depth-first search will therefore evaluate the nodes of the solution tree of Fig. 2.6 in the order shown by the numbering of the nodes.

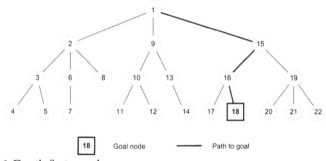

Figure 2.6 Depth first search.

Breadth-first searches
Using a breadth-first search procedure, all the nodes in the solution tree that are at the same level will be tested before testing the nodes at the next level down in the hierarchy. Thus a breadth-first search will evaluate the nodes of the solution tree of Fig. 2.7 in the order shown by the numbering of the nodes.

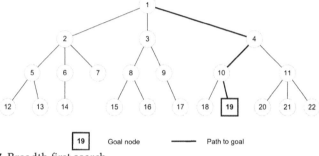

Figure 2.7 Breadth first search.

Learning

In order to support operation in a range of environments and under different operating conditions, systems may well require a learning capability to enable them to adapt their operation according to both circumstances and experience. Additional advantage would be gained if experience could be shared between systems of the same type as this would increase the rate of learning and enhance the ability to respond to unfamiliar environments.

Learning is a complex process and may be broadly categorised into those forms based on discovery or experiment and those based on induction or reasoning. In practice, machine learning strategies have to date tended to emphasise approaches based on either induction or a combination of strategies.

Other techniques employed include the use of generic algorithms to 'breed' solutions from a known starting point, probability based approaches and neural networks. In each case, there is a need to carefully consider the solution space and the training processes involved. It is also important to provide appropriate feedback to reinforce the successful options.

2.4 ARCHITECTURES

Within electronic engineering and software engineering the term architecture is used to describe the organisation and structure of a system and the ways in which the various elements of the system interact. The choice of system architecture is therefore likely to be a critical factor in determining the way in which an intelligent machine operates.

2.4.1 Von Neumann architecture

Named after the mathematician John von Neumann, this architecture describes the configuration of most common computer systems. Also referred to as a random access sequential machine, a von Neumann architecture assumes the existence of a central processor and read-write random access memory as in Fig. 2.8. In operation, the processor sequentially executes instructions stored in memory using data stored elsewhere in memory.

2.4.2 Parallel architecture

Parallel architectures are based around a number of processors sharing access to common memory as in Fig. 2.9. Execution of instructions is allocated to individual processors in turn according to loading, enabling significantly shorter processing times to be achieved.

2.4.3 Distributed and embedded systems architectures

In a distributed architecture, processing power is devolved to individual elements or components within the system, each of which essentially runs its own independent program based on local memory. Communication between the individual elements is then by means of one or other of the systems and communications architectures discussed below. Thus in the camera system of Fig. 2.10, each of the elements; body, lens, flash gun and data back; contains all the processing capability required to enable it to operate as part of the overall

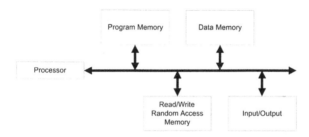

Figure 2.8 Von Neumann architecture.

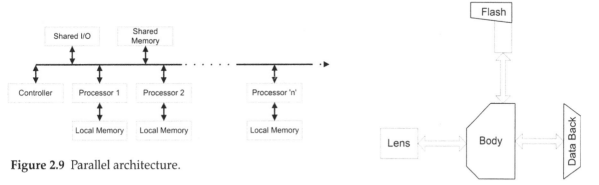

Figure 2.9 Parallel architecture.

Figure 2.10 Camera system.

system together with an appropriate interface to enable communication with the other elements of the system.

2.4.4 Hierarchical architecture

A hierarchical architecture is organised as a tree structure such as that shown in Fig. 2.11 in which each member, with the exception of the topmost member and bottommost members of each branch, are both the parents to the members of the hierarchy in the layer below them and children of a member of the hierarchy in the layer above them. Thus at the top level of the hierarchy sits the complete system while the lowest level of each branch describes one of the individual elements or components that make up the complete system. Information flows up and down the layers in the hierarchy in a strictly ordered manner from parent to children and vice versa.

2.4.5 Network architectures

Networks may take a variety of forms such as those shown in Fig. 2.12. Unlike a hierarchical configuration, all members of a network are essentially equal, though there may be some bias in order to assign priorities to certain nodes. Each node of the network can itself be based around a microprocessor or microcontroller which provides local processing power and hence forms the basis of the distributed and embedded system that is at the heart of many mechatronic systems. Thus, the camera of Fig. 2.10 can be redrawn as a bus architecture in the manner of Fig. 2.13 with the processor in the main body acting as system controller.

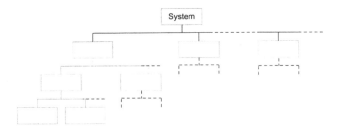

Figure 2.11 Hierarchical architecture.

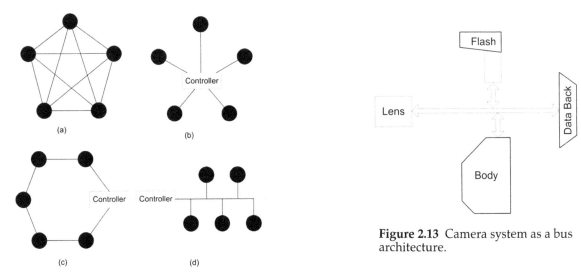

Figure 2.12 Network architectures.

Figure 2.13 Camera system as a bus architecture.

Ring or bus architectures have the advantage over fully-interconnected and radial architectures in terms of the amount of wiring required and are increasingly common in a wide variety of applications and would typically be based around the use of standard network protocols and hardware such as the controller area network bus (CANBus) and Fieldbus. Thus the conventional radial wiring for a vehicle as in Fig. 2.14(a) is increasingly being replaced by the network configuration of Fig. 2.14(b).

2.4.6 Modular architectures

In a modular architecture, each element of the system is a self-contained entity which receives and transmits information about its status via its interface. Each module therefore contains all the necessary processing and intelligence required to enable it to operate as part of the complete system. Communication between modules may be hierarchical or take any of the network forms shown in Fig. 2.12. Thus the camera of Fig. 2.10 is modular in form with each individual lens processor not only being responsible for the operation of the lens but also containing sufficient information to enable it automatically to be identified by the main processor in terms of features such as aperture, focal length and zoom range.

A modular architecture may be chosen for systems which are intended to be reconfigured by changing a module or modules and where 'repair or maintenance by replacement' is intended.

2.4.7 Layered architectures

A layered architecture such as that shown in Fig. 2.15 consists of a number of self-contained elements each of which is associated with its own set of inputs and outputs. Each layer of the system will be responsible for overseeing and controlling different aspects of system behaviour. The final stage of the layered architecture is the arbitration stage which is responsible for resolving conflicts between the outputs of the individual layers and determining the final system outputs.

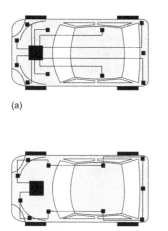

(a)

(b)

Figure 2.14 Vehicle architectures.

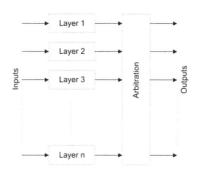

Figure 2.15 Layered architecture.

2.4.8 Subsumption architecture

An approach to the problem of operating in an unstructured environments is that proposed by Rodney Brooks under the general title of a subsumption architecture. Using this architecture, simple, low-level functions are subsumed in turn by higher levels to produce a robust and flexible controller of increasing levels of performance and complexity. Thus, in one of Brooks's early walking robots known as Ghengis, each of the six legs had its own controller but there was no central brain or controller which determined the gait. Instead, each leg attempted to carry out a series of simple motions such as 'lift up', 'move forward', 'put down' and 'move backward' in competition with each of the other legs. Conflicts were then resolved on the basis of simple rules of the form:

'If any one of the right (or left) hand side legs is raised then no other leg on that side can be raised'

and

'If a leg is raised and moving forward then move all legs that are down backward'

The result was that Ghengis rapidly developed a gait of its own which once established for flat ground could be modified by the provision of additional systems. Thus, the introduction of appropriate sensors together with a rule of the form:

'If there is a signal then stop, otherwise proceed'

supports obstacle avoidance. The resulting system and its development can be summarised in the following terms:

- Do the simple things first and learn to do them well.
- Once it is working well then do not change it.
- Add new layers and levels of activity on top of the learned tasks.
- Make the new layer work as effectively and well as the previous layers.
- Repeat the process.

In his 1989 paper with Anita Flyn, 'Fast, cheap and out of control: a robot invasion of the solar system', Brooks proposed to drop onto the surface of Mars a large number of small, solar-powered bulldozers. Though many of these would 'die' on landing and more would fail as a result of exposure to the temperature changes on the Martian surface, the remainder would wander about over the surface governed by a few simple rules to create a level surface. Brooks argues that such an approach would be far more cost effective than trying to position a single large robot such as 'Ambler' developed at Carnegie Mellon University also for use on Mars. This is a six-legged walking machine weighing over 2 tonnes and standing some 6 metres high which used lasers, vision and other sensors to map its environment and plot a path before moving.

2.5 SAFETY

In order to function safely in unstructured environments intelligent systems such as robots, whether static, semi-mobile or mobile, must be provided with an information structure within which they can carry out their function without risk to personnel, to themselves or to other equipment. In an unstructured environment precluding the use of physical barriers, safe operation requires the use of a sensor system capable of monitoring the region around the robot and of detecting the presence of both static and mobile obstacles within that region. In addition, they must be provided with software structures which are themselves inherently safe.

Other safety critical environments such as aircraft or vehicle systems, nuclear reactor and chemical plant control and sub-sea systems are similarly increasingly dependent for their operation on the integration of software and electronics. For instance, an agile combat aircraft such as the Eurofighter Typhoon of Fig. 2.16 is maintained in stable flight by the flight control computers onto which the pilot's commands are superimposed. In vehicle systems, the concept of 'drive-by-wire' in which the direct mechanical link between the driver's steering wheel and the steered wheels is replaced by an electronic link is under development and is likely to appear on the road in the near future.

In both these cases, and in other similar systems, there is a need to ensure high levels of system integrity and safety. Indeed, it may be argued that the resulting systems need to be intrinsically safer than equivalent operator controlled systems in order to gain acceptance. In particular, they require the adoption of a system architecture in which safety is a major feature. This could take the form of the addition of a separate safety system whose role is to independently receive system data and to oversee and sanction system decisions in relation to established safety criteria.

2.6 OPERATING MODES

The development of intelligent systems involving operator interaction such as high-powered manipulators for applications in areas such as forestry, loading derricks, mobile cranes, mining and construction plants will depend on the provision of system architectures able to integrate advanced sensors and safety with appropriate control strategies in relation to operation in unstructured environments. Additionally, these architectures must be designed to ensure the maximum flexibility of operation in relation to operator defined goals while ensuring effective collaboration with the operator.

While the ultimate aim in many instances may be fully autonomous operation, a number of discrete modes of operation can be identified, each of which require the introduction of some level and form of system intelligence. These modes are:

1. *Fully autonomous.* The machine is responsible for its own task planning once the general goal has been defined by the high level strategic processes or the operator.
2. *Operator controlled.* The machine is working under direct operator control with features such as configuration, sensitivity and type of motion selectable by the operator but with the operating envelope defined by the system.
3. *Operator collaborative.* The machine is working under direct operator control but superimposing its own commands and instructions onto those of the operator in order to facilitate performance.

In the first of these modes, the machine is required to interpret the conditions leading to the achievement of the defined goal and must therefore incorporate sufficient knowledge of the task environment in order to enable it to establish the procedures necessary to the achievement of the goal. This requires the presence of appropriate 'knowledge filters' in order to support the hierarchical decomposition of the task environment into the required sub-tasks and intermediate goals. The operation of the intelligent paper roll manipulator described earlier falls into this class of operation following the definition of the position of the delivered rolls by the scanning process.

In the second mode, the machine is relying on operator input for the achievement of the desired goals while at the same time managing the system to ensure that operation is as effective as possible. In this mode the need for the machine to interpret commands is minimised and the level of intelligence required by the machine is set accordingly. An example of this class of system is a fly-by-wire aircraft such as the Airbus A320 of Fig. 2.17. In this case, the flight characteristics of the aircraft are defined and modelled within the control system and pilot commands interpreted in relation to the region of the flight envelope in which the aircraft is currently operating, for instance to prevent excessive climb rates and angles at low speeds.

It is the third of these modes that perhaps places the greatest demands on the ability of the system to integrate its machine based intelligence with that of the human operator. Consider for instance the operation of a telescopic

Figure 2.16 Eurofighter Typhoon.
(*Courtesy of BAe Systems*)

Figure 2.17 Airbus A320.
(*Courtesy of Airbus Industries*)

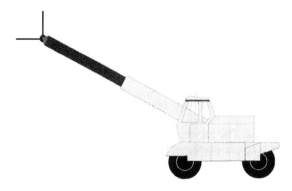

Figure 2.18 Telescopic handler.

handler such as that shown in Fig. 2.18. The reach of such systems means that even with the replacement of the conventional $R\Theta$ control by the more intuitive XY control option, the precise placement of pallets generally requires an assistant to give directions, typically by hand signals or radio, to the operator. If instead, the system was able to recognise the task and to superimpose corrections on the operator's positional control signals as appropriate, then the need for the assistant would be removed. Such interaction does however demand that the system recognise in some way the intent of the operator in issuing command instructions and contain the ability to modify the machine level responses accordingly.

No fully capable system involving the capture of operator intent currently exists with the agile combat aircraft perhaps being the closest realisation to date.

2.7 SUMMARY

The development of intelligent autonomous and semi-autonomous systems capable of operating safely in both structured and unstructured environments involving various levels of operator interaction and collaboration is dependent on the achievement of appropriate levels of machine intelligence as well as on the integration and transfer of functionality between the mechanical, electronic and software domains suggested by mechatronics. In particular, systems must be structured around an architecture which is capable of accommodating a range and variety of systems and operating modes and which is capable of supporting the associated sensing and safety requirements.

The incorporation of machine intelligence to improve operator effectiveness is to a large extent based on the devolution of responsibility for system function to the system itself in a form which is generally transparent to the operator, thereby freeing the operator to concentrate on the higher-level strategic and tactical functions associated with the operation of the system. It is therefore suggested that a mechatronic approach to the design and development of intelligent systems is a major and important factor in achieving the required levels of functionality in such systems.

2.8 CASE STUDY: AN INTELLIGENT ROBOT FOR UNSTRUCTURED ENVIRONMENTS – AUTONOMOUS ROBOTIC EXCAVATOR

Robots are conventionally used within a manufacturing process for the accurate manipulation, positioning and assembly of components and sub-assemblies and for physically arduous or hazardous tasks such as painting and stacking. In the above cases, the environment in which the robot is required to operate is, to some significant degree, structured around the needs and performance of the robot, for instance to prevent human access within the working envelope. In most applications, the operation of robots in manufacturing is primarily concerned with achieving positional accuracy at high levels of repeatability in relation to a highly repetitive and well defined task. In such applications, features such as an ability to exert a controlled force are not a normal operational requirement. This has resulted in an emphasis on control techniques and control strategies which emphasise positional accuracy, repeatability and speed of response onto which features such as collision avoidance, using integrated sensors, and an associated path planning capability have been added.

Robotic applications in unstructured environments, such as those being considered in industries such as construction and forestry or for special applications such as nuclear and chemical plant decommissioning and fire fighting, will be required to become more responsive to their local environment and to assume responsibility for strategic and tactical decisions related to both the task and the environment. Such systems would be required to operate in what is usually a highly unstructured environment within which they must co-exist and collaborate both with other machines and with human operators and workers.

In addition, these categories of task differ significantly from those associated with a production process in that they will in general involve the direct application of a force, for instance in excavation, a long reach and the manipulation of heavy and cumbersome loads. Finally, because of the nature of the task environment, the manipulator arm would typically, but not always, be mounted on a mobile base which itself may well need to be accurately located and oriented in order for the manipulator arm to function effectively.

The automation of the excavation process is a particular example of an intelligent robot system operating in an unstructured environment and one which will be used throughout the course of the text to illustrate and develop the material presented in the individual chapters.

2.8.1 Part 1: problem definition and statement of requirements

The introduction of increased level of factory type automation, including robotic systems, within the civil and construction industries is considered to offer a number of benefits including:

● the removal of site personnel from hazardous locations;
● improved quality and consistency of operation, particularly in areas where there is a shortage of skilled labour;
● increased productivity;
● better planning of work sequences;
● more effective management of the construction process.

Areas for the application of automated systems in construction include the use of automated and robotic plant for the purpose of excavation and trenching. The development of such plant requires that consideration is given to three specific areas as follows:

● the excavation process;
● control systems and strategies;
● site systems.

The excavation process

Figure 2.19 shows the sequence of events associated with the excavation of a trench within which the key element is the ability to:

'Fill the bucket as fast and efficiently as possible'.

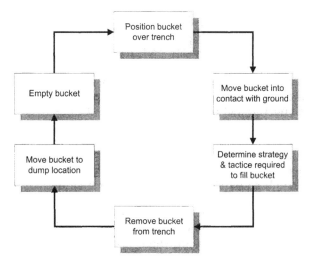

Figure 2.19 Trenching cycle.

Observation of a range of operators ranging from the highly skilled to those with little experience suggests that skilled operators normally control up to three degrees-of-freedom simultaneously with expert operators occasionally operating four degrees-of-freedom simultaneously. It is also clear that while vision is the main source of feedback to the operator, they made use of other performance indicators such as engine noise and the motion of the vehicle in reaching decisions about the sequence of actions they would undertake. In this context, it is worth noting that skilled operators could operate successfully, after relatively few trials, with a chest pack controller enabling them to adopt positions outside the vehicle where they could better observe the motion of the bucket.

Studies of operators indicated that that there were two basic strategies employed for soil removal; namely:

- penetrate and rotate;
- penetrate and drag.

These basic strategies are illustrated in Fig. 2.20 and are summarised in the following sections.

Penetrate and rotate

This strategy is used primarily in soft ground such as sand which can easily be penetrated by the bucket. Using this strategy, the operator would first force the bucket into the ground in a near vertical position. Once penetration to a suitable depth had been achieved, the bucket would be rotated as in Fig. 2.20(a) to fill the bucket. The filled bucket would then be removed from the line of the trench, emptied and the process repeated.

Penetrate and drag

The penetrate and drag strategy is illustrated by Fig. 2.20(b) and is used in harder ground which the bucket cannot easily penetrate. Using this strategy, the operator would first penetrate the ground with the bucket and then adjust the angle of attack of the bucket in the ground to 'plane off' material as the bucket was dragged back towards the body of the excavator. This strategy is much more complex in its demands on the operator, requiring the simultaneous control of more degrees of freedom than the penetrate and rotate strategy.

The major variations associated with the penetrate and drag strategy involved altering the angle of attack of the bucket and the direction in which the bucket was dragged. Investigation of the performance of the machine and, in particular the direction of motion associated with the ability to exert maximum force at any point within the operating envelop of the machine showed that the optimum direction of drag was, in most cases, a line drawn from the current position of the bucket through the boom joint on the machine as shown in Fig. 2.21. Observation confirmed that this was the general line of drag adopted by skilled operators, particularly in difficult ground.

These basic strategies were then modified as required by the operators in response to changes in ground conditions and working environment, for instance in response to the presence of below-ground obstacles.

Obstacle handling

The excavation process is complicated by the presence of obstacles such as boulders and tree roots which are only revealed as the excavation proceeds. Human operators develop their own individual ways of dealing with such obstacles and are readily capable of modifying their approach to adapt to the particular set of conditions encountered. Any autonomous system must

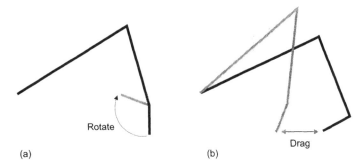

Rotate

(a)

Drag

(b)

Figure 2.20 Excavation strategies.

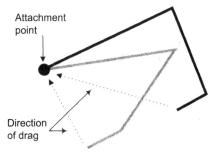

Attachment
point

Direction
of drag

Figure 2.21 Optimum direction of drag.

therefore contain a strategy or strategies for dealing with such obstacles and must, like the human operator, be capable of adjusting these in response to local conditions encountered.

Other factors to be taken into account are described in the following sections.

Control system

Many of the skills and techniques demonstrated by operators are developed through experience to the point where most of their actions are instinctive and not as the result of any conscious thought. The system controller must therefore direct the motion of the excavator both in the ground and through the air in such a way as to achieve a performance comparable with an expert human operator. This requires that, for an autonomous system, the controller must be structured so as to contain the necessary strategic and tactical knowledge associated with the excavation and trenching operations and be capable of integrating that knowledge with the direct servocontrol of the system actuators to achieve the desired motion profile at any instant.

Site systems

The operation of an automated excavator must be integrated within the overall context of site operations and must therefore be capable of integration with site systems at all levels, including other automated items of plant and equipment. Of particular importance is the ability of the excavator to precisely locate and orient itself on-site in order to carry out its task.

A mechatronic design process | 3

3.1 INTRODUCTION

In recent years, products of every type, from the humble toaster to motor vehicles, have become increasingly more complex. Indeed, it can be argued that modern engineering products are typified by their complexity. There are many reasons for this change, but the two key factors are the competitive nature of the global market and the advent of small and inexpensive microprocessors and microcontrollers. By embedding microprocessors, microcontrollers and other related devices such as application specific integrated circuits (ASICs) and digital signal processors (DSPs), it has been possible to dramatically increase the performance of a wide range of engineering systems and has led to the concept of the mechatronic product. Indeed, it is now commonplace to see what were once purely mechanical or electro-mechanical products incorporating microprocessors and the associated software. In many cases, this change has been evolutionary, with software based control and monitoring enhancing functions one by one while in others it has led to the development of entirely new products. However, the penalty is often an extremely complex system that is difficult for any single individual to fully comprehend.

A key consequence of these developments is that mechatronic products are no longer the sole responsibility of the mechanical or electronics designer as two further ingredients have been added in terms of the microprocessor hardware and its associated software. The interfaces between the engineering disciplines are now more crucial since overall performance is dictated by all elements functioning and co-operating with each other. However, the design process adopted by many organisations is still based on the traditional model where only one or two disciplines have to interact. This is not to suggest that this process is incorrect, rather that the emphasis has changed.

Complex products must now be considered as systems made up of contributing elements from the various engineering disciplines that must function together to produce a successful product. The strength of such a systems approach is that it provides a very powerful framework for understanding complexity by allowing different engineering disciplines to be portrayed together. Furthermore, one of the advantages offered by the mechatronic solution is the ability of the product to cope with uncertainty in its operating environment and this increasing importance of the environment also dictates a systems viewpoint.

In Chapter 1, some of the basic concepts and ideas associated with mechatronics and the engineering design process were introduced. The purpose of this chapter is to expand on this introductory material and propose an integrated framework for mechatronic system design. In particular, it concentrates on the process of taking a customer requirement or need and translating it into a feasible and manufacturable design. This process has many names in industry including *integrated product development, product introduction process* or *new product introduction,* but here the term *mechatronic system design* will be used.

Initially, a number of current design models will be discussed together with their strengths and limitations. A review of these models matched against actual practice shows there to be subtle but significant differences. These features are then used to develop a framework that deploys a number of models to overcome the limitations of the basic process type model. The chapter then concludes by considering a mechatronics design strategy based on this framework.

3.2 ENGINEERING DESIGN MODELS

As discussed in Chapter 1, there exist a large number of models that purport to describe the engineering design process. It is worthwhile considering some of these models in more detail since the historical context is important in understanding their strengths and drawbacks. A significant aspect of the range of process models developed to date is the predominance of 'design' models. Engineering design has always been considered a 'difficult' subject largely because it not only involves engineering science but also imagination; something often believed to be the domain of the artist and not of the engineer! Since design is an early activity in the product life cycle, the need to get the design right has been recognised for many years. Accordingly, there has been much effort to determine the 'best' approach to design without inhibiting the creative element necessary for a successful design.

An important group of design models are those which consider the design process; the sequence of events that take place in translating a customer need or requirement into a manufacturing specification. These design process models attempt to provide a framework for the design activity which is not too prescriptive, while giving guidance as to the best approach.

One of the earliest, and perhaps best known, of the design process models is that proposed by French in 1971 and shown in Fig. 3.1. The model outlines the key stages or phases and the information flows between them in a way that provides guidance without over-prescription.

Early design process models have established the basis from which most of the accepted developments have arisen. Certainly through the 1980s, with work such as that by Finkelstein and Finkelstein and into the 1990s with Hales and Pugh among others, the view of the design process as a sequence of stages or phases became well established. Although iteration was an accepted part of the process, it was implicit that these iterations would take place predominantly within the design phases.

One aspect of the model as proposed by French is that it only covers the design process up to the production of working drawings. Pugh addresses this by including manufacture within his model. This gradual recognition that the whole life cycle requires consideration has led to the concepts of concurrent or simultaneous engineering. As outlined in Chapter 1, the core principle of concurrent engineering is that issues of manufacturability should be taken into account from the earliest stages of conceptual design. However in reality the

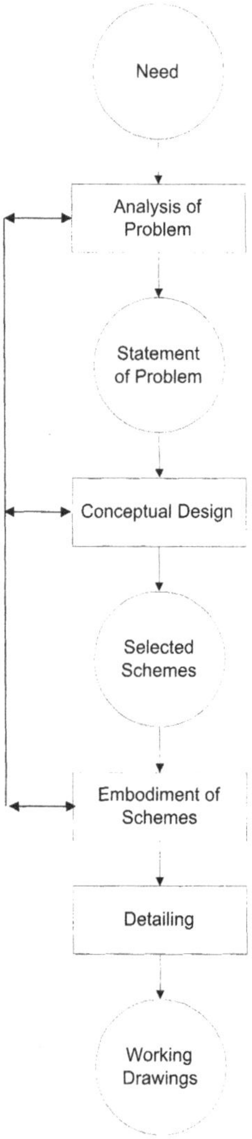

Figure 3.1 French's design model.

process model remains basically linear with activities performed in a series of stages or phases.

While Pugh's design model incorporates manufacture and sales as part of the process, it still does not contain all the aspects of the 'typical' product life cycle. This is not to suggest the model is wrong, just that it emphasises the issues that Pugh wished to consider. This emphasis, achieved by hiding elements of a model, is important as it allows other aspects to be highlighted.

Another classic example of this type of approach is the 'V'-system model shown in Fig. 3.2. Developed by NASA, this model emphasises the fact that the achievement of requirements needs to be supported by verification at each stage or phase in the development process. A further model is that proposed by Matra-BAe Dynamics. Called the 'Information Pipe Model', this is shown in Fig. 3.3 and attempts to focus not on the process but on the maturity of information and the need to consider all the relevant aspects or viewpoints of a problem. The pipe represents information generated on or about the product or system as the design evolves. This information may have many viewpoints as indicated by the various sub-pipes. As progress is made down the pipe from left to right, the information becomes more mature and well defined, starting

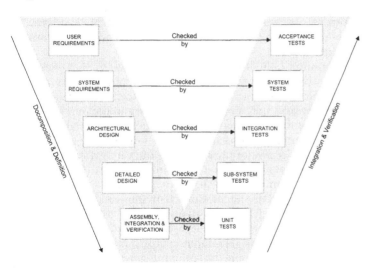

Figure 3.2 The 'V'-system model (adapted from NASA).

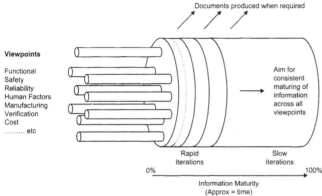

Figure 3.3 Information Pipe Model.

from the left hand end where little is known, to the right-hand end where the system or product is defined.

The strength of this model is the number of features that it captures simultaneously. Firstly, the model shows that information about the product or system should mature consistently over all viewpoints as time progresses. If one aspect is ignored, then unforeseen factors may well emerge at a later stage. Equally, if one aspect is allowed to 'race ahead' it may lead to constraints on the other aspects, resulting in a non-holistic system (i.e. a sub-optimal one). The model also indicates that at the low maturity end of the pipe, rapid iteration across all viewpoints is likely and desirable since making changes at this early stage is not expensive. As the information becomes more mature, then the iterations become less frequent.

The emphasis on information indicates the importance of information management during a project. Good information management will permit the necessary consistent maturity of information and help in ensuring that viewpoints are not neglected. The management of information is also crucial to decision making, and the model further tries to stress the importance of decisions rather than documentation.

Despite its many obvious positive features, this model, like all the others, does not provide the complete picture. In particular it de-emphasises the process view necessary to understanding the basic sequence of events during the life cycle of a product.

Surprisingly, and despite this predominance of models, there has been relatively little research performed into what actually happens during the design, development and manufacture of systems or products. However, some significant work was carried out by Arthur D. Little Corporation in which a selection of US companies with a significant record of innovation permitted researchers to observe their team-based ideas generation and decision-making processes from the beginning of the requirements definition and conceptual design process. From this work, described by Nolan and others, the 'synectics' approach has been promulgated. This places great stress upon the removal of obstacles to creativity by the inculcation of an atmosphere of mutual respect and trust within a team. The aim is to avoid dominance and take away the fear of ridicule from those who might wish to put forward 'way-out' ideas.

Much use is made of ideas generation by association, even to the extent of postulating an empathy with the hypothetical viewpoint of people or entities far removed from the situation. For example, the question could be posed:

'How would the captain of the QE2 view this problem?'

This could even extend as far as creating an empathy with the 'personalities' of inanimate objects such as a fruit in which the colour, springiness and texture of the surface of an orange may have something to 'say' to a requirement for a self-cleaning surface.

To some this approach may seem childish, but perhaps childlike would be a more apt description, since the aim is to break out for a time from the effect of years of educational and task-oriented conditioning.

In the UK, Hales provided some evidence to support the view of the existence of separate stages of design through an empirical study using a participant observer approach where data was collected on a single design project over a period of almost three years. In this study the activities of the participants in the design project were classified, allowing the effort relating to the engineering design process to be extracted. Hales concluded that the 'ideal' engineering design process could be represented by a series of curves defining the effort applied in each phase with time as in Fig. 3.4. This supports the sequential view of the engineering design process that is implied, if not explicitly stated, by the classical prescriptive models.

The work of Hales attempted to match design practice to the various 'theoretical' models of the design process. It required a considerable amount of effort to collect the data; moreover the study was restricted to a single design project. Gardam established an alternative approach to characterising the design process. The phase analysis of Hales is based upon the assumption that processes can be characterised by the activities within them whereas Gardam's approach is based on the relationships between activities and in particular the outputs from these activities. This research provided an additional investigation of the extent to which the classical design process models are followed in practice. Moreover, the study encompassed a number of projects of a multi-disciplinary nature and was able to show that there has been a shift away from the classical design process model, particularly with respect to the mechanical design element.

This shift has occurred relatively recently and therefore could have occurred after Hales conducted his study. Nevertheless the contemporary design process demonstrated a significant overlap between requirements capture and concept design and also between embodiment design and detailed design as suggested by Fig. 3.5. However, there was no corresponding increase in overlap between concept design and embodiment design.

These changes have also coincided with substantial proportionate increases in the effort applied to task clarification and embodiment design while detail design effort has fallen in real terms. In summary, Gardam reports that there has been a move away from a four-phase process to a two phase process and argues that in the case of embodiment-detail design, this is due to the increased use and sophistication of computer based tools. The overlap between task clarification and concept design does not suggest a change in activities, rather that more concurrency is evident.

A study of product introduction processes by Burge and Woodhead has highlighted additional features of the contemporary design process. This study involved modelling the business processes of a number of multi-disciplinary products through structured interviews with staff at different levels within the organisation. Since the study was concerned with business processes and activities outside the boundaries of the 'engineering design process' it was able to identify considerable design activity in response to customer tenders. Figure 3.6 shows a very much simplified model of the business processes found.

Although the companies investigated were all defence contractors, it was apparent that increased design effort was being spent in the bid/feasibility stage since it is during this stage that the major technical, cost and timescale decisions are made. These decisions lead to contractual commitments and thereby define the overall risk of a project. To reduce this risk, increased importance was being placed on requirements capture and concept design.

3.3 A MECHATRONIC DESIGN FRAMEWORK

What is clear from these 'theoretical' and 'empirical' models is that, as suggested in Chapter 1, a mechatronic design process cannot be fully described by

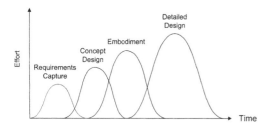

Figure 3.4 Design phases (*After Hales*).

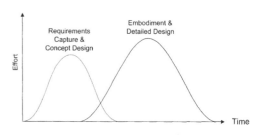

Figure 3.5 Contemporary design process (*After Gardam*).

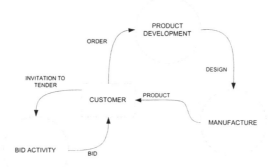

Figure 3.6 Simplified product introduction process.

a single model. In order to understand the mechatronic design process it is necessary to use a family of models to emphasise and explain the different aspects that this highly complex discipline demands.

The process is obviously very important to mechatronic design yet a process model on its own, while it may represent current practice, suffers from being a prescriptive model that implies that if the activities are followed a 'good' product or system will result. This is clearly untrue since there are further important ingredients to consider. Firstly there are the people involved, secondly there are the tools that those people use, and thirdly there is the information generated and lost.

A fourth ingredient, that of management, could also be included. However, it can be argued that many of the activities that are currently considered to be management are in fact an integral part of the process. There currently remains a tendency to separate the engineering and the managerial aspects of a project and to assume that different people should be responsible for these. Accordingly, those staff involved in the technical aspects of design regard some of the activities that are currently classified as management tasks as an imposition on their role and even on their creativity. Project management, risk assessment and reliability are considered by some engineers to be 'somebody else's problem'. This view, although still prevalent amongst many engineers, is manifestly untrue and is a consequence of not integrating these tasks.

One further issue is that of flexibility. Irrespective of how mechatronic design is modelled, the model must be able to adapt, or change, to suit the particular task at hand. To highlight this issue the term framework will be used with the implication that what is proposed within the framework is not rigid but flexible.

The proposed framework is based around the following four *interdependent* absolutes:

- processes;
- people;
- tools;
- information.

These are referred to absolutes because, without these, no system evolution is possible while the term 'interdependent' has been highlighted since it is vital for success. For example, it is recognised that the design of a complex multi-disciplinary product relies heavily on the design team; the diversity of technologies makes it impossible for any single individual to approach the problem. Yet a team cannot function without tools. A group of mechanics given a non-functioning motor vehicle but no tools, would only be able to theorise as to the nature of the problem. With the correct tools these theories can be tested and the problem solved.

Equally a good team must be able to share tools. This does not imply that everybody in the team can use each other's tools, but that there must be some common tools with which all the team are comfortable. The team also requires information, since during all stages of a system life cycle it is information and the maturity of that information that allows for successful progression to the next phase. The interdependence between tools and information is therefore critical. Indeed, it is through the use of common tools that team members share and communicate information.

These four absolutes therefore form the basis of the framework which is shown diagrammatically in Fig. 3.7 from which it is seen that information, or more specifically maturity of information, is at the heart of the framework. It is the output of the various design processes that are supported by tools that in turn are used by people as part of those processes. The 'information arrow' terminating in the people segment of the diagram implies the sharing of information. This is a vital element of the framework since it is only by the generation, maturing and sharing of information that the design of complex multi-disciplinary systems can be accomplished. Indeed, Fig. 3.7 also presents a framework for what is now called the learning organisation.

The process element of Fig. 3.7 can be expanded to provide a more conventional process flow diagram. This is shown in Fig. 3.8 and can be viewed as an extension to the type of model produced by French, but taking into account empirical findings such as those of Gardam and Burge.

The key aspects of the model of Fig. 3.8 are:

- the identification of the typical top level processes and their relationship to the generic life cycle phases in a similar fashion to the conventional design process models, but including the actual practice as observed by Gardam and Burges in particular including the bid/no bid risk assessment;
- the recognition of key outcomes and deliverables and of decision points;
- the inclusion of links to and with customers, partners and sub-contractors.

It is important to note that this model is only an element of the design framework and should not be used in isolation. Its purpose is to support the generation, maturing and sharing of information amongst the people who operate the process and that appropriate tools to support the process are made available.

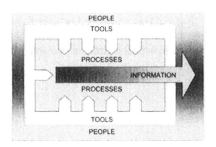

Figure 3.7 The mechatronic design framework.

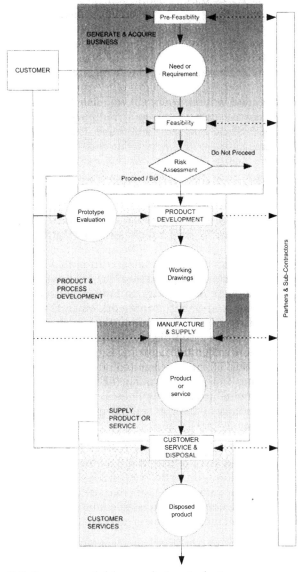

Figure 3.8 Process model for mechatronic design.

3.4 SUMMARY OF PRINCIPLES

A number of design process models have been discussed and their strengths and limitations highlighted. A review of such process models matched against actual practice shows there to be subtle but significant differences. This evidence has led to the formulation of a mechatronic system design framework based upon the four interdependent absolutes of:

- processes;
- people;
- tools;
- information;

and these are shown to be the essential ingredients for a successful mechatronics design process.

3.5 APPROACHES TO DESIGNING MECHATRONIC HARDWARE

The discussion thus far in this chapter has been to introduce high-level approaches observed or employed in the process of design. As such they are not restricted to the mechatronic field but their contribution to it may be very significant. The mechatronic environment is substantially concerned with sensing and actuation systems in which a loads definition, however constructed, leads through a set of procedures, the outcome of which is the achievement of a system comprising hardware and software which achieves the required function and, moreover, whose performance on first start-up holds few surprises because it has been predicted during the process of design.

To even the experienced designer, when viewing the design task at the early conceptual stage, the choice of implementation appears very daunting because of the number of options which may appear to fulfil the requirements. Academic researchers, industrial practitioners and teachers of project management disciplines generally agree that the conceptual stage of design, at which perhaps 5–10% of the total project cost in bringing a product to market is expended, is crucial in determining the project's success.

What the engineering designer needs is to arrive as quickly as possible at a *scheme* which will satisfy the functional requirements of the project and which also meets the non-functional requirements such as cost targets. The functional requirements will include basic parameters which can be set in terms of static or steady state conditions such as force, torque, rotational speed, linear velocity, peak acceleration and also requirements which are most appropriately evaluated by simulation, such as frequency response and stability.

The non-functional requirements may include commercial constraints such as preferred suppliers, component costs in relation to quantities and the manufacturing profile of the organisation. In considering these points, the 'make or buy' decision will be influenced by the availability in-house of suitable manufacturing processes and the appropriate skill base. Concurrent engineering embraces the general principle of considering the functional design requirements, the selection or design of the manufacturing system and the procurement strategy at the same time, or as concurrently as possible.

The process models for mechatronic design which describe an overall sequential process, imply a number of feedback loops where repeated, and ideally rapid, evaluation procedures take place. Here the emerging partial solutions to the problem are tested against the partitioned 'statement of requirements'. This is the core for the conceptual design phase in which several candidate schemes are proposed, developed and evaluated to the level of a scheme. French has defined a scheme as:

> '... an outline of a solution to a design problem, carried to a point where the means of performing each major function has been fixed, as have the spatial and structural relationships of the principal components. A scheme should be sufficiently worked out in detail for it to be possible to supply approximate costs, weights and overall dimensions, and the feasibility should have been assured as far as circumstances allow. A scheme should be relatively explicit about special features or components but need not go into much detail over established practice.'

This statement establishes the close relationship between *function* and *means* and will form the underlying principle on which the following parts of this chapter, its arguments and examples, are based. How a design challenge is approached will be determined by the extent of the constraints which the design authority wishes to place upon the process. For example, the problem statement :

'Find a means of transferring patients to–from hospital trolleys and beds without the Health Authority having to face claims for staff back injuries.'

has fewer visible constraints, at least to begin with, than a statement such as :

'Replace the timing shaft drives in our printing lines with shaft encoders and variable speed motors'.

The process of getting from the original 'client' problem statement to a formal statement of requirements, if it is to be effective in satisfying the client, must involve a thorough and searching stage of determining the client's full and complete requirements, suitably prioritised. This could be described as 'intention elicitation'. When misunderstandings arise, it is often found that this process had not been undertaken sufficiently early or that the design process was well underway while some key areas of the requirement were insufficiently defined.

If the time constraints are sufficiently open, then the statement of requirements should be kept uncompromised by any decision on the means of accomplishing it until both designer and client are satisfied that it is as complete as it may be and both have the same understanding about it. The phrase 'delighting the customer' encapsulates a company's mission to meet the customer's requirement in full, including affordability and timeliness, but with perhaps some pleasant surprises included which the customer had not been able to make explicit at the beginning. Chapter 4 sets out in detail how this can be accomplished by the use of techniques such as viewpoint analysis.

From this point on, the well known procedures for matching scheme embodiments against the design criteria set out in the functional statement of requirements can be followed. This may include rating and weighting techniques provided that a degree of scepticism is retained about the validity of the numerical results. The value of such techniques may lie more in their effect in driving the description of a concept down to a 'kernel' description which can often be usefully be supported by a minimal diagram encapsulating a working principle.

Design according to Dieter, Slocum, and others addressing the processes of machine hardware design, should proceed according to a set of systematic steps, rules and methods in which a number of guidelines should be followed. These guidelines are summarised below together with a commentary as to how they may be applied within the mechatronic environment.

The method should have the attributes described in the following sections.

3.5.1 Be generic and not specific to a particular field

The mechatronic environment is a broad field, comprising energy stores of various types, fluid power and electrical actuators, mechanical transmissions and motion converters, sensors, signal processing and conditioning circuits, control hardware devices, control algorithms and software, all contained within a system topology which provides for the functionality and spatial relationships between the sub-assemblies and parts.

3.5.2 Facilitate the search for optimum solutions

The search for optimum solutions in the mechatronic environment may be very complex. This is especially so, for instance, in the field of mechanical devices such as motion converters comprising variable speed drives that work by friction. The high level working principle of transferring tractive effort through a frictional contact is readily identified though the lower level, and hence more detailed, means of accomplishing this is not. There are many configurations of these devices resulting from a long history of invention and development, seeking improvements in efficiency and durability or inventors navigating around existing patents.

The electronic environment is more orderly as witnessed by the number of books and design guides which give advice on the implementation of well known circuit principles, such as pulse-width modulated (PWM) control for motors.

The task of the designer is therefore to delineate the solution field in which to operate. If it is cast too wide, the task will become very time consuming. Only with the universal adoption of electronic data processing as declared in Section 3.5.7 has it become possible to escape from the confines of textual and diagram-based hard copy, but nevertheless useful and interesting, design guides which were popular especially in the USA in the 1960s . These were often set out in effect as recipes with sections such as 'ten ways of measuring pressure' in the hope or expectation that a designer would find some inspiration, by use, adaptation or combination of the unfocussed solutions offered. It was left to the designer to embrace the working principle and take it forward to a practical embodiment of a scheme.

It is now possible to generate knowledge bases of a size limited only by the effort which organisations are prepared to put in to creating them. These can then be searched algorithmically according to the principles discussed in Chapter 5.

3.5.3 Not constrain inventiveness and experience

The construction of knowledge bases of working principles, especially when the principles are mapped on to means and in effect made synonymous with the means of performing an objective function, may be regarded as being an artificial constraint on the process of design. Indeed, the knowledge base is a distillation of experience and can only be constructed in the first stages by knowledge elicitation from human experts.

The argument may therefore run that to work within a knowledge base is indeed constraining. Where does 'new knowledge' come from? What is invention? These are questions in which a variety of conflicting positions have been taken. In support of the utility of a knowledge base, it is likely that in most organisations the extent of useful knowledge made accessible by a well designed expert system would be greater than any single individual could deploy. This also raises the interesting question, to be considered later, as to how commercial companies view knowledge bases and expert systems.

3.5.4 Be compatible with other disciplines

Mechatronics is above all an integrating approach to design in which the complexities of physical mechanisms generating motion paths and the use of discrete analogue electronic components can both be simplified in the physical embodiment of a product. Embedded algorithms for motion path generation and the use of DSP-based control for industrial electric motors are examples of how this has been achieved. The computer based design tools which will be used in the mechatronic environment will therefore themselves be increasingly domain independent and able to work from a high level statement of requirements expressed in terms of parameters such as load, velocity or bandwidth. Object-oriented (O-O) approaches are used in which generic components can be stored with ports defined in terms of effort and flow variables with the component object containing its embedded transfer function. The assembly of the system representation within the computer then becomes a matter of matching port types. One such system under development, Scheme-builder® Mechatronics, will be described later in the chapter.

3.5.5 Not rely on chance

This ideal is unlikely ever to be realised as even with the use of global means for knowledge transfer via the Internet, there will always be large areas of knowledge which will remain inaccessible to an individual, even if only because of its sheer volume. Within an individual company, or within a co-operating manufacturing association, systems of shared knowledge are feasible covering proven approaches to solving well established design tasks. Organisations such as the British Fluid Power Association (BFPA) co-operating in European bodies such as CETOP have established design principles and dimensional standards to ensure inter-operability of equipment and performance standards to cover parameters such as linearity. The scope for competition then becomes centred on special refinements in performance, efficient design for manufacture and commercial criteria such as product pricing, availability and distributor support.

3.5.6 Facilitate the application of existing solutions to related tasks

With the increasing mobility of technical staff, corporate technological memory is shortening rapidly. Indeed, it has been said that with the process of development of a new aircraft from first requirements specification to flight certification taking on average 15 years, it is unlikely that anyone having a technological role at the start of the project will still have any connection with that role at the end, even if they are still part of the organisation. Tools to enable design recall and re-use are therefore essential if the benefits of prior work are not to be lost.

An important aspect of this is where an 'old' idea was discarded or not pursued further because the technology of the day made it uneconomic or infeasible. An example of this is Torotrak, described in Chapter 12, where an elegant mechanical concept has been made viable by the application of a combination of embedded microprocessor technology, developments in advanced lubrication and surface engineering. This is an example of what Dieter calls the '… application of a new principle to an old problem'.

3.5.7 Be compatible with electronic data processing

The design process is largely based on a search of solution spaces for pre-established practices and design embodiments and finding ways of combining these to produce a new or improved result. A number of ways are available in which this can be done including an expert systems approach based on accumulated and classified knowledge, and an approach based on neural nets or genetic algorithms which progressively refine a solution to meet a set of objective functions.

3.5.8 Be easily taught and learned

It is unrealistic to expect design engineers working in industry to enter the arcane world of the AI specialist. The inference and search techniques employed have therefore to be transparent to the user who needs to be provided with a user interface which allows inputs to be made and outputs presented in familiar language and formats.

3.5.9 Allow objective evaluation

It is the objective in the present developing computer aided engineering (CAE) environments to make 'right first time' a fully achievable goal. This means that within the software it must be possible to represent the hardware prototype in as complete a fashion as possible against all the objective criteria laid down in the statement of requirements. For example, a physical prototype car body with major elements manufactured in fibre-reinforced polymer may represent aspects of aerodynamic performance when subjected to a test in a wind tunnel, but in no way can it represent the response of the structure to physical load inputs including extremes such as crash-worthiness.

Advanced CAE tools can predict performance in a range of such areas with increasing confidence. The need to test varieties of physical prototypes to confirm an optimum design has thereby been very much reduced. Even so, difficulties with the first release of the Mercedes 'A' Class have suggested that extensive testing of complete systems under all conceivable ranges of user input remain essential.

A design tool for the mechatronic environment will be of very limited utility if it can only achieve matching of a system against static criteria. Dynamic simulation will be necessary in most cases involving mobile products and manufacturing systems especially where these comprise high speed processes such as printing, papermaking and textiles.

3.5.10 Encourage creativity and intuition

The question as to whether computer based tools can aid invention is contro-versial. According to Dieter, truly inventive steps are rare and usually occur when the inventor has been struggling with a problem for some time until ' ... a freak occurrence or chance observation provides the needed answer', or ' ... suddenly gains a valuable insight or discovers a new principle that is not related to the problem he is pursuing'.

A number of attempts have been made to facilitate new invention by creat-ing a taxonomy of principles on which existing inventions have been based and providing the designer with interactive guidance to search these for sets on to which the user-defined requirements can be mapped. One such is the 'Invention Machine' based on world-wide patent searches carried out in the former USSR. A more pragmatic view would see the main benefits of com-puter-based design tools towards creativity as being the relief of the drudgery and fatigue associated with searching hard-copy records, standards, databooks and catalogues.

In addition, the increasing capability of CAE tools, now often PC platform-based, allows the designer to immediately make explicit the results of 'three-dimensional (3D) thinking', perhaps a special attribute of successful mechanical design engineers. The downstream software compatibility with tools such as AutoCad and ProEngineer now allows the design engineer to enter a continuum of applications from the 3D representation to the automatic generation of 2D piece part manufacturing drawings if required together with bills of materials, and links to finite element mesh programmes for stress, thermal and fluid flow analysis. Thus the designer should have more time to think creatively and eval-uate the potential results of design decisions before a commitment is made.

3.5.11 Reflect modern management thinking

The commercial time constraints acting on the design process in a highly com-petitive manufacturing environment have accelerated the move from hierarchi-cal to matrix management structures but have also increased the a pressure on functional specialists to broaden their scope. One of the problems with a matrix organisation such as that of Fig. 3.9 is the competition for specialist design resources which often exists between the managers of the individual project teams, the resolution of which may occupy much time at executive director level.

In contrast, the 'ship' or hierarchical organisation for project management of Fig. 3.10 is structured such that a dedicated project team contains much more of the detailed technical resources needed to complete its task. Some intermediate variations on these structures are found, and the principal points in their applic-ability are briefly summarised in the following sections.

The project matrix system

The project matrix system has good attributes:

- It allows resources to be more easily re-allocated by senior management on conclusion of a project.
- It enhances the acquisition and retention of specialist expertise within the functional groups and makes this readily available as a service to the project managers.
- It facilitates design re-use and standardisation.

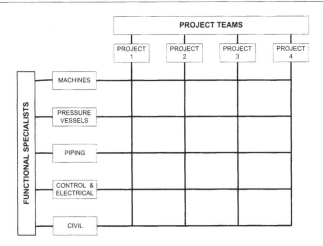

Figure 3.9 Functional matrix project management structure.

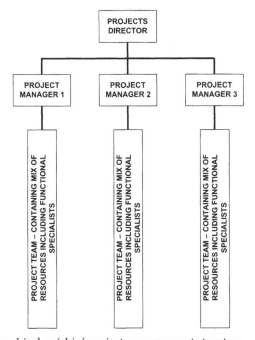

Figure 3.10 Hierarchical or 'ship' project management structure.

● It strengthens supply chain links to manufacturers of commonly used lines of equipment.

Possible difficulties are:

● potential for compartmentation between the functional groups and 'over-the-wall' attitudes akin to the pre-concurrent engineering design/manufacture dichotomy;
● potential for conflicts of authority between functional group managers and project managers over resource allocation, e.g. insufficient has been provided or it is not of the right technological balance or quality;

- the competition for resource is more complex, involving at least three interest groups, the project managers, the functional group heads and the corporate managers who have to hold the ring somehow and allocate overall resources when they may not have the necessary technological insight to make optimal judgements;
- perceived loss of authority by project managers/engineers and conflicts between functional group heads and project engineers over preferred technologies or equipment suppliers.

The hierarchical or 'ship' system

This system has recognised good attributes:

- The larger, more self-contained, team generates an esprit de corps, hence the 'ship' analogy, which may improve the chance of the particular project meeting its specification, time and cost objectives.
- The authority of the project manager/engineer is perceived to be increased, thereby in the best circumstances calling upon and successfully deploying traditional charismatic leadership skills.
- Reduces the 'fight for resources' to a two-cornered affair between project managers and senior level or board management.

Possible difficulties are:

- Core expertise in specialist areas is acquired in an idiosyncratic manner and is transferred to succeeding projects in a fragmented way as teams are split and re-formed. In a 'lumpy' product development or capital programme, the expertise may be lost altogether if the organisation has a hire-and-fire culture.
- The possibility for technological innovation may be inhibited. Unless there is a strong R&D function well aligned with overall company objectives, the transfer of expertise and cross-fertilisation which may occur when a new project team is formed from parts of several previous ones may be dependent upon how the multi-faceted lower level technological conflicts are resolved and how long this takes.
- The capture of re-usable design information becomes far more difficult.
- The breadth of the supplier base may increase unmanageably according to the individual preferences of strong project managers. Thus the commercial benefits of standardisation are less attainable.

The matrix organisation is observed to perform well in environments such as the chemical or petroleum industry where the technological focus for development is on the process chemistry rather than the physical means of performing the operations. The latter tend to follow well-established routes, hence the division into functional specialist groups to carry out the design tasks. As the primary clients have increasingly 'de-layered' their organisations in the 1980s and 1990s, so the functional group structures have been reflected more and more in the contractors and consultancy group companies which serve them. It is now the contractors who have to face the problems of 'lumpy' demand which is handled by the payment of premium rates for less secure employment and which may also be offset by the phasing of the international growth – recession cycles. The hierarchical system is observed to be in use more with companies involved in longer project timescales with a greater degree of engineering innovation, such as the automotive and aircraft industries.

In any organisation which is committed to new product development, and which intends to remain competitive, the question needs to addressed as to what are the most appropriate procedural tools for the organisation to use in response to the challenge of its market sector. This is the field of systems engineering, Fig. 3.11, which overlaps with the purview of mechatronics at the more detailed implementation level. Evidently there must be an overlap between the two, but some boundary may be apparent at which the functional specification is taken into, or invoked by, a set of procedures or an information-based tool which will assist the designer in finding the best implementation to satisfy the design requirement.

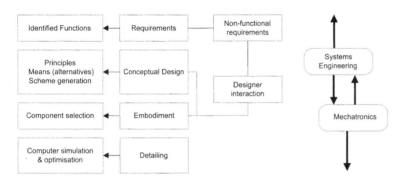

Figure 3.11 Systems engineering and mechatronics.

Over the whole spectrum of types of organisation for project engineering and management, the need for soundly based information, based on proven historical precedents or underlying physical principles is very evident. A number of developments are underway of information based tools to provide a top-to-bottom guide to the design engineer of which an example is 'Schemebuilder® Mechatronics'. This methodology aims to promote design reuse, automatic recording of the design steps and decisions made and to be a rapid prototyping tool for the automatic generation of dynamic simulation models of the complete system concept. The purpose is to aid product innovation and the diversification process of finding new system applications for existing and new products. This infrastructure of Schemebuilder Computer Aided Conceptual Design (CACD) can be used to build expert systems for design fields which are mature and can be considered to rely very heavily on design reuse rather than innovation.

In these cases a finite number of functional requirements can be identified in order to produce a satisfactory design. Hydraulic and pneumatic systems have been successfully researched in this context and the prototype 'Schemebuilder Expert: Fluid Power Module', whose development is explained in more detail in Chapter 12, provides the designer with a choice of functional requirements which have been validated with industrial experts. This approach does aid both design re-use, best practice in design and concurrent engineering. Although such expert modules provide a fast track to a conceptual design with a high level of design re-use, they do not allow the designer freedom to invent where new solutions are needed.

Instead, the expert knowledge in the form of *design principles* (rules) acts as a form of advice to the designer while allowing the freedom to create new principles and search through a database of past solutions. In this way the designer can reuse an existing solution which may be advised by the expert or create a novel solution by the use of fundamental understanding of the expert advice.

This approach, 'Schemebuilder® Mechatronics' uses a sophisticated object-oriented software infrastructure, which integrates CLIPS, Visual C++, MATLAB/SIMULINK and a web browser. This infrastructure comprises a design methodology consisting of functional decomposition represented in the working principles and alternatives for seeking design solutions. The means of performing functions has been integrated with the knowledge of components stored in an expert system for control system design. At the component level, the objects are stored with attributes comprising their transfer functions and port types, thus ensuring that the assembly of schemes can only be accomplished in accordance with underlying fundamental physical principles. Within the approach, the aim is to balance the confidence which is delivered by the expert system approach together with a freedom to explore potential new solutions.

The expert system for controller design uses design principles in the form of rules based on root-loci and non-linear design methods and with industrial expert experience to design linear and non-linear control systems for single input single output (SISO) systems. This controller design approach is object-oriented and is synergistic with Schemebuilder® Mechatronics object-oriented modelling of mechatronic systems using MATLAB/SIMULINK. The object-oriented modelling tool in SIMULINK uses new definitions of port types and a novel bandwidth modelling approach to overcome the requirement for complex symbolic analysis which often conflicts with the requirements to model discontinuous behaviour. This behaviour exists in mechatronic systems through physical effects such as power saturation in power control and backlash in transmission systems.

Schemebuilder® Mechatronics is a tool for interactive design and provides the user with a graphical user interface (GUI) consisting of a scheme building site, incorporating a function means tree, expert advice and help using the web browser and a knowledge base editor to allow the designer to enter past solution knowledge in the form of new working principles. It also allows a designer to move from concept to rapid prototyping by automatically building a dynamic simulation model which uses a mechatronic library of means and components in SIMULINK. This facility assists the designer in evaluating the relative performance of different schemes.

Having generated a non-linear dynamic performance model of the scheme, Schemebuilder® Mechatronics allows the designer to use optimisation techniques to fine tune the performance. Lancaster University Engineering Design Centre has focused on applying mechatronics to control and servomechanism design, flight control systems and heating, ventilation and air-conditioning (HVAC) systems. The development of the mechatronics and expert elements of Fig. 3.12 have involved industrial experts to validate the expert system and give comments on usability.

Schemebuilder® Mechatronics has been used in an industrial case study to improve the accuracy and repeatability of a precision servo-pneumatically actuated glass forming machine, which has resulted in the field implementation of a novel control algorithm in servo-pneumatics, an application not traditionally regarded as suitable for the control of high speed proportional precision motion with significant inertia loads. Figure 3.13 shows how such systems can be reverse-engineered within Schemebuilder® in order to abstract

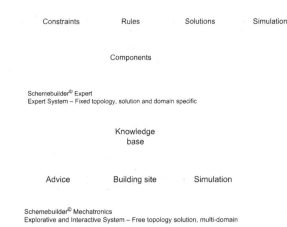

Figure 3.12 Schemebuilder® structure.

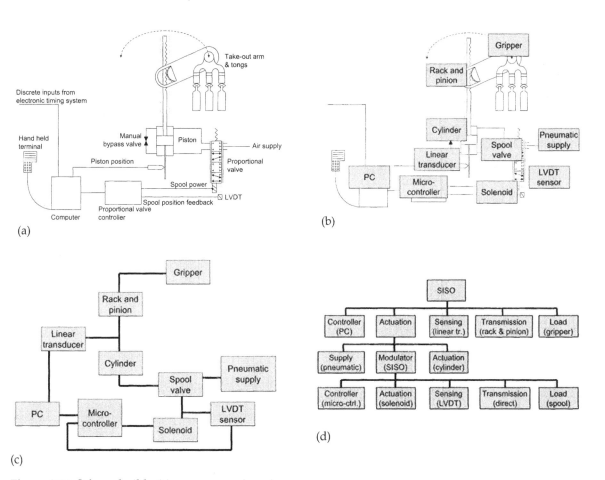

Figure 3.13 Schemebuilder® in reverse engineering.

their existing working principles and find new solutions by a process of trial and replacement.

Thus the structured, i.e. object-oriented, methodologies traditionally used in software engineering to provide a top-down approach, are very applicable to conceptual system design involving requirements capture, functional modelling and component selection. The ability of this method to hand-shake efficiently with object-oriented approaches for system modelling allows a virtual prototype computer model to be rapidly produced ready for system performance assessment.

In addition, the simulation of a process allows the impact of component choices on a system's performance to be assessed and is very useful in concurrent engineering to allow rapid feedback to and from the customer regarding technical performance and cost issues. Normally, modelling is an expensive exercise and Schemebuilder® allows the system models to developed using a library of component models, enabling the overall system models to be automatically generated, with the designer needing only limited knowledge of the simulation language, for example SIMULINK. Moreover, such tools can provide a cost effective new way of working between equipment supplier and systems manufacturers, especially if the information can be exchanged via the Internet using hypertext mark-up language (HTML) format. Within this development outline perhaps lie several of the principles on which future steps in practical expert system and AI aids to the design engineer can be based.

Requirements interpretation for mechatronic systems

4

4.1 INTRODUCTION

An important part of any design process is the need to properly understand and interpret customer requirements. Failure to gain a full understanding of customer requirements will lead to an unsatisfactory product or system, irrespective of the excellence of its design or manufacture. Moreover, errors committed by not accurately interpreting customer needs are typically not identified until the product or system is with the customer, resulting in their dissatisfaction.

There are several inter-related reasons why such misunderstandings occur. These include:

- hidden requirements or requirements perceived as 'obvious' by the customer not verbalised or otherwise expressed;
- ambiguous requirements;
- lack of consideration of whole life cycle aspects;
- limited attention to the 'total environment' in which the product operates where 'total environment' is used to indicate a systems engineering view which covers such aspects as human factors, supply chain, maintenance, logistics and so forth;
- the lack of suitable tools to address these issues.

Mechatronic products and systems present a further challenge because of their complex and multidisciplinary nature which increases the need for accurate requirements interpretation. Also, in mechatronic system design there are usually several groups of people who have a vested interest in the final product or system. There can be up to four categories of these stakeholders:

- the sponsor seeking the development of the system or product;
- the end users;
- the technical experts who have an input to the design or are responsible for technical evaluation;
- the financial controller responsible for the release of funds.

In this chapter a framework and tool set for the requirements interpretation of mechatronic products is suggested and outlined. The tool set comprises a number of simple yet powerful techniques that allow engineers and other team members to capture and expand a set of requirements. The diagrammatic nature of some of the tools provides an excellent communication method among the design team members, and where appropriate to the customer.

The chapter will also show the position of requirements interpretation in the mechatronic systems design framework presented in Chapter 3. Through consideration of life cycle aspects and the nature of customer requirements, it is suggested that current approaches to requirements analysis are generally not sufficient to fully interpret customer requirements and that the lack of a structured and systems approach, particularly in the case of complex multidisciplinary situations, leads to an increased risk of product failure. Requirements interpretation is therefore more than just about understanding customers needs, but is also concerned with allowing the design team to reach a common understanding.

These features are used to develop a framework for requirements interpretation that is based around the four interdependent absolutes of people, process, tools and information introduced in Chapter 3.

Central to the approach is the use of a range of tools that support the consideration of the requirements from a number of different angles, permitting a clearer, more consistent and unambiguous understanding of the requirements together with methods for communicating within a team and to the customer. These tools are:

● viewpoint analysis;
● functional analysis and modelling;
● sensitivity and failure mode analysis;
● textual analysis.

Viewpoint analysis and functional analysis and modelling have their origins in software engineering where the importance of requirements capture has been recognised for some considerable time. There are many 'software related' tools available; however, the two considered here are simple to use and apply equally to engineering systems as to software systems. Their main strength lies in the use of diagrammatic methods to simplify and convey structure, allowing for greater clarity which in turn aids consistency and reduces ambiguity.

Viewpoint analysis in particular provides a method by which the various stakeholders and their issues can be captured and structured to improve completeness and is particularly useful in identifying whole life cycle issues. Specifically, it provides a means by which the views of the stakeholders can be captured and structured, improving the completeness of the requirement. Functional analysis and modelling is particularly powerful in that it can identify missing or hidden customer requirements. It also allows for the development of a system model that can form the basis for further design development.

Sensitivity and failure mode analysis, while not new of itself, is here applied at an earlier stage in the overall design and development process than is conventionally the case. The benefit arises from the early identification of both technical and project risk, together with an indication of potential trade-off studies that can arise through the presence of conflicting requirements.

Conventional 'textual analysis' is included since the familiarity of engineers to this approach is an important factor. Indeed, users must feel comfortable with any tool if it is to be successful.

4.2 CUSTOMER REQUIREMENTS AND THE PRODUCT LIFE CYCLE

As discussed in Chapter 1, people have been designing, developing and manufacturing products for a considerable period of time. Despite all this experience, products still fail to meet customer expectation and frequently cost more than planned, something which has been exacerbated as products have become more complex. While there are many reasons for this situation, there is no doubt that the failure to fully understand the customer requirements is a significant factor.

Most design process models such as those of French, Hales and Pugh discussed in Chapter 3 and product life cycle models such as that given in ISO 9000 show the analysis of customer needs to be one of the earliest activities. It follows that errors committed during these early stages will, if unidentified, be successively built upon in subsequent stages as suggested by Fig. 4.1.

This obvious statement becomes critical when it is realised where or when such errors are discovered. Evidence suggests that these early errors are often difficult to identify and Fig. 4.2 shows a plot of the phases or stages of product development against the logarithmic of cost of error removal and indicates that the earlier an error is identified, the less costly it is to put right. However, errors committed early in the development process are often only identified at much later stages, as suggested by the dotted lines in Fig. 4.2. Indeed, it is frequently found that errors committed during market analysis and requirements capture and interpretation are only identified when the product reaches the customer.

Figure 4.1 Errors in the life cycle.

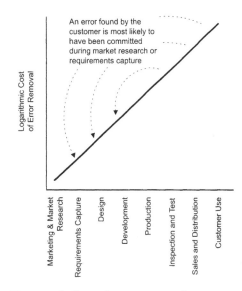

Figure 4.2 Cost of error removal.

Two main questions arise from these observations:

- Why are errors committed during these early stages of product or system development?
- How is it possible to determine that an error has been committed during these early stages, if it can only be identified much later?

The answer to the first question includes factors identified earlier as contributing to misunderstandings, namely:

- hidden requirements or requirements perceived as 'obvious' by the customer not verbalised or otherwise expressed;

- ambiguous requirements;
- lack of consideration of whole life cycle aspects;
- limited attention to the total environment in which the product operates.

The solution to the second question is to note that the outcomes of Fig. 4.2 are based on current traditional approaches to product development, and it is these processes themselves together with a lack of suitable support tools and combined with increased product complexity that are significant causes of error.

People are also critical. Pugh has likened the outcome of the early stages, the design specification, to a ' ... mantle or cloak that envelops the subsequent stages in design ...'. Far too often, designers put the cloak on too early in their rush to realise a product which is later found not to meet customer requirements.

In summary, failures in fully understanding customer requirements are hard to identify because of the lack of suitable processes and tools to support requirements interpretation. However, before considering such a process and tool set, it is necessary to understand the nature of customer requirements and the current approaches to capturing these.

4.3 CUSTOMER REQUIREMENTS

Requirements are defined as constituting a specific need or want. Typically they are written documents and as such are often structured in a manner that can easily lead to mis-interpretation of the customers' requirements. Moreover, the resulting document will frequently not contain all the required information nor place the information given in an order of preference. Unstated requirements, or a misunderstanding of the relative importance of the requirements, is often quoted as a major reason why a particular requirement may not be satisfied.

The need for a product will, in general, arise from one of two principal sources, the producer or the customer. Where a product is intended to meet the demands of a *broad market*, the producer will normally generate the requirement. Broad markets are typified by mass produced products such as motor vehicles and consumer goods, therefore attempting to fully satisfy every prospective or possible customer is likely to be difficult, if not impossible. Indeed, a compromise requirement that satisfies the majority of customers is usually the best that can be achieved. However, the best compromise must always be sought. Companies that sell in broad markets normally establish a marketing group or team to determine customer requirements. This group of people should:

- determine the need for a product;
- accurately define the market sector and demand, since doing so will determine the grade, quantity, price and timing for the product;
- accurately determine customers requirements by reviewing previous products and market needs including the identification of any unstated expectations or biases held by prospective customers;
- communicate customer requirements clearly and accurately to the design team.

The last point results in the generation of a formal statement of the customer requirements covering items such as:

- the function or purpose of the product;
- the required performance characteristics, including environmental and usage conditions, reliability expectations, disposal, maintenance requirements and cost;
- aesthetic characteristics including style, colour, shape and form together with human factors;
- all applicable standards and statutory regulations;
- packaging;
- quality requirements.

This customer requirement, sometimes called a product brief, is the input to the feasibility stage of the mechatronic system design process model.

Where the product is to be supplied to satisfy a particular contract there would normally be a single customer and therefore a *narrow market*, hence the customer will usually write the requirement. Narrow market products are typified by products such as military equipment, power stations and specialist items of plant. The resulting formal statement of customer requirements should however cover the same points as identified for a broad market product.

Pugh identifies another dimension to customer requirements with the idea of static and dynamic concept designs. This argument suggests that some products are conceptually static at the system level and quotes the motor car as an example with the Model T Ford of 1908 being conceptually identical to a Ford of the 1990s. There are also products that are conceptually dynamic. An example here is the recent shift from single-hull car ferries to multi-hull designs. An important feature of the argument is that there is inevitably an element of dynamic design at the sub-system level in what are conceptually static products. These points reflect back to customer requirements in that they too will reflect the static and dynamic nature of the product.

These two categories of requirements can be combined to form the two-dimensional classification model shown in Fig. 4.3. This also serves to indicate that products that fall into the dynamic category will tend over time to migrate to the static category. Indeed, the use of a compact disc or CD player as an example of a dynamic design may now be incorrect since the basic concept has become well established. Equally, the static condition is not permanent and there are examples where a conceptual 'jump' can introduce a new concept for a previously static design. The example of multi-hull car ferries previously referred to earlier is a typical case of such a jump as for many decades the concept of a car ferry comprised a single hull and would have been considered as static in design terms. Such conceptual jumps normally come about as a result of the development of new materials, new manufacturing processes, new technologies or improved engineering analysis of the basic design problem.

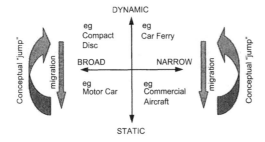

Figure 4.3 Two-dimensional requirement model.

This categorisation is very crude but does serve to show the extremes. In all categories, however, a common feature is that, irrespective of how they are generated, the outcome is the same, a set of written requirements.

It is these written requirements that are passed to the design engineers to interpret by what French calls 'analysis of problem' or in current terminology 'requirements capture'. The definition provided by French of this activity is still correct, but he also states that: 'The analysis of the problem is a small but important part of the overall process'. This view is not entirely supported by the designers of complex products and an engineering director of a major international company has said that:

> ' ... with out doubt our biggest problem is making sure we understand what the customer wants, and also making sure the customer understands what they want'.

For engineering products a common method of attempting to capture requirements is through the textual analysis of the requirements document leading to a cardinal point specification. While it can be argued that this approach has been demonstrated to be successful, the increased complexity of products and the commercial risks associated with failure suggest that new approaches to requirements capture are necessary. Indeed, while textual analysis still has a role to play it has a number of drawbacks including:

- Information is gathered in a relatively haphazard manner.
- There is little or no analysis of the information collected.
- The use of natural language is ambiguous.
- The final product specification is unstructured and textual.
- Such structure that does exist does not clearly show the interrelationships between system elements.
- There is little or no customer feedback, thus the specification is not verified.

This can result in one of two possible situations arising:

- If the design specification does not reflect the customers' requirements then the WRONG system will be designed.
- If the design specification is ambiguous, then the design WILL NOT meet the intended customer requirements.

The result in each case is that the design engineers generate concepts and final designs that do not reflect what the customer actually desires. The proper elicitation of the customer requirements is necessary in order to develop a design specification that is:

- clear;
- consistent;
- complete;
- unambiguous;

and is therefore essential to the mechatronic design systems process.

When considering the requirements for a complex system it is important to note that there exist a number of different types of requirement:

- operational requirements which define the major purpose of the product or system, that is why the system is required in the first place.
- functional requirements which specify what the system has to do in order to satisfy the operational requirements.

Wash and Go
Requirement for intelligent washing machine
The marketing department has identified a small but potentially prof-
itable need for an intelligent washing machine that is capable of
autonomous operation.

This machine will replace the existing 'top-of-the-range' model. It will
be of standard size and take a standard load. The machine will be capable of determining the load make up and fabric type and of determining
the 'best' wash cycle. It will detect mixed loads and where necessary
inform the user of any changes. The cost must be within 15% of existing
machine prices. The style must be distinctive. The machine will use
domestic water and currently available detergents.

Standard electricity supplies will provide the power source. It will
operate at appropriate temperatures and wash cycles most suitable for
the fabric type, which will be determined by the machine. The user will
have the facility to check the wash cycle and override the machine decision. The machine will rinse and spin-dry the clothing as appropriate to
load and type.

- non-functional requirements which define system constraints or modifying
  influences on the system. There are various sub-sets of non-functional
  requirements including:
  (a) performance requirements that define how a particular function must
      be performed;
  (b) system requirements that define modifying influences which affect the
      whole system. These typically include:
      (i) safety;
      (ii) conformance to standards;
      (iii) reliability;
      (iv) human factors;
      (v) cost;
      (vi) size and weight;
  (c) implementation requirements which set out how a system is to be built.

In practice, these various types of requirement usually follow the pattern that
for each written functional requirement there is an associated statement about
the corresponding performance and implementation requirements. However,
the constraining non-functional requirements generally reflect the current
state of knowledge and therefore may be present in great detail or may even
be absent. Quite often several functional requirements are served by one non-
functional requirement and visa versa. As a consequence, the resulting requirements document is often confused and unstructured. Moreover, any analysis
that follows results in a design specification that is also confused and unstructured.

This situation is illustrated diagrammatically in Fig. 4.4 which illustrates the
confused and unstructured nature of a textual requirements document. Visi-
bility is poor and operational requirements are not separate from and placed
before the functional requirements.

To illustrate the type of requirements document often faced by design
engineers, consider the above statement of requirements for an intelligent

washing machine that will be used as an illustrative example throughout the remainder of the chapter.

It is apparent from this simple but typical statement of requirements that:

● There is no explicit operational requirement.
● Functional and non-functional requirements are stated haphazardly.
● There are both functional and non-functional requirements missing.

To overcome the problems with the traditional requirements model, it is necessary to establish a new model in which the relationships are clearly defined before analysis, so that the analysis is forced by the model. Functional requirements define what the system is required to do and must be in turn be driven by the operational requirements. The implementation must then be driven by the functional requirements and finally, the non-functional requirements that modify the system must be broken down into those that modify the functional requirements and those that modify the implementation requirements. This structured and visible set of requirements is shown diagramatically in Fig. 4.5.

In the structured model of Fig. 4.5, function is the master requirements set. Thus the starting point of any analysis must be to separate the functional and non-functional requirements. Moreover, methods must be found that allow the relationships between the various requirement sets to be both clear and visible.

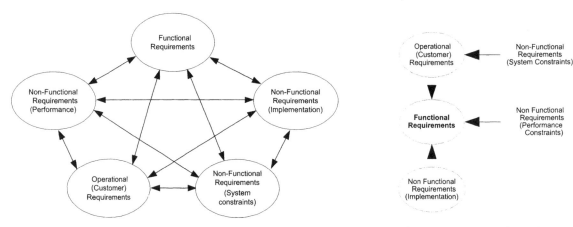

Figure 4.4 Relationships between requirements in a textual requirements document.

Figure 4.5 Structured requirements model.

4.4 CUSTOMER REQUIREMENTS AND THE MECHATRONIC SYSTEM DESIGN FRAMEWORK

The preceding sections have discussed the nature of customer requirements in order to arrive at a structured requirements model. It has also been argued that present approaches to understanding customer requirements and translating these into practical design solutions are not always appropriate for mechatronic products or systems and it is therefore necessary to consider the position of requirements interpretation within the context of the mechatronic systems design framework presented in Chapter 3.

The first stage in developing this framework is the feasibility stage or phase.

The aim of this stage is to provide sufficient evidence to allow an assessment of the risk of proceeding with the bid or design project to be made. It is absolutely fundamental to an organisation that whether bidding against a customer tender or responding to an internally generated design brief, that the risk to the company is determined and understood. The level of this risk can be determined by:

- gaining a clear understanding of customer requirements;
- developing conceptual designs that will satisfy those requirements;
- evaluating the conceptual designs to arrive at the preferred solution.

A development of the mechatronic system design process model to indicate the processes in the feasibility phase is shown in Fig. 4.6 in which the overlapping boxes are used to indicate concurrency. Figure 4.6 is essentially an activity model and is a necessary, but not sufficient, structure in the realisation of good engineering design and due account must also be taken of the balance of people, tools and information. Although many aspects of people, tools and information can be considered in relation to the various activities shown in Fig. 4.6, there are issues that do not fit into this prescriptive model.

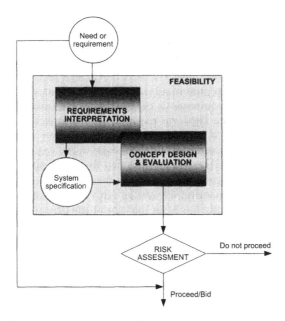

Figure 4.6 Feasibility phase process model.

Perhaps the most important of these issues is that of the people who will undertake the feasibility stage studies. Since it is implied that the system is multi-disciplinary, the team must comprise personnel with all the necessary engineering skills. Although the full scope of the product is likely to be unknown, it is usually straightforward to identify from the requirements a provisional operational requirement. This can then be used to formulate a feasibility team.

For example, consideration of the requirements statement for the intelligent washing machine suggests that a provisional operational requirement might be:

'To identify load make up and wash clothes cleanly.'

The requirement for autonomous operation implies mechanical, electronics and software design and therefore engineers with appropriate backgrounds should form part of the team. Some management is also necessary and could well be provided by one of the engineering team members. To ensure a concurrent engineering approach, there is no reason why manufacturing, production and quality engineers should not also be part of the team. Indeed, they may add dimensions to the feasibility study that the design engineers may not.

It is possible that at the early stage of the process there will be no significant involvement of commercial staff such as marketing personnel in the technical design team. Cooper, however, states that:

'non-engineering [sic] – must become part of, and involved with, the design teams of the future …'.

This view is wholeheartedly supported, but such people can be involved in the technical aspects of the study only if appropriate tools to support communication are available and are used. It is not in general difficult to justify the inclusion of various other staff members in the team but the project, and hence the size of the team, will generally be constrained financially. Team selection is therefore the first decision of the feasibility study and one which allows the processes of Fig. 4.6 to be undertaken.

Within the overall feasibility process, two sub-processes are identified: (a) requirements interpretation and (b) conceptual design and evaluation. It should be noted that the term 'requirements interpretation' is used here in preference to the more common 'requirements capture' since it implies expanding on the customer's requirements. The term 'requirements capture' is used when there is no extension or expansion of the customer's needs taking place.

The formal link between these sub-processes is the requirements or system specification, but it must be remembered that the overlapping boxes in Fig. 4.6 imply iteration and concurrency. Much has been said about conceptual design and evaluation in engineering. However, with the exception of the software industry, surprisingly little has been said about the process of, or the tools for, requirements interpretation. In consequence, it is perhaps the least understood and least recognised aspect of any design or feasibility study. Yet it is obviously the foundation on which any design is based and many products have failed to satisfy the customer because customer requirements were not properly understood. This is not to place blame solely with the engineers as customers are generally equally culpable. The remainder of this chapter will therefore concentrate on this omission by introducing processes and tools to support this important activity.

4.5 REQUIREMENTS INTERPRETATION

Requirements interpretation is defined as the activity of understanding what a customer requires and of expanding any statements made by the customer to produce a specification that can be taken forward to a product or system that will meet customer requirements. While this is certainly true, it has another very important use, that of allowing the feasibility team to reach a common understanding.

Every customer requirement will contain ambiguity, if only because natural language is used to describe the requirements, but also because the customer is not in general entirely clear as to how to express the requirements or is not even certain of what they require.

It has been suggested that ambiguity is the main cause of requirements not being met in software design. If that is the case, what chance does a complex multi-disciplinary product or system have! The approach proposed to overcome this problem is to use a range of tools that allow consideration of the requirements from a number of different perspectives. This permits for a clearer, more consistent and unambiguous understanding of the requirements, together with methods for communicating within a team and to the customer. As has already been stated, these tools are:

- viewpoint analysis;
- functional analysis and modelling;
- sensitivity and failure mode analysis;
- textual analysis.

Although the current emphasis is increasingly away from prescriptive approaches to design, some form of temporal framework must be used. The designers have to start somewhere but must recognise that iteration is highly likely and essential. It is therefore suggested that the requirements interpretation should begin with a textual analysis, concurrently proceed with viewpoint analysis and functional modelling before conducting a sensitivity and failure mode analysis. Iteration is important and the overall requirements interpretation process is implied by Fig. 4.7 with each tool supporting the other in the maturing of information about customer requirements.

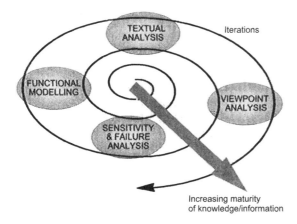

Figure 4.7 Iterative requirement interpretation process.

4.5.1 Textual analysis

Textual analysis is concerned with a line-by-line assessment of the requirements document with the purpose of interpreting, expanding and clarifying the individual requirements. It is also common to attempt to structure the requirements through appropriate numbering systems to produce what is frequently referred to as a cardinal point specification. The final outcome of a textual analysis is a specification suitable for further analysis or for conceptual design. However, the process of arriving at this stage is also very important since textual analysis is itself an iterative process where areas of ambiguity are identified and discussed with the customer for clarification.

In the approach proposed, textual analysis is not used in isolation but in association with other tools. A common approach to textual analysis is to start by separating out the various requirements. For example, in the case of the intelligent washing machine the basic statement comprises a number of combined requirements that can be separated into the following set of statements:

- It will replace the existing top of the range model.
- It will be of standard size.
- It will take a standard load.
- It will be capable of autonomously determining the make up of the load.
- It will be capable of autonomously determining the fabric type.
- It will be capable of autonomously determining the 'best' wash cycle.
- It will detect mixed loads.
- It will, where necessary, inform the user of any changes.
- It will cost within 15% of existing machine prices.
- It will have a distinctive style.
- It will use domestic water supplies.
- It will use currently available detergents.
- It will use standard domestic electricity supplies.
- It will operate at the temperature most suitable for the fabric type.
- It will operate the wash cycles most suitable for the fabric type.
- It will be possible for the user to check the wash cycle at any time.
- It will be possible for the user to override the machine's decisions.
- It will rinse the clothing.
- It will spin-dry the clothing.

These statements can be ordered in a number of ways. The structured requirements model of Fig. 4.5 provides excellent separation. For example:

1. Functional requirements:
    1.1 It will be capable of determining the load make up.
    1.2 It will be capable of determining the fabric type.
    1.3 It will be capable of determining the 'best' wash cycle.
    1.4 It will detect mixed loads.
    1.5 It will, where necessary, inform the user of any changes.
    1.6 It will be possible for the user to check the wash cycle at any time.
    1.7 It will be possible for the user to override the machine's decisions.
    1.8 It will rinse the clothing.
    1.9 It will spin-dry the clothing.
2. Non-functional requirements:
    2.1 System requirements
        2.1.1 It will replace the existing top of the range model.
        2.1.2 It will be of standard size.
        2.1.3 It will take a standard load.
        2.1.4 It will cost within 15% of existing machine prices.
        2.1.5 It will have a distinctive style.
        2.1.6 It will use currently available detergents.
    2.2 Implementation requirements
        2.2.1 It will use standard electricity supplies.
        2.2.2 It will operate at appropriate temperatures most suitable for the fabric type.

2.2.3 It will operate appropriate wash cycles most suitable for the fabric type.

This separation allows for the identification of ambiguous and 'missing' requirements such as:

1. Missing functional requirements:
   (a) loading method?
   (b) detergent determination and loading?
2. Missing constraints:
   (a) system
      (i)   reliability?
      (ii) safety?
      (iii) legislation?
   (b) performance
      (i)   cycle times?
      (ii) cleanliness?

Such lists may not be complete at this stage but they are asking key questions about the requirements. An obvious area for concern is the lack of information about the loading method which results in the questions:

- Is the machine to be a front or top loader?
- Are there other possible loading methods?

These questions arise from a classic form of requirement ambiguity. Firstly, the requirement may be absent because the customer wishes other options to be considered. Alternatively, the customer may have considered the requirement so obvious that it does not need stating. In either case, the customer should be encouraged to state what is required since the danger of not doing so is the generation of concepts that will not meet their needs.

As questions such as these are answered, the structured textual specification can be developed. However, despite structuring the specification, it is still a textual document and as such it remains difficult to see the true structural relationships between different aspects of the requirements. This clarification can be achieved through the use of viewpoint analysis together with functional analysis and modelling.

4.5.2 Viewpoint analysis

Viewpoint analysis is not new and is based upon the common sense perspective that, given a complex situation, only a partial understanding will be obtained from any single point of view. It follows that a more complete understanding will be obtained if several viewpoints are taken.

Consider a commercial aircraft, the perspective of the maintenance engineer will be significantly different to that of the pilot. For the maintenance engineer, ease of access and line replaceable units would signify a good design, whereas the pilot is primarily concerned with flying the aircraft and hence in its stability and control characteristics. From an overall systems viewpoint both perspectives are important.

Viewpoints are not restricted to human perspectives and the product itself also has a perspective, as might Government in the form of standards and legislation. While all these considerations are obvious and based on common sense, it is easy for other viewpoints to be missed through design teams being too technology biased or through pressures on time. It is therefore essential that viewpoint analysis is conducted within some form of structured framework and that suggested is adapted from that developed as part of the CORE (Controlled Requirements Expression) method.

The approach adopted is to separate functional from non-functional requirements and hence to show both the relationships between the functional requirements and how the non-functional requirements impact on them. This is achieved by defining two differing types of viewpoints as follows:

- functional viewpoints which support the logical partitioning of the system into modules that transform information. There are two types of functional viewpoint;
  (a) *bounding viewpoints* which provide an external view of the system; i.e. how the system looks as seen by the outside world;
  (b) *defining viewpoints* which present an internal view of the system and which are used to describe the internal functions of the system;
- non-functional viewpoints which represent groups of requirements that modify or constrain the functional requirements of the system.

The non-functional viewpoints are *not* transformers of information, instead they modify parts of the system. Notice also that implementation issues such as 'push button user interface' are not included: the method is only concerned with requirements.

The first stage in the process is to determine the functional and non-functional viewpoints by attempting to write down all possible viewpoints in the form of a *viewpoint bubble diagram*, and then to decide which of these are functional and which non-functional. However, more important is that this activity is performed as a team and is where all the various viewpoints from different team members and stakeholders are collected. A key point to note is that some of the viewpoints will come from customer requirements and other will come from experience. It must also be remembered that customer requirements are unlikely to be complete and 'missing' viewpoints will need to be introduced as necessary. It is also probable that any brainstorming process will generate viewpoints that are outside the scope of the problem. These should be identified and omitted.

Consider the example of the intelligent washing machine, the given set of requirements resulted in the viewpoint bubble diagram of Fig. 4.8 which contains a great deal of information that is not contained within the original requirements document. For example, viewpoints on maintenance and reliability are included that were not expressed in the original statement but which obviously must be considered. In effect, viewpoint analysis attempts to control the desire of design engineers to 'design without full knowledge' by forcing them to consider the problem first.

The next step in the process is to partition the diagram into functional and non-functional viewpoints as in Fig. 4.9 after which the non-functional viewpoints are temporarily discarded and consideration is given to logically structuring the functional viewpoints. The aim of this structuring is to develop a logical viewpoint hierarchy in order to split the analysis task into a number of specification levels and thus be in control of the amount of detail at any one level.

The first step in structuring the functional viewpoints is to identify the bounding and defining viewpoints as shown in Fig. 4.10 after which the viewpoints are grouped together as in Fig. 4.11 such that at any one level there is a maximum of five to seven viewpoints.

The penultimate stage is then to convert the structured viewpoint bubble diagram into a viewpoint structure chart of the form shown in Fig. 4.12. In practice it is found that the bounding viewpoints occupy one level only in this chart.

The final step in viewpoint analysis is the structuring of the non-functional viewpoints. The requirements model discussed earlier indicated that a good specification would result only if the hierarchy of functional requirements was 'matched' with a hierarchy of non-functional requirements. To satisfy this condition it is necessary to structure the non-functional viewpoints as closely as possible to the structure of the functional viewpoints.

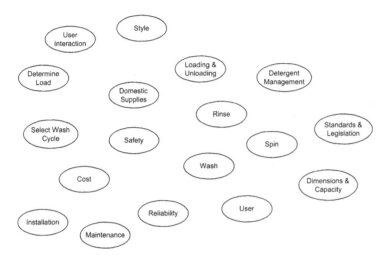

Figure 4.8 Viewpoint bubble diagram for washing machine.

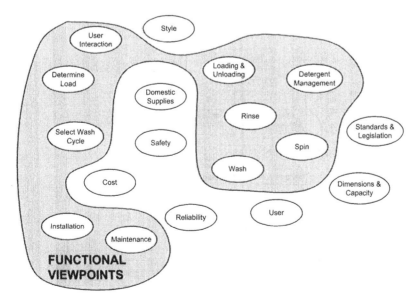

Figure 4.9 Separated viewpoint bubble diagram for washing machine.

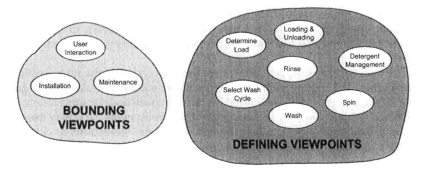

Figure 4.10 Separation of functional viewpoints for washing machine.

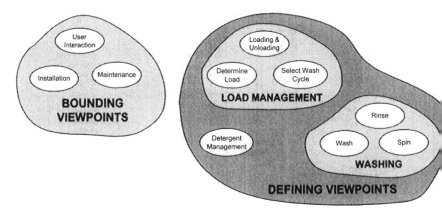

Figure 4.11 Grouping of functional viewpoints for washing machine.

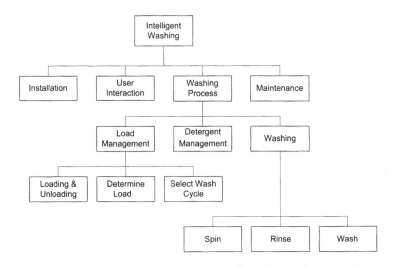

Figure 4.12 Functional viewpoint structure chart for washing machine.

For each 'level' of non-functional viewpoint it is also necessary to specify the corresponding constraints. These are developed from the customer requirements document together with other sources of information such as national and international standards. It is common that a customer will miss or omit important non-functional requirements in drawing up their initial statement of requirements.

The outcome of viewpoint analysis is a diagrammatic representation of relationships between functional elements in the system/ product together with the constraints that affect them, allowing each to be considered individually and in detail. Moreover, viewpoint analysis is a simple tool that can be used by a whole team, irrespective of the individual team member's background. This team approach results in the sharing of information to force a common understanding of customer requirements.

4.5.3 Functional analysis and modelling

Developing a functional model of customer requirements is a very powerful method for understanding and clarifying those requirements and the interrelations between them. Moreover, a good functional model of the 'problem' can also form the basis for conceptual design since as design decisions are made the model can evolve to reflect them. Since the prime concern is with mechatronic systems, the modelling method must satisfy a number of technological considerations. In particular it must:

- be sufficiently abstract to capture both physical and information characteristics;
- be able to include human and other non-technical forms of interaction;
- present the information in a clear and simple fashion.

There are many modelling techniques for systems, several of which have emerged from the software industry such as CORE, SADT, SSADM, VDM and IDEF. There are also more complete approaches to modelling such as Checkland's soft systems methodology. Most of these techniques provide the necessary tools to construct a model of the system under consideration. The important point is that the model replaces the need for a human to commit to memory the details of a system. Hence any ambiguity can be removed and consistency and completeness achieved.

However, most of these techniques require a high degree of expertise in order to gain the most from them. Since the concern in requirements capture and analysis is with design engineers and other non-technical contributors, any modelling method must be simple to use and such that all the team members feel comfortable with its application and use. The method described here is that of DeMarco and was originally developed for the software industry but is universally applicable to any system irrespective of its constitution.

When used as a requirements capture tool, the DeMarco method has a strong relationship with the need to structure the requirements. The use of diagrammatic methods assists in separating the functional from the non-functional requirements and makes these highly visible, permitting omissions and errors to be readily identified.

The approach has limitations, but when used within a framework such as that under discussion provides a unique method for modelling the mechatronic system throughout its design cycle and of providing an interface to other tools for risk, reliability and cost analysis among others.

Modelling concepts

Before discussing the detail of the method, it is important to reflect upon some very important modelling concepts. These are discussed in the following sections.

Diagramming

Natural languages are ambiguous and the use of textual methods to convey complex information will often lead to errors due to the inability of text based descriptions to convey structure. Diagrams on the other hand are very good at conveying structure.

Information hiding

Most systems are too complex to represent on a single diagram. However, humans are able to understand any degree of complexity provided it is presented in small 'chunks' together with a 'map' of how the chunks are to be brought together. When combined with diagramming, information hiding allows complex systems to be represented in a logical hierarchical fashion. The highest levels of the resulting family of diagrams omit the fine detail and concentrate on the essential information at that level. The information that is hidden at the higher level is then revealed in lower-level diagrams. This approach has many names including top-down, modular and abstraction. The latter name however also has another very important meaning.

Abstraction

Although sometimes used imply information hiding, abstraction is used here to support consideration of different physical characteristics using the same diagramming conventions. For example, a voltage signal or hydraulic flow can be abstracted as a information flow on a diagram using the same conventions as a user command.

Structured analysis

The aim of structured analysis is to produce a readable, maintainable, complete, consistent, and unambiguous model of an existing or planned system of any size. In order to achieve this, the three tools described below are used.

Dataflow diagram

The dataflow diagram (DFD) is a network representation of the system which portrays the system in terms of its component processes or activities together with the information flows between those processes. The basic elements that make up a dataflow diagram together with their graphical conventions are shown in Fig. 4.13 and are:

- the dataflow, a vector indicating a well defined information flow;
- the process which transforms incoming dataflow(s) into outgoing dataflow(s);
- terminator(s) which act as a source or sink of data external to the system;
- datafile(s) which serve as a temporary repository of data.

The role of the DFD is that of presenting a description of the system from the viewpoint of the data which flows through it rather than from the viewpoint of the system. Indeed, it de-emphasises the flow of control within the system.

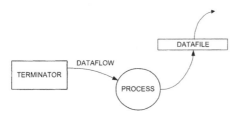

Figure 4.13 Dataflow diagram conventions.

Procedural information is defined in the process specification. The important point is that most system inefficiencies and ineffectiveness occurs at interfaces between processes. The DFD highlights all the interfaces and therefore the areas where problems will occur, or are likely to occur.

A process is an abstract concept that is concerned with describing the functions that a system has to perform. The DFD shows the various functional requirements of a system and how they are related. As the DFD is concerned with the flow of information it can be used just as effectively to model a set of manual processes, and the interfaces between them, as it can a set of automated processes. The dataflows can also be abstract in that they can represent physical qualities such as electrical supplies or fluid flows as well as information conveyed electronically or by paperwork. It is the ability of the DFD to show these various mixtures that make it a powerful tool.

Large systems require that a 'top-down' treatment is used to produce a hierarchical set of DFDs. Thus, a single DFD is used to represent the whole system at a level of detail which can be realistically represented on one diagram. The 'processes' of this top-level DFD can then be considered as sub-systems. Each of these sub-systems is then taken in turn and decomposed into further processes to form a lower-level DFD. These lower levels can incorporate other, simpler sub-systems or individual processes. This activity of successive decomposition is continued down to the 'primitive' level where processes cannot usefully be further decomposed. They are then judged to be small enough to be rigorously described with a process specification.

Structured analysis provides a number of rules for constructing DFDs which includes a simple numbering system and rules for ensuring consistency between diagrams. It also introduces a very important feature in that the top of the set is a single diagram comprising a single process, the system, showing all the incoming and outgoing dataflows to that system. This top level diagram is called the context diagram and defines the boundary of the system. The context diagram fulfils two important tasks. Firstly, it reflects the operational requirements of the system and secondly it shows how the system interacts with its environment.

Data dictionary

The data dictionary (DD) is used to specify precisely what is meant by, that is what comprises, every dataflow and datafile on every DFD. In particular, it specifies the data elements which are contained in each dataflow or each datafile. Just as the DFD effects a top-down decomposition of the processes within the system, the DD effects a top-down decomposition of the data.

At the highest levels, the dataflows or datafiles are defined as groups of subordinate items consisting of one or more data items that can or cannot be subdivided. This decomposition of dataflows into data items is continued until no further decomposition can usefully be achieved, at which point 'data' elements are defined. The dataflows and datafiles represented on DFDs and defined in the DD can be abstract in nature. For example, there is no reason why a piece of steel bar entering a machine tool cannot be considered as a dataflow and the machine tool as a process. Equally, the machined part coming off the machine tool constitutes an outgoing dataflow.

The DD makes use of a restricted set of simple relational operators which can be used in combination to construct more complex operators that allow for the definition of any possible dictionary entry. The set comprises the following operators:

IS EQUIVALENT TO
AND
EITHER-OR
ITERATIONS OF
OPTIONAL

An example of a DD definition would be:

Telephone_directory IS EQUIVALENT TO:
ITERATIONS OF:
EITHER: Business_Name
OR: Personal_Surname
AND: OPTIONAL: Initials
AND: Address
AND: Telephone number

However, it is clear that documenting the definition in this form is tedious, hence a more concise notation is actually used in the DD. This is:

=   means   IS EQUIVALENT TO
+   means   AND
/   means   OR
[] means   EITHER-OR (that is: make a selection of the contents of the brackets)
{} means   ITERATIONS OF the item enclosed
() means   OPTIONAL use of the item enclosed

Thus, for the telephone directory example:

Telephone_Directory = {[Business_Name/Personal_Surname + (Initials)] + Address + Telephone Number}

Process specification

The process specification (PS) specifies the component processes. The DFDs are successively decomposed to greater levels of detail until a 'primitive' process is characterised and each and every one of these primitive processes is then described by a PS, which defines the transformation that converts incoming dataflows into outgoing dataflows. The PS defines the policy governing transformation but not the method of implementation and is also where any non-functional requirements such as constraints, are introduced. Thus the non-functional requirements are related in a structured way to the functional requirements.

The PS is documented using structured English or some other suitable language; normal English is too ambiguous and should never be used. Structured

English is a specification language that makes use of a limited (English) vocabulary and limited syntax. The vocabulary consists of:

● imperative English language verbs;
● terms defined in the data dictionary;
● reserved words for logic formulation.

The syntax of a statement is limited to:

● procedural sentence;
● closed-end decision construct;
● closed-end repetition construct;

or combinations of these. Structured English is therefore very much like a high-level programming language.

Together, the DFD, DD and PS allow the construction of a model that is consistent, complete, and unambiguous. By the very nature of these tools, the model also serves as a descriptive document for the system as suggested by Fig. 4.14.

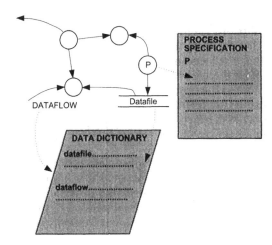

Figure 4.14 A DeMarco model.

The simplicity of the method means that all members of a design team can use it. Moreover, it provides a common language that shows functionality and structure simultaneously through the use of DFDs. This sharing of a common view is vital when attempting to interpret customer requirements, even to the extent of being able to refer back to the customer and 'walk' them through a model of their own requirements.

In many respects, it is evident from the above that structured analysis is really nothing more than a language, a communication tool, but one that does not suffer from the ambiguities of natural language. This, however, is achieved by the use of a strict notation, and consequently means that the model does not cover every aspect of a system. For example, the timing of various processes may be important, DeMarco makes no attempt to address this though there are extensions such as those proposed by Ward and Mellor which are intended to cater for a temporal arrangement of processes.

Indeed, the method only purports to highlight the flow of information through a system. However, this should not be underestimated since most systems are inefficient, not because of timing considerations, but because of poorly defined information flows and interfaces.

To illustrate the use of the method, consider again the example of the intelligent washing machine. Figure 4.15 shows a context diagram for the system based on the given set of requirements. The central process is entitled to reflect the operational requirement and in this example reflects the whole life cycle of the system. This may or may not be appropriate dependent upon the system being modelled and it is possible to have several variants of the context diagram to show different views of the system to different audiences.

A common trait of many engineers is not to consider issues such as installation and maintenance until the design is essentially complete or in the detailed stage. The modelling technique does not necessarily force these views to be taken into account, particularly if the design team comprises just engineers. Yet they can be captured and the interaction with other parts of the system clearly seen, particularly if as was suggested the 'feasibility team' includes non-engineering members.

Figure 4.16 shows the decomposition of the process in the context diagram (CD) into three separate processes. It is possible at this stage to include more detail by having more processes, but this is a decision that is made by the team. The key point is that the team must all be in agreement that this diagram is understood and that it reflects and interprets the requirements. Interpretation of the requirements is therefore an important aspect of functional modelling. There is no statement in the original requirements about installation or maintenance; but it must be there and must be included in the model if a complete picture is to be generated. Indeed, it is these interpretative features that makes the modelling tool so powerful as it forces design teams to explicitly include unstated customer requirements.

Note also that the method captures to some extent the structure of the viewpoint analysis. Indeed, viewpoint analysis should be used to assist in construction of the DFDs and the two methods should be cross-referenced to ensure consistency. Equally, textual analysis should be used to provide assistance in identifying and structuring the non-functional requirements.

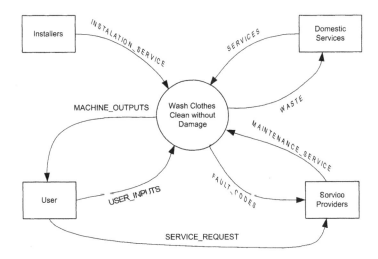

Figure 4.15 Context diagram for washing machine.

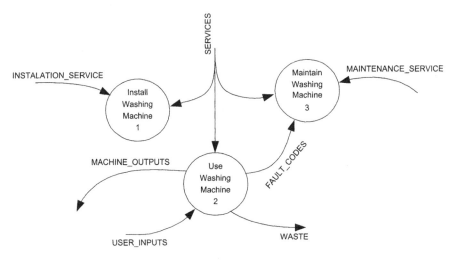

Figure 4.16 Level 0 dataflow diagram for washing machine.

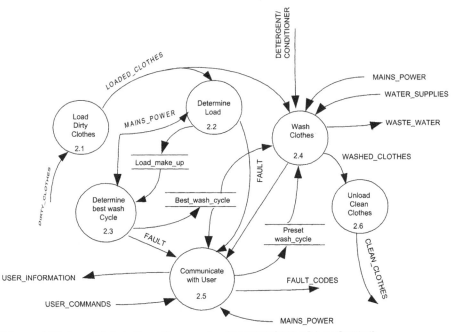

Figure 4.17 Level 1 dataflow diagram for USE WASHING MACHINE process.

To continue the modelling and analysis, each of the processes identified in Fig. 4.16 can be further decomposed into their own 'child' DFD. For example, Fig. 4.17 shows the decomposition of the 'USE WASHING MACHINE' process. The decomposition process can be continued to lower levels providing more detail of the functionality of the intelligent washing machine system.

The diagram of Fig. 4.17 also illustrates some further important benefits of this modelling approach. Firstly, it allows the design team to consider unstated functions such as 'LOAD DIRTY CLOTHES'. It could be argued that this is a trivial function, but the point is that it is often trivial functions that are overlooked. This particular function is also interesting since it shows how the modelling method is used to capture user interaction with the machine. It therefore captures the human–machine interface and ensures that ergonomic matters are considered early in the design process.

Note also that in the diagrams of Figs 4.15–4.17, no decisions as to implementation have been made, but that the modelling exercise has generated significantly more information than was contained in the original customer requirement.

The DFD is only part of the modelling tool set. To complete the model it is necessary to define the dataflows and datafiles and also the functional processes. The DFDs not only decomposes the functional processes, but also the dataflows between them. Thus to 'read' between DFDs it is necessary consult the DD. For example,

USER_INPUTS = USER_COMMANDS + [DETERGENT / (CONDITIONER)] + DIRTY_CLOTHES

The DD is also used to capture some of the non-functional requirements, for example:

DIRTY_CLOTHES = Up to 2.5 kg of fabric items of various types and finishes

Although the non-functional requirements expressed above were not part of the original requirement, it is easy to see how they may be included in a structured fashion.

DeMarco originally proposed that only the lowest level processes need to be defined with a PS. This is logical since a definition of the lowest level or 'primitive' process completely defines that system. However, it is suggested that higher-level process definitions can perform an additional task beyond that originally considered by DeMarco. The use of DFDs almost automatically separates out the functional and non-functional requirements and, as already indicated, the DD is used to capture in a structured fashion some of the non-functional requirements. Other non-functional requirements are captured in the PSs but adopting DeMarco's convention of only specifying the lowest level processes will not capture all of these, particularly the non-functional system requirements such as overall cost. Indeed, the requirement that the cost of the intelligent washing machine should be within 15% of existing devices cannot sensibly be assigned to any of the primitive processes. However, this could be included as part of a higher-level process specification.

An example of a PS might look like

PROCESS SPECIFICATION
PROCESS 2.1 COMMUNICATE WITH USER

WHILE MAINS_POWER ON

   DO THE FOLLOWING
     Generate USER_MESSAGE appropriate to BEST_WASH_CYCLE
     Display USER_MESSAGE
     await USER_COMMAND
     determine appropriate action
       CASE 1 USER_COMMAND = GO
        start wash cycle using BEST_WASH_CYCLE
       CASE 2 USER_COMMAND = MANUAL_OVERRIDE
        start wash cycle using MANUAL_OVERRIDE

END CASE
UNTIL MAINS_POWER OFF

In summary, structured analysis is a simple but powerful modelling tool that can be used to capture and interpret customer requirements. Producing a diagrammatic model of the system separates the functional and non-function requirements of a system and its simplicity is such that it is an ideal tool for use in multi-disciplinary teams and allows the team members to share a common perception of a problem.

4.5.4 Sensitivity analysis

Requirements interpretation using viewpoint analysis, functional modelling and analysis and textual analysis separates functional and non-functional requirements. It also provides for a logical structuring of the functions and supports a shared understanding between the members of the feasibility team. While this partitioning of functionality may not reflect the final implementation, it is important to start the task of assessing the degree of interaction between the functions early on in the design process. Although methods such as functional modelling and analysis do not explicitly show the time-order between functional processes, it is clear that the performance of one process will impact on all the other processes with which it interacts. Gaining an appreciation of this interaction at an early stage is important because it provides:

- an early understanding of possible trade-offs;
- an identification of key processes with regard to overall system performance;
- identification of 'sensitive information';
- identification of potential system failure modes;
- guidance when generating and selecting conceptual solutions.

The common approach to sensitivity analysis is to generate concept solutions and then to assess their worth through parametric and sensitivity studies. These studies should still be performed, but an assessment at the earliest possible stage can lead to rapid maturing of the conceptual solution.

An understanding how one element impacts on another is provided through the work of Dr Genichi Taguchi. As indicated in Chapter 1, his ideas relate to the achievement of product quality, but the concepts can be extended to the design process.

Taguchi's loss function

Taguchi's method is a methodology for improving quality by eliminating variability at appropriate stages during product design and production and has gained acceptance in many different industries. The cornerstone of the approach is the concept of the loss function and it is this that has wide reaching implications.

Taguchi argues that any product that does not meet its target specification will either impart a loss to the company or a loss to society. It is reasoned that a loss is always incurred when a product's functional quality characteristic, normally denoted by y, deviates from its target value, normally denoted by m, regardless of how small is that deviation. The quality loss is therefore zero only when $y = m$. From this the loss function $L(y)$ can be defined such that:

$$L(y) = k(y-m)^2 = k\Delta^2 \qquad (4.1)$$

The term $(y-m)$ is the deviation from the target value and so the loss A due to a deviation is:

$$A = k\Delta^2 \qquad (4.2)$$

where $\Delta = (y-m)$. Thus, the loss is quadratic and dependent upon the constant k. An example of a loss function is shown in Fig. 4.18.

Every component in a product will have its own characteristic loss function, but the 'degree of loss' is dependent upon the constant k. Thus the sensitivity is determined by the value of k as shown by Fig. 4.19.

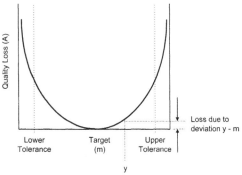

Figure 4.18 The Taguchi loss function.

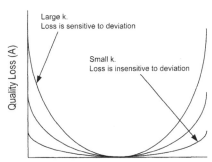

Figure 4.19 Sensitivity of the loss function.

Taguchi's argument is that products should be made robust to these various sources of loss by design and the selection of appropriate production methods. For design a three-stage process is specified as described in the following sections.

Systems design
The fundamental design and engineering concepts are established.

Parameter design
Target values are determined and the sensitivity of the design established by determination of the loss functions for each component. Components should be chosen such that the loss function is robust in which case a large deviation away from the target value will only impart a small loss. If some components are particularly sensitive then they must be appropriately toleranced.

Tolerance design
Design tolerances are determined from the results of the sensitivity analysis. Robust components do not need a tight tolerance, but sensitive ones do. The aim is to identify from their loss functions which are the sensitive components and to determine the appropriate tolerance limits for these.

The procedures described above will result in a design that has been appropriately toleranced. It is then important to use, design or develop production processes that are capable of meeting the required tolerances.

Taguchi recognised that many companies do not know what their production processes can achieve and to overcome this issue he advocated the design of experiments to determine the performance of individual production processes and to identify the parameters that affect that performance. The experiments use statistical techniques to determine the relative contributions to process performance of any influencing factors. For example, when paint spraying there are several potential influencing factors such as ambient temperature and pressure, humidity, spray pressure, nozzle size and variability. An experiment can be designed which will allow the relative contribution of each of these to the quality of the finished item to be determined. In most cases it is likely that only one or two items will be significant contributors and once known these can be precisely controlled to ensure that the product is manufactured at its target value.

Taguchi and requirements interpretation

The implication from the above is that Taguchi's approach is applied only once the basic concept design has been agreed. However, recognising that all parameters in a system are subject to target values and that functionality will be affected through the variation of input quantities about their respective target values, means that even during requirements interpretation an assessment of critical aspects can be made.

Consider the 'USE WASHING MACHINE' process of Fig. 4.17 in which subprocess 2.4 'WASH CLOTHES' has associated with it a number of inputs and outputs. All inputs will have some nominal target value and variations about that value will obviously affect the performance of the complete process. This could be regarded as stating the obvious; however, it is the obvious that is often overlooked. The questions that need to be asked at this stage include:

● What are the likely variations in the inputs?
● How sensitive is the process to these variations?
● Which are the critical inputs?

There is also a 'knock-on' effect since the functional processes do not exist or operate in isolation. This gives rise to another set of questions:

● Which inputs affect which outputs?
● How will variations in the inputs affect the variation of the related outputs?

By attempting to answer these questions, knowledge can be generated about which inputs have significant variation and which processes are critical to the overall operation of the system, providing an early assessment of technical risk.

The approach adopted on the above analysis is therefore primarily one of simple common sense. In the case of complex systems a logical framework is necessary in order that the approach is structured and logical. Figure 4.20 outlines the methodology as applied to the assessment of the sensitivity of a set of requirements. Two approaches are used to determine the sensitivity as described in the following sections.

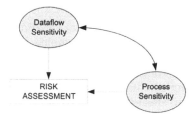

Figure 4.20 Sensitivity analysis.

Dataflow sensitivity

Dataflow sensitivity analysis is a simple tabular method for assessing the relative effect of dataflow variation on the performance of a system. The technique is to identify and list, from the functional model, the data-flows and datafiles and their receiving processes. Typically this is done only for the primitive processes since more than one process can receive the same dataflow. For each case when a process receives a dataflow the effect of variation in the dataflow on that process is judged using a simple three point scale for which:

⊕   means *very sensitive*
O   means *sensitive*
Δ   means *insensitive*

The tabular layout is then as shown in Table 4.1. This tabular form is very useful in identifying those processes that are sensitive to variations in input dataflows and the contents of datafiles. These can then be highlight when considering concept solutions. However, the table does not immediately show how these sensitivities permeate through the system. Hence, a process sensitive to a certain input could result in an output having increased variation. This output could then form the input to another process and so on, ultimately affecting the overall performance of the system.

Table 4.1 Dataflow sensitivity analysis table

Dataflow	Receiving process	Sensitivity
x	1.1	O
y	1.2	Δ
w	2.1.1	Δ
u	2.1.2	⊕
etc.		

To identify these possible relationships, the results from the table are used to annotate the DFDs so that potential causal threads can be clearly seen. To illustrate this consider the 'USE WASHING MACHINE' process of Fig. 4.17. For the sake of brevity it is assumed that this is the lowest level diagram. The corresponding sensitivity table is then as given by Table 4.2.

If the data from this analysis is transferred back on to the dataflow diagram, the result is as shown in Fig. 4.19 in which the most sensitive information flows are shown by the heavy lines. Thus from Fig. 4.21 it can be seen that there is a critical thread between the DIRTY_CLOTHES, through LOAD_MAKE_UP, BEST_WASH_CYCLE to WASHED_CLOTHES which is the output of the system. Thus the causal effect chain has been identified, which in this case backs up common sense.

Process sensitivity

Dataflow sensitivity examines the effect of variation in dataflows on functionality. However, while this does indicate whole system effects, there are other possible situations that have to be considered. One is the failure of an individual function or process. Although the precise nature of the failure is likely to be unknown until its implementation is decided, it is still often possible to carry out a failure mode and effects analysis (FMEA).

Typically, a FMEA is conducted once the product or system design is

Table 4.2 Dataflow sensitivity analysis table for USE WASHING MACHINE process

Dataflow	Receiving process	Sensitivity
DIRTY_CLOTHES	2.1	⊕
LOADED_CLOTHES	2.2, 2.4	⊕
MAINS_POWER	2.2,2.3,2.4,2.5	Δ
LOAD_MAKE_UP	2.3	⊕
FAULT	2.5	Δ
USER_INFORMATION	USER	Δ
USER_COMMANDS	2.5	O
FAULT_CODES	MAINTAINERS	Δ
PRESET_WASH_CYCLE	2.4	Δ
BEST_WASH_CYCLE	2.4	⊕
DETERGENT/CONDITIONER	2.4	O
WATER_SUPPLIES	2.4	O
WASTE_WATER	DOMESTIC SERVICES	Δ
WASHED_CLOTHES	2.6	⊕
CLEAN_CLOTHES	USER	⊕

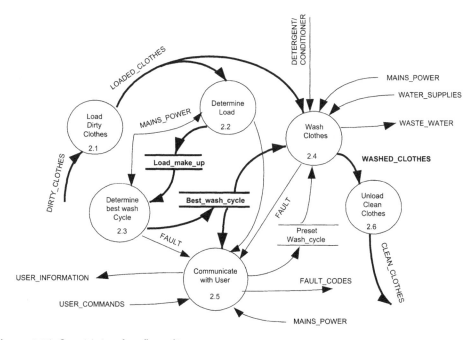

Figure 4.21 Sensitivity dataflow diagram.

complete. However, the principles can be applied to all the primitive functions of a functional model. The approach is to identify the failure modes for each function and then assign the standard FMEA ratings to probability of occurrence, severity of occurrence, and probability of detection. By then calculating the criticality index, or risk priority number, the sensitive functions can be identified. Figure 4.22 shows part of an FMEA table for the 'USE WASHING MACHINE' process.

Viewpoint: Function	P_o	Probability of Occurrence	Criticality Index	Date		No:	
System: Washing Machine	S	Severity of Occurrence	$C_i = P_o \cdot S \cdot P_d$	Author:		Issue	
Subsystem:	P_d	Probability of Detection		Checked:			

| FUNCTION | CHARACTERISTICS OF FAILURE | | INDEX | | | | | COMMENTS |
	MODE	EFFECTS	CAUSE	P_o	S	P_d	C_i	
LOAD DIRTY CLOTHES	OVERLOAD	POOR WASH	USER ERROR	6	6	3	108	
	UNDERLOAD	POOR WASH	USER ERROR	4	4	3	48	
	EXTREME MIX OF LOAD	POSSIBLE COLOUR RUN	EXTREMES NOT IDENTIFIED	4	8	3	96	
		FABRIC SHRINK	EXTREMES NOT IDENTIFIED	4	8	3	96	
DETERMINE LOAD MAKE UP	INCORRECT LOAD MAKE UP INDENTIFIED	POOR WASH POSSIBLE COLOUR RUN FABRIC SHRINK	ITEM OF CLOTHING NOT INDENTIFIED	4	8	3	96	

Figure 4.22 Part FMEA for the USE WASHING MACHINE process.

Risk assessment

Dataflow and process sensitivity analysis provide evidence for an overall assessment of risk. While this is certainly not precise, it does provide an early indication of where potential difficulties might lie. Of particular importance are:

● sensitive dataflows that are constrained by some non-functional requirement;
● processes that have particular performance requirements;
● overall system requirements due to functional failures modes.

Both analysis methods also allow some indication of causal links between dataflows and processes, and may even indicate complex multi-cause situations. They also provide information for activities such as concept selection and trade off studies. Moreover, the method allows all of the team to contribute; indeed team members from service or maintenance are extremely valuable in conducting FMEAs. This sharing of knowledge to reach a common understanding is vital to successful mechatronic system design.

4.6 SUMMARY OR BEYOND REQUIREMENTS INTERPRETATION

This chapter has concentrated on understanding customer requirements. This is the perhaps the least understood and least recognised aspect of any design or feasibility study yet it is obviously the foundation on which any design is based. The outcome of the processes and tools described are a structured functional model of the required mechatronic system together with a provisional assessment of the technical risk associated with that system. Moreover, the use of the tools considered allows feasibility or design teams to capture different viewpoints that lead to a sharing of knowledge and information.

Tools such as functional modelling are by their very nature compatible with other systematic design methods. In particular, function-means analysis is a natural choice which can build upon the functional decomposition obtained from functional modelling and viewpoint analysis. The approaches presented are also compatible with and enhance the capability of other requirements capture tools such as quality function deployment.

4.7 CASE STUDY: AUTONOMOUS ROBOTIC EXCAVATOR

The defining operational requirement for the autonomous and robotic excavator is that it should:

'Fill the bucket as quickly as possible.'

A basic statement of requirements is then provided by the following statements:

- It should be capable of working in any type and conditions of soil.
- It should excavate with an accuracy of 50 mm.
- It should achieve a speed of autonomous operation comparable with that of an average operator in all conditions.
- It should be autonomously capable of dealing with different types of underground obstacle as appropriate.
- It should be capable of both autonomous operation and of various levels of operator controlled operation.
- It should be capable of locating and orienting itself on-site and have a self mobile capability under certain conditions, i.e. to provide incremental movement along a trench line.
- It should support a range of operator interface structures ranging from conventional joysticks to off-line programmed commands as appropriate.
- It should be capable of integrating its operation with that of other site systems.
- It should be supported by an appropriate simulation and modelling capability.

Figure 4.23 shows a viewpoint bubble diagram for the excavator while Fig. 4.24 shows the associated CD.

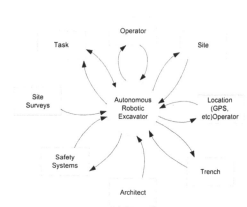

Figure 4.23 Viewpoint bubble diagram for autonomous robotic excavator.

Figure 4.24 Context diagram for autonomous robotic excavator.

5 | Artificial intelligence

5.1 INTRODUCTION

An introduction to some of the fundamental ideas of artificial intelligence (AI) was given in Chapter 2. The current chapter expands on those ideas to provide an overview of the techniques of AI and to give guidance on the considerable amount of jargon that is used in what is a huge and expanding research field. However, the main aim of the chapter is to concentrate on those aspects of AI that are particularly relevant to the development of intelligent machines that operate in real-time in the real world. This implies that such machines must make intelligent decisions based on sensor data from the environment, and effect some change either to their own state or the world around them by means of actuation as a consequence. There is currently a large base of active AI research, largely centred in university computer science departments, where the emphasis is more concerned with the solution of closed logic problems, such as formal proofs, within the computer itself. Such work has only marginal relevance to the type of intelligent systems that are considered here, and indeed it can be argued that such reasoning plays a fairly insignificant role in the way that humans operate effectively in a complex and unstructured world.

Additionally, in recent years there has been considerable development in 'biologically inspired' techniques such as artificial neural networks and fuzzy logic. These however are not regarded as coming within the definition of mainstream or classical AI and so are considered separately in Chapter 6. An exception is the technique known as genetic algorithms which is described in this chapter.

The task of developing a practical intelligent system, which can carry out useful work in the real world, must begin by collecting existing relevant knowledge about the field of operation or particular *domain*. This is known as *knowledge acquisition* and is covered in Section 5.3. The knowledge must then be presented in a form which is suitable for manipulation by a computer. This is the *knowledge representation* stage. Several different techniques have emerged for knowledge representation and the bulk of the chapter is devoted to describing the most common approaches. The whole process of collecting the knowledge and presenting it in a suitable form is referred to as *knowledge engineering*. The *knowledge engineer* acts as a broker between the domain experts, who may know nothing about AI, and the AI programmers, who may know nothing about the domain.

5.2 AI PROGRAMMING VERSUS 'CONVENTIONAL' PROGRAMMING

Although this may seem a gross over-simplification, the vast bulk of computer programming is concerned with manipulating numbers according to the rules of arithmetic. Typically, under the 'dataflow' paradigm a program takes a set of numbers from an input device or file, subjects them to some kind of arithmetic transformation according to a sequential procedure of functions, and then sends the 'answer' to an output device. The bulk of the knowledge required to programme using this 'conventional' approach is referred to as *procedural knowledge* or *imperative knowledge*, i.e. it is centred around commands and actions. The procedure defines *how* a problem is solved. The numbers are the *data* on which the procedure operates. Clearly conventional programs, such as the word-processor with which this book is written, can also manipulate strings of characters, but in general no account is taken of the actual meaning of the words.

Artificial intelligence programming manipulates knowledge. The knowledge is often represented in the form of statements and rules which are manipulated according to the rules of logic in order to achieve *goals* with real-world concepts being replaced by symbols. Much of the knowledge used in AI programming declares truths about the world and is often referred to as *declarative knowledge*.

The resulting rules declare 'what' the problem is rather than 'how' it will be solved. The statements, rules and goals are the *data* on which the AI system operates. In conventional programming a procedure may of course contain rules of the form:

IF (condition TRUE) THEN (do something)

but these play a different role when compared to the rules in an AI system.

In most AI programs, the order in which the rules are listed will not affect the outcome, whereas in conventional programs the rule sequence is an inherent part of the procedure. There is unfortunately an extensive, and sometimes confusing, list of terms used to describe the AI approach to programming. These include:

- knowledge-based programming;
- rule-based programming;
- symbolic programming;
- logic programming.

These approaches should not be confused with the term expert system which describes a particular type of end product.

In practise, both conventional and AI programs will use elements of both procedural and declarative knowledge. However, the AI approach is biased towards the declarative and the conventional approach to the procedural. A large intelligent mechatronic system may combine modules, some of which are AI programs and some of which are conventional programs. The aim is always to use the most appropriate tool for the job.

5.3 KNOWLEDGE ACQUISITION

It cannot be assumed that the knowledge engineer knows anything about the domain for which they are building an AI system which may require obscure and subtle knowledge in the field of say medicine, metallurgy or food science. Knowledge acquisition is the process that the knowledge engineer uses to capture the requisite knowledge and is often regarded as a major bottleneck in the development of intelligent systems. There are available six main sources of knowledge:

- experts in the field;
- literature;
- case studies;
- theoretical studies;
- behavioural modelling;
- emulation.

Each of these sources are briefly considered in the following sections.

5.3.1 Knowledge acquisition from experts in the field

The process of obtaining relevant knowledge from human experts is known as *knowledge elicitation*. Burton *et al.* describe and compare four basic techniques. Of these, *formal interview* and *protocol analysis* are most appropriate for eliciting procedural knowledge while *goal decomposition* and *multi-dimensional analysis* are primarily used to elicit declarative knowledge. It is however important to note that most knowledge acquisition tasks will require a combination of techniques.

Formal interview

This is the standard knowledge acquisition technique and consists of the knowledge engineer holding formal structured meetings with the domain expert. There are three basic steps to the approach, which may take place over several interviews:

1. The expert provides a general overview of the task in their own words.
2. The knowledge engineer then seeks confirmation of specific rules and procedures in order to impose some structure on the knowledge in the form of a preliminary model. The ways in which the knowledge engineer can intervene are restricted to certain probes.
3. As the knowledge emerges, the knowledge engineer will put hypothetical problems to the expert in order to test the consistency and completeness of the model.

A useful extension to the formal interview is the *questionnaire*, which must be carefully designed if it is to be of real value. A questionnaire can be sent to a wider audience of experts than would be reachable by personal interview, and it is particularly suitable for clarifying an accepted approach to a specific contentious point. Questionnaires are also useful for guarding against the danger of over-reliance on one expert whose views may be deviant or unrepresentative of the wider expert community.

Protocol analysis

This is a behaviourist approach in which the knowledge engineer passively observes the expert actually solving problems. The expert is encouraged to verbalise the reasons for particular actions during the procedure. The knowledge engineer makes notes and then post-processes the information to induce rules which describe the surface behaviour. These rules can then be put to the expert for comment. It is particularly important that the knowledge engineer should pick up deviations from routine behaviour, and to obtain an explanation for these deviations.

This can be a useful means of eliciting subtle manual skills from non-professional experts where the knowledge may be subconscious and not possible to verbalise. For example, it is not easy for any human to explain how they walk or ride a bicycle in terms which would be of any use to the programming of a robot. The analysis of video recordings may also be useful in this context.

Goal decomposition

One form of this technique is the *laddered grid*. This is a pencil and paper method in which a complex problem space can be explored by building up a network of relationships. For example, supposing the aim is to gather knowledge about medical techniques. The engineer may first insert a random entry into the problem space by suggesting 'antibiotics'. The expert is then asked to expand up, down and across the hierarchy of techniques by means of probes such as:

What forms of treatment, other than drugs, exist?
What are the main classes of antibiotic?
What alternative families of drugs are there?

Multi-dimensional analysis

Commonly referred to as a *card sort*, this technique involves the creation of a stack of cards where each card is labelled with a particular concept from the domain to be explored. The expert then has to sort these cards into classification groups, identify what each group has in common and then indicate the relationships between the groups. In this way a hierarchy of concepts is built up. The expert may be asked to intuitively give a factor which indicates the 'strength' of the relationship between groups. Some statistical analysis of the results may therefore be required.

5.3.2 Knowledge acquisition from literature

For professional diagnostic-type activities, such as the identification of diseases or geological specimens, there exists a large body of literature which can form a good starting point for a knowledge base, although an expert is likely to see the result as naive. For machinery or process control there may be available operator instruction manuals which can enable the knowledge engineer to gain some insight before confronting the expert.

5.3.3 Knowledge acquisition from case studies

A useful technique is to acquire information by a systematic analysis of previous case histories in the relevant domain. This of course presupposes that adequate records have been maintained, but the widespread use of computer-based record keeping means that this is an increasingly viable option. It is a good technique for 'project' based activities where a bundle of documents describes the entire life-cycle of an activity. A systematic analysis of such documents can often reveal knowledge that is not apparent to individuals and would otherwise be lost to the business.

5.3.4 Knowledge acquisition from theoretical studies

The distinction between procedural and declarative knowledge has already been made. It is also necessary to distinguish *surface knowledge* from *deep knowledge*. Where possible, humans try to avoid difficult reasoning based on first principles. They do this by devising simple rules, or *heuristics*, that work in certain situations *most* of the time. This is surface knowledge. An example of such a heuristic is:

'If the car engine does not turn over, the battery is flat.'

However, if the battery is replaced and the engine still does not turn over, it is necessary to think more fundamentally about the problem. This may require knowledge about the car wiring, the starter motor or even the basic principles of electricity. This is deep knowledge.

There is an obvious danger that some of the techniques described for knowledge elicitation will reveal only the surface knowledge of a domain. A particular problem is that, for instance, a machine operator may not be aware of the deep knowledge that underlies the working of their machine. If the machine is to be automated it is important to know what limits its performance. For example, is it limited by human reaction time or is it limited by the power of the motor?

Questions like this mean that the knowledge acquisition process may need widening to include experts in related fields so that a good theoretical understanding of the domain can be achieved.

5.3.5 Knowledge acquisition from behavioural modelling

One way to obtain knowledge about a particular domain is to build a model which demonstrates the phenomena of interest and then to exercise the model to see how it behaves. Models can be: mathematical, either algebraic or statistical; physical, often to scale; or increasingly computer based. A computer model enables a *simulation* of various scenarios to be tested. This is particularly valuable for learning about hazardous environments such as those associated with space, sub-sea or nuclear operations. A model can be used to test behaviour which is beyond the limits of human capacity such as the control of unstable fly-by-wire aircraft.

Models are usually, and quite deliberately, simplifications of the real world. This can be beneficial because it enables a particular phenomenon to be isolated and properly understood without other factors complicating the situation and obscuring the knowledge. The danger of course is that the user misunderstands the simplifications within the model and uses it beyond its useful limits.

5.3.6 Knowledge acquisition from emulation

If the aim is to build a bricklaying robot, one way for the knowledge engineer to learn about the process is to take a course in bricklaying. This approach tends to be frowned upon by the AI community on the basis that the knowledge engineer will be only an amateur and not a true expert. They argue that the knowledge engineer should remain independent, analytical and detached. This may be true in some domains, but on the other hand, it can be helpful in familiarising the knowledge engineer with the difficulties and vocabulary of a subject.

5.4 KNOWLEDGE REPRESENTATION

Having captured the knowledge, the next stage is to represent it. Knowledge can be represented in many forms, and two methods that are familiar to humans are natural language, as in a text book, and pictures, as in a wall chart. However, neither of these methods are appropriate representations if the aim is subsequent manipulation of the knowledge by a computer program.

The two principal requirements for a representation *scheme* are that it must contain adequate *semantics* and *syntax*. Semantics is concerned with *meaning* and clearly the representation must be rich enough to retain as much of the knowledge as possible. This means that it may be necessary not just to store a list of facts, but to also be aware of the relationship between those facts in some kind of relational hierarchy. Syntax is concerned with the rules and structure of the representation, thus the rules of grammar are the syntax of natural language. The syntax should permit the easy storage, manipulation and retrieval of the knowledge and also the creation of new knowledge from existing knowledge.

A good example of an appropriate knowledge representation scheme is the London Underground Map. Designed in 1931 by Beck it was a radical departure from what is normally thought of as a map, and yet it contains an enormous amount of meaningful knowledge (semantics) in an easily understood format (syntax). The breakthrough was to realise what is, and what is not, important for a passenger to navigate an underground network. The key requirements are to clearly distinguish the different lines, indicate the sequence of stations and show the points of connection between the lines. The real distances between stations, the actual route of the tracks and the relationship to surface objects are irrelevant to the required understanding. Beck's solution matched the requirements exactly, and he rejected the irrelevant information in order to make the map clearer. The challenge for AI is to find such effective representations that are suitable for computer manipulation.

Because AI applications are so varied, there is no single representation scheme that is ideal for all cases. Clearly, an expert system containing knowledge about 'metal fixings' would require a different representation to an intelligent controller for a mobile robot. A key consideration therefore is the selection of the most appropriate representation scheme. Complex intelligent systems may adopt a different representation for particular program modules, the whole possibly being linked by an object-oriented structure. The remainder of the chapter describes the following commonly used representation schemes and gives an indication of when their use is appropriate:

- logic methods;
- associative networks;
- frames;
- production systems;
- case-based reasoning.

The first two of these are most suited to the representation of declarative knowledge whereas the remainder are hybrid techniques that can be used for both declarative and procedural knowledge. In general, adopting any of these techniques implies the use of a dedicated computer package or programming environment. These will enforce a rigid syntax on the knowledge and in return will provide the facilities for storage, manipulation and presentation of that knowledge. Such a program is often referred to as a *shell* because it contains all the functions for manipulating the knowledge but is initially empty of any domain facts.

5.4.1 Logic methods

The application of logic is at the root of all science, mathematics and western philosophy and is essentially concerned with reasoning about interconnected facts which can be either TRUE of FALSE. There are numerous logic representations that have been developed to support reasoning about particular types of knowledge. Two of the most general and commonly used are *propositional logic* and *predicate logic* and these are explained in more detail in the following sections.

Although these are sufficient to tackle any logic problem, philosophers have found that logics with a more specific syntax are advantageous when working in specific areas. They enable a briefer and more coherent representation of the problem. Examples of more specialised logics include:

- temporal logic, i.e. reasoning about time and the order of events;
- modal logic, i.e. reasoning about possibility;
- epistemic logic, i.e. reasoning about knowledge and beliefs;
- deontic logic, i.e. reasoning about moral obligations and ethics.

Clearly the above have little relevance for existing mechatronic machines and systems, but they may become useful for future intelligent robots.

Propositional logic

Propositional logic, or propositional calculus, is the simplest form of logic consisting of a series of statements, or propositions, which can either be TRUE or FALSE. The common syntax is that each proposition is included in parenthesis so that it can be clearly distinguished from other propositions, e.g.

(a car has an engine)
(an engine needs fuel)

The power of propositional logic is that, as well as being able to check for a known fact in a list, *new knowledge* can be generated, e.g.

(a car needs fuel)

Although the above example is trivial, in a real application the technique enables a non-expert to interrogate thousands of propositions about an unfamiliar topic. A simple proposition, such as one of those above, is known as an *axiom*. It is also sometimes referred to as *well-formed formula* or *wff*. As well as stating facts about the 'permanent' world they can also express the current state of a changing situation, e.g.

(the fuel tank is over half full)

It can also be convenient, as in algebra, to replace a particular proposition by a symbol, e.g.

(the engine is running) could be replaced by **A**

More complex facts can be expressed by linking propositions with *connectives* such as ANDs, ORs and NOTs. Unfortunately the symbols used to represent the connectives vary depending upon the background of the user. Table 5.1 compares the different conventions. It should be noted that the 'OR' is more fully called the 'inclusive OR', and the statement is TRUE when either A or B or both are TRUE. In translating statements into English it can often be replaced by 'otherwise'.

Table 5.1 Logic connective symbol conventions

	Engineer	*Mathematician*	*Scientist*
A AND B	**A • B**	**A & B**	**A ∧ B**
A OR B	**A + B**	**A ∨ B**	**A ∨ B**
NOT B	**B̄**	**~B**	**¬B**
A implies B		**A → B**	**A ⊃ B**

Here, the mathematician's convention will be used. More complex relation-ships can now be expressed as follows:

(the engine is running) → ((the ignition is on) & (the car has fuel))

or

A → (B & C)

Notice that, as in arithmetic, extra parenthesis can be added to remove ambi-guity and add clarity.

The formula above is known as a *rule of inference*. The fact that **A** is TRUE implies that both **B** and **C** must also be TRUE, but not vice versa. Also, if **A** is FALSE, then **B** and **C** can be either TRUE or FALSE without contradicting the above statement. It is possible to simplify the range of connectives required to express statements. For example, **A → B** can always be re-written as (~ **A ∨ B**).

A truth table can be constructed which tabulates all possible values (TRUE or FALSE) for each proposition and the effect that this has on the truth of the compound statement. Table 5.2 is then the truth table for the statement **A → (B & C)**. Readers familiar with binary digital logic and Boolean algebra will find Table 5.2 straightforward but others should take the time to understand each line of the table. If in doubt refer back to the meaning of the symbols thus:

A means 'the engine is running'

B means 'the ignition is on'

C means 'the car has fuel'

and so forth. Note also that the column for ~ **A ∨ (B & C)** contains the same truth values as that for **A → (B & C)**, thus indicating that the statements are equivalent. In English both these statements can be read as:

B and **C** can have any value so long as **A** is FALSE, otherwise they must both be TRUE

It must also be remembered that although the manipulation of formulae and the construction of truth tables is dependent only upon the syntax of proposi-tional logic, it is the user that actually supplies the semantics or meaning for the statements, e.g.

P & ~ Q

means

'James loves curry but hates pizza'

when

P = James loves curry

and

Q = James loves pizza

Table 5.2 Truth table for **A** → (**B** & **C**)

A	B	C	A → (B & C)	~A ∨ (B & C)
f	f	f	t	t
f	f	t	t	t
f	t	f	t	t
f	t	t	t	t
t	f	f	f	f
t	f	t	f	f
t	t	f	f	f
t	t	t	t	t

First-order predicate logic

In propositional logic, each proposition or axiom is a complete entity which cannot be decomposed into smaller parts. This means that a large number of individual propositions may be required to define complex problems. Predicate logic, or predicate calculus, is more expressive because it enables logic statements to be broken down into a *predicate* and one or more *arguments* or *terms*. Consider the following sentence:

A car needs wheels.

The predicate part of the sentence generally contains a verb and could in this case be considered to be 'needs wheels'. It is therefore convenient to connect these words together with hyphens to indicate that they must be taken together. The sentence therefore becomes:

car needs-wheels

in which 'car' is the argument or term. Also, a convention known as *prefix notation* is widely used which means that the predicate is clearly identified by putting it first. Thus:

needs-wheels car

Alternatively 'needs' could be made the predicate. In this case the predicate requires both a subject and an object when:

needs (car, wheels)

Here, the use of the predicate 'needs' implies that the first argument must be the subject. The use of parenthesis is also a convention (i.e. part of the syntax). It is also possible to write:

needs (car, wheels, brakes, engine)

Remember that it is the user that supplies the meaning. The words themselves are only convenient symbols and should be defined in other formulae. It is even possible to write:

needs (car,*x*,*y*,*z*)

The usual connectives can also be used:

moving (car) → has (car, wheels, engine)

when the above would be read:

'If a car is moving it implies the car has wheels and an engine.'

A powerful aspect of first-order predicate logic is that the arguments may take on variable values. Thus it is possible to replace 'car' by the variable *x* which is used to represent all vehicles when:

moving (*x*) → has (*x*, wheels, engine)

The value of the variable, *x*, can be qualified by the *universal quantifier*, ∀, which means that it is TRUE in *all* cases. Thus the statement:

∀ *x* moving (*x*) has (*x*, wheels)

means:

'All vehicles that move have wheels.'

provided that *x* has been defined as being the class of vehicles. This statement of course may, or may not, be TRUE.

The variable *x* can also be qualified by the *existential quantifier*, ∃, which means that it is TRUE in at least one case. Thus the statement:

∃ *x* moving (*x*) has (*x*, wheels)

reads:

'There is a moving vehicle that has wheels.'

Although predicates can be represented by symbols, such as replacing 'moving' by **M**, these cannot take on variable values in the way that arguments can. This requires second order predicate logic.

It is now possible to show how logic can represent knowledge and be manipulated to produce new facts. Consider the following example in which the logic is shown in both natural language and in predicate logic form. Lines numbered one to nine are *given*. Lines five and six, containing →, are *rules of inference*. The remainder being facts or axioms. Lines ten to fourteen are *generated* from the first nine, the numbers at the end showing which lines were used as the basis of the new knowledge.

Given

1. is-car (Ferrari, Ford) A Ferrari and a Ford are both cars
2. drives (John, Ferrari) John drives a Ferrari
3. drives (Susan, Ford) Susan drives a Ford
4. is (*x*, car) let *x* represent a car
5. ∀ *x* moving (*x*) → If any car moves it
   has (*x*, fuel) implies it has fuel
6. ∀ *x* moving (*x*) & A moving car is always
   ~ moving(*y*) → going faster than a
   going-faster (*x*, *y*) stationary one
7. ∃ *x* moving (*x*) At least one car is moving
8. ~ has (Ferrari, fuel) The Ferrari has no fuel
9. has (Ford, fuel) The Ford has fuel

Generated

10. drives (John, car)	John drives a car	
		(1 &2)
11. drives (Susan, car)	Susan drives a car	
		(1 &3)
12. ~moving (Ferrari)	The Ferrari is not moving	
		(1,8, 4 & 5)
13. moving (Ford)	The Ford is moving	
		(1,9, 4, 5, 12 & 7)
14. going-faster (Ford, Ferrari)	The Ford is going faster than the Ferrari	
		(12, 13, 4, & 6)

It is worth noting what *cannot* be deduced from the above. It cannot, for example, be said that:

'Susan is going faster than John.'

because it has not been stated that a driver goes at the same speed as the car. Neither can it be said that *all* Fords go faster than all Ferraris. In short, it is not permissible to make any assumptions beyond the strict facts provided. In this case it is only possible to deduce facts about two specific cars and their owners. The range over which the knowledge is assumed to be relevant is known as the *universe of discourse*.

It is also worth noting that the order in which the given statements are made has no effect on the outcome. This highlights the difference between this approach and conventional procedural programming.

Forward and backward chaining

New facts can be arrived at in two ways. Consider line thirteen above:

'The Ford is moving.'

The first strategy that may be used to obtain this fact is to look for all the formulae that contain a reference to 'Ford' as an argument. In other words to ask the question:

'What are all the things that a Ford can do?'

Thus lines one and nine together indicate that a Ford is a car and that it has fuel. The next step is to look for all the formulae with 'car' as an argument, and so on. Eventually, the fact of interest, or *goal*, will emerge. This is *forward chaining*, and it consists of using facts and rules to obtain new knowledge by *deduction*.

The alternative strategy is to start with the goal, and ask the question:

'Can the Ford move?'

Thus line five indicates that only cars with fuel can move. Line nine then indicates that the Ford has fuel, and so on. This is *backward chaining*, which works backwards from the goal to see if any facts support its truth. In general, if too many irrelevant conclusions can be generated by the data, it is better to use backward chaining

A good way to illustrate the difference between forward and backward chaining is to consider how a doctor might diagnose an illness. Suppose a child reports spots and a raised temperature, this enables the doctor to deduce that the patient is probably suffering from either measles, chickenpox or scarlet fever. The doctor would therefore ask for further symptoms. If the patient revealed that the spots started behind the ears and were preceded by a cough, this would provide the doctor with enough facts to deduce measles. This is forward chaining.

On the other hand, if the doctor was aware of a recent measles outbreak in town, as soon as they heard the word 'spots' they could ask:

'Did they start behind the ears, and have you had a cough?'

If the patient answers 'Yes' then the diagnosis of measles is confirmed by backward chaining. Of course, in reality the doctor is likely to use a subtle combination of both forward and backward chaining in order to home-in on the diagnosis.

Creating a logic-based program

Clearly, a realistic logic-based program might contain thousands of facts and rules and there are currently available two basic options for creating such a program. It could either be programmed using one of the dedicated AI computer languages such as Prolog or LISP, or exploit one of the many expert system shells such as ART or KEE, although such shells can be expensive.

The language LISP is a general and well established AI language, but it has an extensive learning curve for inexperienced users. Prolog is much more special-purpose and is designed specifically for logic programming. A Prolog fact is expressed in almost exactly the same syntax as the first order predicate logic examples in this book and the Prolog programming environment provides the facilities for querying the knowledge base.

Expert system shells such as ART and KEE support several knowledge representations in addition to logical reasoning and also provide for a variety of search strategies such as forward and backward chaining. Such a shell is likely to be written in either Prolog or LISP and, as such, they are higher-level programming environments than either language. Simpler shells are available, some of which are in the public domain.

5.4.2 Associative networks

Associative or *semantic* networks are essentially a graphical means of presenting predicate logic statements and should *not* be confused with *neural networks* which are dealt with elsewhere. They are a compact and easily understood method of representation.

Figure 5.1 shows part of the knowledge from the car and driver example presented in the form of an associative network. In Fig. 5.1, the arguments of the predicate logic statements are shown as *objects* which form the *nodes* of the network. The objects are connected by *directed arcs*, or arrowed lines, that indicate the nature of the *association* or relationship between the objects. the associations are closely related to the predicates of the logic representation. The use of 'inst' is short for 'instance of' and means that a 'Ford' is a particular instance of the class 'car'. If a class is itself a member of a super-class the relation 'is-a' is used, thus, 'car' is-a 'vehicle'. Similarly, 'engine' and 'wheels' could be represented as 'part-of' the class of 'car'. Advantages of associative networks are:

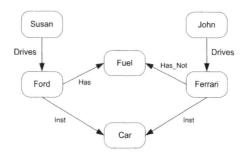

Figure 5.1 Associative network.

- They are easy to understand.
- They lead naturally to an *object-oriented* view of the knowledge.
- All the facts relating to a particular node can be easily seen.
- The 'big picture' can be simplified by representing the networks in layered form. i.e. all the 'parts-of' a car can be shown on a separate diagram in order to keep the top-level diagram simple.

Disadvantages include:

- It is not easy to represent the logic connectives such as ANDs and ORs. The syntax has been extended to cover this, but it ceases to be as illustrative.
- It is easy to think that the network contains more meaning than it actually does. It can contain no more than the equivalent logic representation.
- There is a lack of agreed syntax and software tools.

5.4.3 Frames

Frames can be used to represent both declarative and procedural knowledge and are most appropriate for situations in which the knowledge is commonly associated in predictable 'large chunks'. Thus a frame for a 'car' might contain fixed information concerning its make, engine size, colour and so on together with variable information such as current owner, mileage and date of last service. This variable information is contained in *slots*, which can be thought of as windows that contain changing values. Frames can be constructed from an *archetype frame* or *scheme* which contains the required 'blank' fields and slots, together with default values that can be used in the absence of better information.

Slot values can be given upper and lower limits and can be updated by various means. They can take their value from other connected frames such as an 'owner' frame or a 'mechanic' frame. In a real-time application, they could be connected to an 'instrument' frame which is in turn connected to actual sensors. Another approach is to employ a *demon*. This is a procedure or algorithm which can calculate a value based, for instance, on the information contained in other slots. Once the slots of a frame are filled-in it becomes *instantiated*, i.e. a representation of a particular instance of, for example, a car.

A simplified example of a frame is shown in Fig. 5.2. The data above the line is the fixed information and that below the line the variables in slots. Note how the service details can be stored as a procedure. If the current mileage is not known it could be generated by a demon that uses the time elapsed between the last service and the current date to estimate a value.

The idea of *inheritance* is strong with frames. This means that once a frame has been constructed for a particular model of a car, all other cars of the same model will inherit many of the same characteristics. The acquisition of new knowledge is therefore made relatively easy. The challenge for frames is the updating of the slot values, known as *matching*. This is made more difficult because the slot values can be obtained from a large variety of sources;

The parallels with the object-oriented paradigm are obvious and the technique is also related to *case-based reasoning* (CBR) which is dealt with later.

5.4.4 Production systems

Production systems are one of the most useful and simple techniques of knowledge representation and are particularly appropriate for mechatronic systems because they are primarily concerned with sequencing actions. The term *production system* is however rather confusing and should not be confused with a manufacturing system.

Alternative names are *rule-based system* or *inference system*. Although production systems are regarded as a hybrid technique for representing both declarative and procedural knowledge, they are biased towards the procedural. Production systems can, like logic systems, be used to manipulate knowledge to form an expert system; however they can also be used for real-time control and even task planning. Although development software is available for production systems, they are one of the few techniques that are simple enough to program from scratch without too much difficulty. A production system consists of three main components:

- the production rule memory;
- the working memory;
- the inference engine.

An optional component for real-time control is a set of external actuators and sensors. Figure 5.3 shows how the components are connected.

The production rule memory

The production rule memory contains a series of rules of the following form:

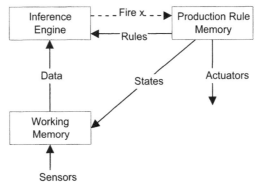

Figure 5.2 A frame for a car.   **Figure 5.3** The components of a production system.

IF (condition TRUE) THEN (perform action)

The conditional part of the rule is a logical formula that can be either TRUE or FALSE. Both the condition and the action can be compound statements using connectives such as AND. The rules can either change the real world by sending a signal to an actuator or they can change the data held in the working memory. The list of rules represent the permanent knowledge in the system and are not normally changed, although it is possible to envisage a *learning* system where certain conditions or actions are tuned to optimise performance, or even new rules being added automatically. Thus a typical rule in the programmer for an automatic washing machine at the start of its cycle might be:

IF (stage = filling) & (temperature > 40) & (water-level = full)
    THEN replace(filling, washing) & switch-on(wash-motor)

This states that if the machine is currently in the 'filling' stage and the water temperature is above 40°C and the water level is full, then change to the washing stage and switch on the appropriate motor. The instruction to switch on the motor is obviously a real-world action whereas the change from filling to washing is an instruction to change information held in the working memory. The actual syntax of the rule obviously depends upon the programming language being used.

The working memory

The working memory contains knowledge that can change. It contains current facts about the world and the state of the process. The working memory must contain all the facts necessary to check the truth of the left-hand-side of all the rules contained in the production rule memory. The actual values can be changed either by sensors in the real world or by production rules. Thus in the washing machine example the working memory must contain an indication of the current stage in the cycle; filling, washing, rinsing and so forth, the water temperature and a water level indicator. Other rules may require further information such as elapsed time.

The inference engine

The inference engine or rule interpreter is the active part of the program which cycles through the rules and decides which rule to invoke or *fire*. The steps are as follows:

1. For the entire list of rules in the production rule memory read the condition part.
2. Check each one against the facts in the working memory to see if it is TRUE.
3. Note which rules are TRUE.
4. If more than one rule is TRUE, carry out some form of conflict resolution in order to select a **single** rule.
5. Fire that rule.
6. Repeat the cycle.

It is essential that only one rule is allowed to fire on each cycle through the production rule memory. This ensures that there is some level of predictability or determinism in the system, which is particularly important for safety-related systems. The system is normally allowed to cycle through the rules as quickly as the processor will permit. The approach described here is *data driven* and hence is a form of forward chaining; however other implementations can utilise backward chaining or *goal-directed reasoning*.

Conflict resolution

If the conditions of more than one rule are found to be TRUE, there must be some means of determining which action is executed. The candidates are collected in the *conflict set* and some form of *conflict resolution mechanism* performed. One or more of the following strategies is possible for this purpose.

1. *Prioritising of rules*. Each rule is assigned a priority, for instance from one to ten. The rule with the highest priority is then the one which is selected from the conflict set for execution. If there is more than one rule with the same high priority, then one of the other conflict resolution methods must be applied. A variant of this approach is to prioritise the rule *conditions* rather than the rules themselves. Hence any rule involving a safety sensor condition would be given a high priority.
2. *Ordering of rules*. The rules are listed in the order of priority. Therefore the first rule in which the conditions are TRUE becomes the rule to be fired. This approach makes the inference engine very simple, but the effective control of very large rule-bases becomes difficult to manage. AI purists can object to the fact that the order in which the rules are placed becomes part of the control procedure. The knowledge engineer must place the more important rules higher on the list and this can sometimes be counter-intuitive. For example, a rule to check that a process is complete must come first, otherwise it may never be encountered as a rule higher up the list will be fired first. It is also important to remember that, after firing a rule, the inference engine must always start at the top of the rule list, and *not* the one after the previously fired rule.
3. *Specificity*. If one rule has, for instance, three conditions on the left-hand-side which are TRUE, and another rule has the same three conditions plus one or more additional conditions which are also TRUE, then the rule with the largest number of TRUE conditions is fired. The justification for this is that the longer rule is more specific to the particular solution. A variant of this approach is simply to give the highest priority to the rule with the longest list of conditions.
4. *Elimination of previous fired rule*. A strategy used in some expert system type applications is to discard a rule from the conflict set if it has previously just been fired. This is *not* an appropriate strategy for control applications as the desired objective is often to carry on doing the same thing until conditions change.
5. *Random choice*. As a last resort, if none of the other techniques have resolved the conflict, it is possible to pick a rule at random. Clearly this should be avoided in a safety related control system, where a high degree of determinism is required.

Some further points on production systems

The following points should be noted.

● Whereas ANDs are acceptable in production rules, ORs should be avoided. The use of OR connectives make predicting the behaviour of a system much more difficult. It is generally possible to avoid their use by simply adding more rules.

- The washing machine referred to above is an example of a *finite state machine*. This is a system which can only exist in a limited number of states, e.g. filling, washing, rinsing spinning and so forth. It is clearly pointless to cycle through all the rules for washing if it is known that the machine is in the filling stage. Performance can therefore be improved if a *branching-rule* is invoked. Thus once it is established that the machine is 'filling', only those rules relevant to 'filling' are considered. The last act of a particular stage is then to inform the working memory that the state is changed. The branching-rule must be read on each cycle to check for the change and invoke a different set of rules. In effect a branching-rule is a *meta-rule* or a rule about rules!

- It has been stated that the working memory can take values from external sensors. It can also take values from other independent processors. Thus a production system can be part of a *parallel* or *concurrent* system. In this case the working memory has become a means of exchanging data between concurrent processes and can be referred to as a *blackboard* or *whiteboard* and openly displays the latest information, making it available to any interested component of the system.

5.4.5 Case-based reasoning

The case-based reasoning (CBR) approach to knowledge representation emerged as recently as the 1980s and is still an active research area. It constitutes a possible basis for the really intelligent robots of the future that need to perform complex tasks in unstructured environments. Such robots may act as kitchen assistants, garden labourers or car mechanics.

It is argued that common situations arise in which events follow a fairly predictable pattern and the CBR concept is that the way to behave in a particular situation is to refer back to an earlier similar experience or *case* by referring to a *case library* which is a repository for previous experiences. Certainly, most of normal human life consists of engaging in familiar and predictable scenarios such as getting dressed, cleaning ones teeth, eating breakfast, travelling to work, eating in a restaurant, going to the bank or to the library. In each of these situations individuals behave in a certain way and have certain expectations concerning the events that will unfold.

Humans also have the ability to perform such actions with relatively little active thinking and might in such circumstances be said to be on a form of 'autopilot'. We all recognise the problems when we wish to deviate slightly from a recognised behaviour pattern. For example, we may wish to deliver a package on our way home from work. However, the routine of driving home is so well established that we arrive home with the package still on the car seat. The point being that, if the bulk of behaviour is merely a repeat of earlier events then there is no need to engage in fresh reasoning, planning and optimisation each time such an event is repeated. The same logic would apply to intelligent robot behaviour.

Of course not all events are familiar. The first time humans engage in an *unfamiliar* experience, for instance making a parachute jump, they are more wary, cautious and alert. However, this experience is then added to our case library for future reference. It is also true that even familiar experiences do not turn out *exactly* as expected, so that some adaptation from the norm is required.

Spectacular deviations from the norm become separate cases and are remembered, e.g. running out of petrol on the way to work. The CBR approach to determining behaviour or achieving a goal is thus:

1. Assess the situation.
2. Find a similar previous experience.
3. Use the previous experience to understand the current situation better.
4. Adapt previous actions to cope with the new situation.
5. If the new case turns out to be significantly different to the old, then remember the new case and also what was wrong with the old approach.
6. Apply feedback and analysis to the situation. Study what went wrong and what worked well. Hence repair gaps in the remembered cases.

Cases

There are two types of case. Firstly, the *prototypical* or *normative* case describes what would be expected to happen in a normal event. These are also sometimes referred to as *schemata* and, when related to everyday events, as *scripts*. Thus a script for a visit to an expensive restaurant will invariably involve actions such as:

- booking a table;
- travelling to the restaurant;
- being shown to a table;
- selecting from a menu;
- eating and drinking;
- paying;
- leaving.

These essential elements, or *snippets*, can be depended upon in virtually all restaurant cases. A script for a visit to a burger bar would be very different. The prototypical cases are generally used first in CBR as they provide a baseline against which the current situation can be compared.

Secondly, there are *specialised* cases which contain specific knowledge tied to a particular context. For example, a specific restaurant may offer a fixed menu, or the meal may be part of a wedding celebration and involve no payment. Clearly if the system were to include, say, six alternative cases for different types of restaurant there is likely to be a lot of duplication in stored knowledge. One approach to reducing this inefficiency is to adopt *memory organisation packets* or MOPs. A MOP describes a generic situation in a distributed manner.

Every case thus consists of contextualised knowledge that indicates the steps necessary to achieve a goal. It thus has two parts:

1. the lesson it teaches;
2. the context in which it is used.

The second of these is known as the *matching problem* and is achieved by some means of *indexing* the cases. This is one of the most difficult issues in CBR and is still an active research area. One approach is to use certain keywords or features, extracted from a checklist, that are known to be predictive of different behaviour patterns.

CBR applications

CBR can be used for problems involving planning, diagnosis or design. Some examples of current work in these categories is given in Table 5.3 in which the name in parentheses is the name of the research project or package. In addition several generic tools or shells are available.

Table 5.3 Case-based reasoning packages

CBR and planning	CBR and diagnosis	CBR and design
Robot navigation (ACBARR, ROUTER)	Car purchase advice (BROADWAY)	Architectural design (ARCHIE)
Land warfare battle planning (Battle Planner)	Computer operating system failures (CASCADE)	Electronic circuit design (BOGART)
Machinery failure recovery (CABER)	Heart failure diagnosis (CASEY)	Design of mechanical devices (CADET)
Manufacturing scheduling (CABINS)	Car repair (CELIA)	Structural design of buildings (CADSYN)
Meal planning (JULIA)	Computer repair (The Compaq SMART system)	Cooking recipes (CHEF)
Resource planning (MEDIATOR)	Legal consequences (GREBE)	Autoclave loading (CLVIER)
Telephone traffic management (NETTRAC)	Lung tumour diagnosis (IVY, MEDIC)	Landscape architecture (CYCLOPS)
Industrial relations (PERSUADER)	Audiology (PROTOS)	Design of physical devices (KRITIK)
Common-sense planning (PLEXUS)	Disaster analysis (SWALE/ACCETER)	Organisational change advisor (ORCA)
Automated manufacturing (PRIAR)		
Bank message routing (PRISM)		
Operation of everyday devices (SCAVENGER)		

Summary of CBR

We can sum up CBR in the following points.

- Rather than build up solutions from individual facts or axioms, as in conventional knowledge based systems, CBR uses ready-made responses as a starting point. This can be very quick for routine situations.
- CBR can provide solutions to problems which are not fully understood or where information is incomplete. It is therefore a *behaviourist* approach.
- CBR provides a means of learning from past mistakes.
- Whereas CBR is compatible with the way humans reason about problems, there are formidable problems in broad and unstructured domains where there is the danger of being too dependent on old cases when they are not appropriate.

5.5 SEARCHING FOR SOLUTIONS

This section is not concerned with how knowledge is represented, but with how it is retrieved. There are certain classes of problem where this presents significant difficulties, such as:

- game-playing problems such as chess programs where the aim is to select the best move from many possible options in order to achieve the *goal* of winning the game.
- planning and scheduling problems where the aim is to determine the sequence in which a complex series of events, or *sub-goals* , is ordered, to achieve a primary goal. Examples include the scheduling of events on a construction site or determining the machining and robot assembly sequence of a complex manufactured product.

- navigation problems where the aim is to move from a current position to a goal position, by the most efficient route. An example is the well-known 'travelling salesman' problem, which is concerned with finding the best route between a given set of towns.

The common factor in all the above is the large number of possible options or choices to be made on the route to the goal. Consider the following example of a navigation problem:

'Figure 5.4 shows a symbolic map representing two towns connected by a motorway. The aim is to work out the best route from the current position to the goal.'

A human would probably solve the problem by a subtle combination of forward and backward chaining. Firstly, they might use forward chaining to find the best route from the current position to the motorway. Then they might use backward chaining to work out which is the nearest motorway junction to the goal. The rest is easy.

Various AI search techniques suitable for advising a human or instructing an autonomous delivery vehicle will be considered. Clearly at each road junction there is a choice to be made and, where two roads cross, this amounts to 'go straight-on', 'turn left' or 'turn right'. These choices are known as *operators* and in this case applying an operator leads the search to another junction. The choices for a journey which involves encountering three such junctions are shown in Fig. 5.5.

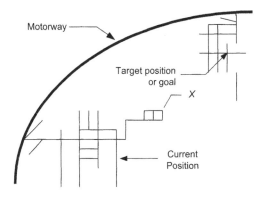

Figure 5.4 Navigation map.

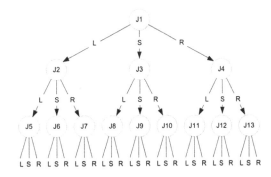

Figure 5.5 Tree-structured search space.

If Fig. 5.5 was extended to include every road junction it would represent the *search space* or *world state space*. This search space is often tree-shaped and each junction would then be referred to as a *node* in the search space. A high-level node can be referred to as a *parent* and the nodes connected to it in the level below are referred to as *children*. It is clear that the number of choices increases exponentially with each *level*. In fact, for j nodes with m choices per node, the total number of choices, N, is given by:

$$N = m^j \qquad (5.1)$$

If $j = 20$ and $m = 3$ then $N = 3.5$ billion choices!

This is known as a *combinatorial explosion* and finding an effective solution faced with such a wide choice is the essence of the 'search' problem. Another difficulty is that a junction can be approached from different directions and might therefore appear in the search space more than once.

5.5.1 Blind search

This crude method consists of wandering around the search space until the goal is encountered. Clearly some means of identifying the goal is required but otherwise no useful information is used to guide the search. There are two basic search strategies that can be employed, *depth first* and *breadth first* which were first outlined in Chapter 2.

Depth first search

Depth first searching consists of following a single route from the current position, down through the levels until the goal is encountered. Thus a depth first search would follow a path like the one shown in bold in Fig. 5.6.

If a dead end is reached, the search starts again at the top and follows a different route down through the search space. With large problems this can be very inefficient. Consider for example the navigation problem of Fig. 5.4. If the first route selected involves turning the wrong way on the motorway then this search could end up wandering the streets of the wrong town!

For this reason, depth first searches normally have a constraining *depth limit*. Thus an arbitrary limit of perhaps twenty junctions is imposed when, if the goal has not been found, that route is abandoned and another one started from the top level. This modification brings its own problem however, as reaching the goal may, in fact, require a minimum of twenty two junctions to be passed. This means that with a depth limit of twenty, the goal can never be found. Thus a depth first search with a level limit cannot guarantee to produce a result.

Once the goal has been found, it is highly unlikely that the first blind search has found the 'best' route. It thus becomes necessary to repeat the process until the goal has been reached by several routes and to have some means of evaluating the different solutions. The simplest way is to take the route with the minimum number of junctions.

The advantages of the depth first search are that it is simple to program and that in computational terms it requires little memory for storage of previous data, although it does require a method of ensuring that routes are not repeated.

Breadth first search

A breadth first search explores all the nodes at one level before proceeding to the level below. It therefore proceeds on a broad front as shown in Fig. 5.7. A breadth first search is guaranteed to produce a result and the first result obtained will involve the minimum number of junctions. Compared to depth first search it is, on average, quicker.

The disadvantage of a breadth first search is that it has high computer memory requirements. All information concerning all junctions must be accumulated so that the search can return to a junction and eventually extend it to the next level.

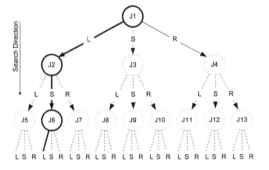

Figure 5.6 Depth first search.

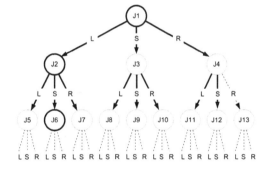

Figure 5.7 Breadth first search.

Iterative-deepening search

An iterative-deepening search is a hybrid blind search that combines the speed of breadth first search with the reduced memory requirements of depth first. It consists of firstly making repeated depth first searches to a shallow level and then gradually increasing the depth until the goal is found. Because no record of past nodes is remembered it requires the same node to be revisited many times, but in certain circumstances this can still be more efficient than either pure depth first or breadth first searches. The selection of the best blind search technique ultimately depends upon the nature of the search space. Table 5.4 gives a rough indication as to when each technique is appropriate.

Table 5.4 Selection of blind search technique

Blind search technique	Problem search space
Depth first	Finite space with many branches but few levels
Breadth first	Few branches but many levels

5.5.2 Heuristic or 'best-first' search

Blind search took no account of any structural information that may be contained in the search space. If such information exists then it is easy to improve performance. This is generally accomplished by adding a common-sense rule or heuristic to help determine the route through the search space. As well as making the route to a solution quicker, such techniques usually ensure that the resulting solution is optimised to some degree. They are often known as *best-first searches*. A simple heuristic to help with a navigation problem might be:

'Take the route whose direction points closest to the goal.'

In the problem shown in Fig. 5.4, the above heuristic would in fact lead to point *X*, and this particular problem would be best solved by introducing a sub-goal to:

'Find the nearest motorway junction first.'

Hill climbing

A similar approach is known as *hill-climbing*. In a hill-climbing search each of the children nodes adjacent to the current position is visited and an *evaluation function* applied to each of them. The evaluation function may for instance be, say, the inverse of the estimated distance to the goal. If this evaluation function value was plotted for all nodes in the search space, the junction nearest the goal would have the highest value and the one furthest away would have the lowest. A contour plot of such values would result in a three-dimensional surface with the goal at the top of a hill.

The hill-climbing search chooses, from among the children, that node which has the highest function value. In a basic hill-climbing search only the best value is remembered. This reduces memory requirements but increases the likelihood of homing-in on a local maximum such as point X in Fig. 5.4. No facility exists for backtracking, and there is no point in returning to the start point and repeating the search, as the same result would ensue.

The A* algorithm

There are numerous variations on basic hill-climbing, all with the aim of improving the speed of convergence and producing a more optimal solution. The increased sophistication is generally at the expense of the requirement for more memory. One variation is the so-called *A* algorithm*. With this technique, the evaluation function consists of the sum of two parts. The first part is the actual 'cost' of the route from the start point to the current node. The second part is the estimated 'cost' from the current node to the goal. Minimising the sum of the two parts thus gives the best current estimate of the optimum route. This is then combined with a variant of depth first search where the most likely routes are explored concurrently. Because more than one route is explored at the same time, this can be thought of as either a 'depth first search over a broad directed front', or a 'breadth first search over a deep directed front'. In other words, like iterative deepening search, it is a hybrid between depth first and breadth first search.

In practical terms, for the navigation problem shown in Fig. 5.4, each road junction could be numbered and its x, y global co-ordinates stored. As the search progresses the evaluation function for each junction is also remembered. The first term could consist of the sum of all the straight line distances between the previously encountered junctions, and the second term could be the straight line distance from the current junction to the goal. If the application was a robot delivery vehicle, the importance of the robot having an internal representation of the map, or *world model*, is apparent. It enables the robot to carry out a search strategy to determine an appropriate route before actually moving. In other words the robot can carry out a series of simulated or virtual journeys. A vehicle without a world model could still reach the goal provided it was fitted with some kind of positioning system such as a satellite global positioning system (GPS). However, it would have to actually visit each prospective child node in order to determine its evaluation function.

5.5.3 Genetic algorithms

The techniques described above are best suited to finding solutions in tree-like search spaces, but many complex design and planning problems are not like that. Rather than finding a route through a problem space which contains a large number of similar choices, these problems are more concerned with sequencing different operations or the effect of changing design parameters. One technique that can be used to resolve such issues is *genetic algorithms*. This technique is *biologically inspired*, and borrows most of its vocabulary from Darwinian evolutionary theory. However, the parallels should not be interpreted too literally.

As an example, consider a robotic machining and assembly cell for a complex sub-assembly. There may be hundreds of individual operations involved that can take place in various orders. Some operations may be achievable in different ways, and this may produce conflict between, say, speed and quality. Clearly there may be millions of different ways in which the individual operations can be combined to produce the best compromise. The genetic algorithm approach to such problems is as follows.

Step 1: create a population

The problem must be broken down into relevant variables, each of which can be represented as a binary number. Typical variables could be; the speed of a particular cutting process, the method of clamping the workpiece or the order in which two events are carried out. For example:

speed of cut	0110	i.e. sixteen possible options
machine bottom	0	a Boolean
first		variable – either do it or not

Each of these binary numbers is referred to as a *chromosome* and the chromosomes are then strung together to form the *genotype* of a particular *individual*, for instance:

0110001100111010101011000110 ....

This individual may contain hundreds of individual binary numbers and represents one possible solution to the problem. Clearly, by knowing the size of each chromosome and order in which they are connected, it is possible to decode the string and convert it back into a series of work instructions.

A *population* of individuals is first created, usually by randomly selecting 0s and 1s for each individual genotype. A difficulty is that all the resulting individuals must be *valid* solutions to the problem, although some, of course, will be better solutions than others. This is achieved by careful selection of the variables, which may be valid *ratios* between two parameters rather than the parameter values themselves. For a simple problem, the *population size* may be of the order of 50 individuals whereas a more complex problem may use two hundred.

Step 2: test for fitness

All members of the population must be evaluated to assess their relative fitness for achieving the given task and the criterion used to judge the best solution must be decided. This can be a single criterion such as 'minimum cost', 'maximum quality', or a compound combination of several such factors. The fitness of a particular individual is evaluated by applying an *evaluation function* to each individual in turn. Most literature tends to trivialise this stage which for real problems can be difficult and very computer intensive. For example, an early genetic algorithm application was the optimisation of the shape of a complex turbine blade. The only way to evaluate the fitness of each shape was to carry out a sophisticated computer simulation which must be repeated for each individual.

Step 3: breeder selection

The proportion of the population that will be allowed to *breed* is then selected. This is known as the *generation gap* and is typically around 40% of the total population. Simply selecting the fittest 40% of the population is regarded as too crude and runs the risk of losing genetic material currently contained within less successful individuals that may prove to be valuable for future *generations*. A common selection technique is *weighted roulette* in which the whole population is represented on a virtual roulette wheel, but the fittest individuals have more 'slots' and so are more likely to be selected. The wheel is spun once for each breeder required and the *mates* are usually paired off at random.

Step 4: mating

Each mating pair produces two *progeny* or children. Common mating mechanisms include *crossover* and *mutation*. *Single point crossover* consists of randomly selecting a point in the genotype of each mating pair and swapping the front-end of one of the pair with the front-end of the other to create two new individuals. This mating strategy is illustrated in Fig. 5.8

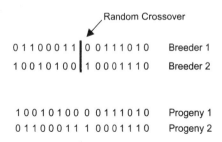

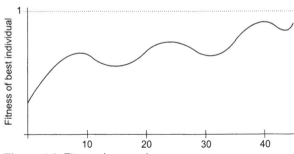

Figure 5.8 Genetic algorithms – breeding using single point random crossover.

Figure 5.9 Fitness/generation curve.

Mutation consists of the random flipping of one, or possibly more, binary digits in any individual according to odds which are determined by the *mutation rate*. This may lie between, say, 1 in 100 to 1 in 1000 digits. The purpose of mutation is to introduce new genetic material into the population. All of the new individuals created must, of course, be valid solutions to the problem.

Step 5: dispersal

The progeny are now returned to the population. However the population generally remains a constant size and some individuals must therefore be culled from the population. Various strategies exist for this, such as random selection or 'not the fittest', also referred to as *elitism*.

Step 6: generation monitoring

The process now returns to step two and the breeding cycle is repeated for a number of generations. The fitness of the best individual is constantly monitored and the process halted when no significant improvement occurs over a predefined number of generations. A typical graph of fitness against number of generations is shown in Fig. 5.9.

The genetic algorithm approach is subject to many variations. For example the individual population may be *seeded* with chromosomes that look favourable at the outset. The basic aim is to achieve a reasonable compromise between speed of convergence and maintenance of *genetic diversity*. The problem of homing-in on a local peak of fitness remains, and there is no guarantee that the fittest solution is a true optimum.

5.6 CASE STUDY: THE USE OF ARTIFICIAL INTELLIGENCE FOR AN AUTOMATED ROBOTIC EXCAVATOR

The autonomous excavator system known as LUCIE (Lancaster University Computerised Intelligent Excavator) was introduced in Chapters 2 and 4 when the basic requirements for such an excavator were considered. This section considers the AI approach adopted in the development of LUCIE to carry out its task of digging a trench.

Although the full LUCIE system consists of several distinct software modules, one module in particular can be considered to be based on AI principles. This module is known as the 'activities manager' and its function is to achieve the goal of excavating a trench to a specified depth in variable soil types. The *input* to the activities manager consists of:

1. data to describe the desired trench;
2. data to inform the excavator of its current position in terms of six global degrees of freedom;
3. some indication of the current progress being made in moving the excavator bucket through the ground.

The *output* from the activities manager consists of a demand to move either the excavator bucket or the vehicle tracks at a specified velocity. This demand is then processed by another module, the 'low-level controller' that sends the appropriate signals to the electro-hydraulic valves

The activities manager must therefore emulate the skills of an expert operator in order to achieve the desired goal. To do this it must contain knowledge about basic digging strategies, as well as tactical knowledge about how to deal with tricky real-time problems such as very hard ground or boulders in the path of the bucket.

5.6.1 Knowledge acquisition

Several of the techniques listed in Section 5.3 were employed to establish the knowledge base.

Experts in the field

Visits were made to sites where excavation was underway. The JCB excavator factory was also visited, where their expert demonstrator drivers were made available. All the drivers were interviewed and on the whole were very co-operative, but largely unable to conceptualise their actions with; 'I just do it!' being a common response. A more fruitful approach was to observe the drivers in action with the aim of:

1. discovering their basic approach to the various stages of the digging cycle;
2. noting under what circumstances they deviate from their basic approach, and then asking them specifically to explain this later.

A video recording was also made and this proved valuable for subsequent detailed analysis.

The predominant reaction to these visits was the realisation of just how good a really expert driver is, and consequently how difficult it will be to compete using a robot system. The driver's control over the excavator arm was both smooth and fast. Indeed, the degree of control was as though the excavator was an extension of their body and they invariably moved the bucket along the shortest possible path. It was however noted that a human operator slowed down significantly as they approached the target depth because they do not want to risk over-digging. At this point one or two *banksmen* must repeatedly test the depth of the excavation with a *traveller* lined up against pre-surveyed *profiles*.

Examples of the types of rule acquired are:

'In soft ground use a continuous "penetrate and scoop" action to fill the bucket.'
'In hard ground use a "penetrate, drag and scoop" action to fill the bucket.'

Literature

A literature survey revealed three types of relevant publication. The first type tended to be of academic origin and was either concerned with excavator kinematics and control or trajectory planning. On the whole they provided little expert knowledge of the strategic or tactical type required.

The second source or information is trade literature and training manuals. These proved to be more fruitful and descriptions such as that given below as the means of coping with problems such as removing a boulder were obtained:

'To dislodge a boulder, dig down on both its far side and its near side. Then place the bucket just beyond it with the bucket teeth inclined at a steep angle. Apply boom-down pressure and move the bucket lever back and forth rapidly. When the teeth have penetrated to the under side, crowd and curl the bucket to loosen the boulder.'

The Construction Industry Training Board (CITB) training notes also provided a description of current practice while some subtle advice for improving efficiency was given in a magazine:

'Never raise the bucket higher than necessary when swinging towards the spoil heap. Instead of trying to dump on top of the pile, dump through the pile, pushing the dirt back as the bucket is uncurled and replacing it with the dirt in the bucket.'

The third type of relevant literature is that produced by excavator manufacturers themselves. Thus, Caterpillar provide technical data on average excavator cycle times that provide useful targets for automation. Thus for a 20 tonne excavator operating in hard clay at 2 m depth the times shown in Table 5.5 are estimated.

Table 5.5 Excavator cycle times

Activity	Time (seconds)
Load bucket	5.5
Swing loaded	5.0
Dump bucket	2.0
Swing empty	3.5
Total cycle time	16.0

Behavioural modelling

An early decision was made to construct a fully operational fifth-scale model of an excavator arm. The structure was based on drawings supplied by JCB Excavators. Hydraulic cylinders were used as the means of actuation, and it was intended that it should be driven and controlled by the same hardware and software as a full-sized excavator. As can be seen from Fig. 5.10 it was provided with a sand box for trial excavations.

The model proved extremely useful for developing strategies for both the low-level controller and the activities manager. Its advantages were:

1. The development work could be carried out under laboratory conditions without the problems of inclement weather.
2. The model was easily fenced off and overall caused less concern over operational safety. It enabled confidence to be gained before transferring the technology to the full-sized excavator.
3. Many trial holes could be dug without concern for the environment or time lost in back-filling them.
4. Compared to a computer model, the fifth-scale model was probably no more expensive and took no longer to build.
5. It proved to be a good demonstration and marketing tool to attract further interest and funding for the research.

Figure 5.10 Fifth-scale model of excavator arm operating in sand box.

The model was used to develop X-Y resolved motion control, 'Forceball' control, teach-and-repeat operation, and full autonomous excavation. For autonomous digging a strategy was developed to remove obstructions such as a half-brick. The technology was eventually successfully transferred to the full-sized excavator with relatively few problems.

Emulation

For safety reasons all individuals connected with LUCIE were required to complete a basic one-day excavator training course. One individual also attended a CITB one-week course. These courses provided good insights into the basic excavation process and also exposed people to an expert driver. The CITB course also provided a good set of training notes.

Theoretical studies

At the outset it was considered important to understand both the theoretical basis for soil cutting; forces involved, optimum bucket angles and so forth, and the theoretical limits to forces and movements that a hydraulic excavator can provide. This was considered to be a parallel to previous work done on metal cutting and machining.

The theory of soil cutting
The aim is to obtain a relationship between the established soil mechanics strength parameters; shear strength (C_u), friction angle (ϕ) and plasticity index (I_p) and optimum bucket angles and cutting forces. Some relevant theoretical work on soil cutting is reported in the literature, but most of the more rigorous work originates from the former Soviet Union with the most appropriate work being that by Zelenin. The required excavation force is related to an empirical penetration test on the soil. A standard 2.5 kg hammer is dropped through 400 mm onto a 1 cm² circular rod. The number of impacts, C, to drive the rod 100 mm into the soil is the fundamental measure of soil resistance. Zelenin does show an empirical graphical relationship between C and the established soil strength parameters. For cohesive soils, this can be expressed algebraically as:

$$C = \frac{3.9C_u}{I_p} \qquad (5.2)$$

where C_u = soil undrained shear strength (kN/m²), I_p = plasticity index.

Because excavation takes place relatively quickly, most soils, other than dry sands and gravels, will behave as though undrained, and hence the friction angle, ϕ, is irrelevant. The dragging excavation force, P_k (N), for the buckets of mechanical shovels with teeth is then given by:

$$P_k = 10Ch^{1.35}(1+2.6l)(1+0.0075\alpha)z \qquad (5.3)$$

where h = depth of cut (cm), l = width of bucket (m), α = angle of bucket relative to ground (degrees), z = reduction factor for teeth ~ 0.65.

Typical values for LUCIE digging in firm clay are:

C_u = 100 kN/m²
I_p = 30
h = 15 cm
l = 0.60 m
α = 30°

The 5 cm depth of cut, h, is adequate to fill the bucket with a 2 m long drag. Substitution in the above equation gives:

$$P_k = 10.25 \text{ kN}$$

Excavator kinematics and forces
Although it is easy to plot the outer limits of an excavator's working envelope, it must be realised that kinematic limitations on joint angles means that the bucket *cannot* be moved in any desired direction within that envelope. For example, at half the arm reach, the bucket cannot penetrate the ground in a vertically downward direction. It is obviously important to understand these limitations before attempting to programme the bucket movements.

As regards forces, there is a highly non-linear relationship between the maximum force that can be applied to the ground and the arm joint angles. A program, 'DIGVECTA', was written that produces a graphical representation of available bucket directions and forces throughout the working envelope. The program takes into account:

- the three arm hydraulic rams (boom, dipper and bucket) reaching their force and displacement limits;
- the vertical equilibrium of the excavator (lifting);
- the horizontal equilibrium of the excavator (sliding).

The program firstly assumes the bucket is in the fully curled position. The bucket tip is then systematically moved to grid points within the excavator's operating envelope, and the maximum force available to move it in a specified direction is calculated. The bucket is then uncurled in 20° increments and the process repeated until the bucket is in the fully open position. Results are produced in tabular form and also plotted on-screen in the form of vectors. The locus of the vectors thus represents both the range of possible movement and the magnitude of available digging force. Figure 5.11 shows a sample output for LUCIE when the bucket angle relative to the dig direction is 30°, i.e. dragging.

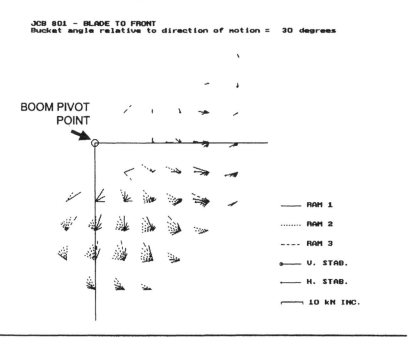

Figure 5.11 Output of DIGVECTA for **dragging.**

Points to note from Fig. 5.11 are:

1. Much higher penetration forces are available when the direction of movement is inclined towards the cab compared with when the motion is vertical.
2. When dragging, the highest forces are available when digging towards the boom pivot point. This ties in with expert drivers' unconscious behaviour when trying to remove boulders.

3. The maximum dragging force over much of the envelope is of the order of 10 kN, which is similar to the value obtained from the soil cutting theory in order to fill the bucket with a single drag in firm clay. This implies that with harder material a shallower dig must be made, and that consequently multiple drags may be necessary in order to fill the bucket.
4. Other than general points, such as those above, it was concluded that the theoretical studies are of limited use for digging automation and should not be part of the real-time decision making. This is largely because the soil is generally highly variable and its properties not known to the required degree. Hence an adaptive approach based on real-time performance feedback, as with human operators, is preferred.

5.6.2 Knowledge representation

A useful intermediate stage between knowledge capture and formal programming was to represent the digging process in the form of a finite state transition diagram. This is a graphical technique consisting of boxes that contain the actions that the system is performing when in a particular state. Arrows connect the boxes to indicate the sequence of actions that make up a particular task. Labels on the arrows indicate the circumstances that must occur to trigger the transition from one state to another.

Such diagrams can be produced at various levels by expanding individual boxes. Thus one diagram can be produced for the whole digging process whereas another could be concerned only with, say, the process of dragging the bucket. Figure 5.12 shows an example of an intermediate finite state transition diagram which deals with the actions that make up a digging cycle.

When it comes to programming, representation of the digging knowledge within the activities manager is by means of a modified Production System. A series of rules of the form:

IF (condition TRUE) THEN (perform action)

have been derived from the knowledge acquisition stage. Features include:

1. The rules contain logical ANDs but not ORs, as this facilitates checking the determinacy of the system and is hence an important safety feature.
2. The system is a 'finite state machine' and consequently it is only performing certain actions at particular points in its operating cycle. This means that there is no point in reading all the behaviour rules all of the time. Thus. the rules on 'penetrating the soil' are irrelevant when the excavator is in the 'going to dump' stage of its cycle. Meta-rules or branching rules have therefore been introduced to reduce the number of rules considered at any time and hence improve performance.
3. The rules occur in the order of descending priority. This means that the first rule encountered whose conditions are TRUE is fired. As suggested earlier it can be argued that this goes against the spirit of an AI program because the rules cease to be just data and become an important part of the functional structure of the program. It also makes it more difficult to add new rules. However placing the rules in random order and giving each rule an explicit priority is inherently just the same. Keeping the rules in order means that the inference engine reduces to a simple loop in the program.

The production rules in pseudo-code used for a basic 'penetrate' and 'drag' are

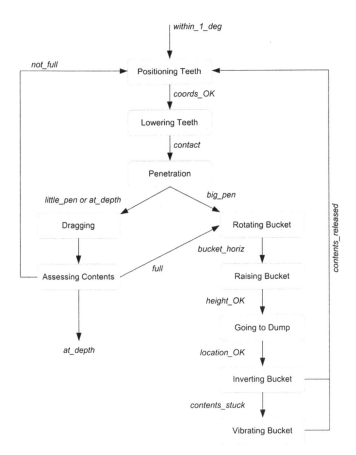

Figure 5.12 Finite state machine diagram for 'digging within reach'.

given in Table 5.6. The left-hand side indicates the conditional IF part of the rule and the right-hand-side shows the action to be performed if TRUE.

The working memory must contain all the data needed to check the conditional part of all the rules. In the latest version of LUCIE the three processors – activities manager, low-level controller and safety manager communicate via CANbus. The data therefore comes from a variety of sources including:

- the other processors via the CANbus;
- sensors connected direct to the activities manager processor via serial ports;
- a high-level planning program;
- rules in the production system itself

The working memory occupies a dedicated part of the processor memory and is served by a portion of programme whose job it is to keep it up-to-date.

Table 5.6 Production rules pseudo-code

Penetrate

Y≥ required_depth	Remember penetration_point
	X_vel=0
	Y_vel=0
	Penetrate=OFF
	Drag=ON
Tilt_Angle > 5 degrees	Remember penetration_point
(hard ground)	X_vel=0
	Y_vel=0
	Penetrate=OFF
	Drag=ON
Penetration > Pen_Depth	Remember penetration_point
(soft ground)	X_vel-0
	Y_vel=0
	Penetrate-OFF
	Scoop=ON
At position	X_vel=cos(pen_angle)
	Y_vel=sin (pen_angle)
Penetrate=ON	Extend arm horizontally to a target point at 75%
No penetration_point exists (implies a new dig)	reach Set bucket_angle to pen_angle
Penetrate=ON	Move to penetration point
Penetration_point exists (implies previous cycle)	Set bucket_angle to pen_angle

Drag (X_vel=–X_vel_start)

X < x_min	X_vel=0
	Y_vel=0
	Drag=OFF
	Scoop=ON
Contents > Bucket_Full	X_vel=0
	Y_vel=0
	Drag=OFF
	Scoop=ON
Y > Required_Depth	Y_vel=0
X(n)–X(n–1) > X_inc	Record Y
(This records the cutting profile)	
Bucket started moving	Y_vel=Y_vel+Y_vel_up_inc
X_vel_actual > X_vel_max	
(If moving too quickly go down)	
Bucket started moving	Y_vel=Y_vel_–Y_vel_up_inc
X_vel_actual < X_vel_min	
(If moving too slowly come up)	
Drag=ON	X_vel=X_vel_start
	Y_vel=0

Neural networks and fuzzy systems | 6

Humans are capable of reasoning using incomplete, uncertain, and noisy data and of developing effective solutions based on their experience and understanding of the problem domain. While AI and expert systems have been produced which attempt to reproduce the ways in which a human expert would reach a solution, these are primarily based on digital computers and often require complex program structures in order to arrive at a solution. Such systems also lack flexibility and, as discussed in Chapters 2 and 5, are incapable of operating outside the restricted problem domain for which they are constructed.

The capability of neural networks and fuzzy systems to operate in a more 'human-like' manner in reaching a conclusion has led to increasing interest since the 1960s in their potential application in a wide range of system environments. A neural network, or neural computing, attempts to reproduce electronically through the use of artificial 'neurons' the functions of the human brain. The resulting network consists of the parallel combination of a large number of simple processing elements, essentially summing junctions, which instead of being programmed, are trained how to carry out a particular task. This ability of a neural network to learn and to modify its response as a result of training has been a major factor in their development and even simple networks are capable of carrying out quite complex functions given proper training. Applications of neural networks include speech recognition, image processing and analysis, processing of sensor data, classification of sonar and seismic data, financial forecasting and adaptive control.

There are few absolutes in the human world. Instead, humans work with approximate and imprecise concepts such as 'normally' and 'in the region of' which cannot be accommodated within the syntax of a conventional computer language. Fuzzy systems and the associated fuzzy logic were developed from work by Lodfi Zadeh in the 1960s and enable the manipulation of 'human-like' concepts and data. As such, they bear a close relationship with expert systems, and indeed it will be seen that the construction of a set of fuzzy rules has a lot in common with the knowledge engineering process associated with the creation of a conventional expert system. Applications of fuzzy systems include control, image recognition, system analysis and process monitoring.

In this chapter the basic principles of neural networks and fuzzy systems are introduced along with examples of their application to a range of engineering problems.

6.1 NEURAL NETWORKS

The human brain is a massively interconnected structure containing some 10^{10} of its basic processing elements or *neurons*, each of which is connected to approximately 10^4 other neurons. Of the brain's neurons, the great majority are what are referred to as *interneurons* which carry out what are essentially local processing operations with input and output connections of the order of 100 microns in length. The remaining neurons are used to provide the connections between the regions of the brain, to the various sensory organs and to muscle groups. One result of this massive interconnection, both locally and between the different parts of the brain, is that although different parts of the brain may assume primary responsibility for certain functions, this knowledge is also distributed to and is accessible from other parts of the brain. Thus if part of the brain is lost, it is in many instances possible through a process of relearning for other sectors of the brain to take over the functions previously primarily associated with that part of the brain that has been lost.

An individual neuron is shown in diagrammatic form in Fig. 6.1 and consists of a main body or *soma* attached to which are a series of long, at least at the scale of the cell, irregular filaments or *dendrites* which act as the inputs. The output from the soma is normally provided by the *axon*; though in the case of some interneurons, the dendrites may provide both input and output functions. Axons terminate in a connector known as a *synapse* which links the axon with a dendrite from another neuron. For any individual neuron, each dendrite may be linked with a large number of axons while its output axon may connect to a large number of dendrites to create the massively interconnected network already referred to.

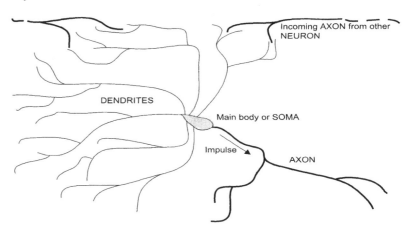

Figure 6.1 The basic structure of a biological neuron.

When a neuron fires it generates a voltage pulse, referred to as an action potential, of around 1 ms duration on the axon. Should the potential at a synapse reach its threshold potential as a result of the arrival of one or more action potentials, the synapse releases chemicals known as *neurotransmitters* which diffuse across the gap between the synapse and the associated dendrite and activates receptors on the dendrite. The activation of these receptors results in a flow of ions along the dendrite which, depending on the activation mode of the synapse, will result in either a positive or negative change in potential on the dendrite being transmitted to the soma.

The soma itself acts to in some way 'sum' the signals received via the dendrites. If this input sum reaches a threshold activation level, the neuron will be activated and will fire, generating an output action potential on its axon which can then be transmitted to other neurons. If the activation level of the soma is not reached then the neuron will remain inactive.

6.2 ARTIFICIAL NEURAL NETWORKS

Artificial neural networks mimic the human brain in that they are constructed from a large number of highly interconnected, simple processing units or neurons which behave like biological neurons in that their output is derived from the weighted sum of their inputs. The primary characteristic of such networks is that they are not programmed, instead they 'learn by example' as a result of presenting them with sets of training data during the course of which process the weights are adjusted according to the learning rules adopted to achieve the desired level of performance.

As a result, artificial neural networks have gradually evolved to applications in such diverse areas as neuroscience, psychology, mathematics, engineering and management. Table 6.1 shows the gradual evolution of neural networks.

Table 6.1 The development of neural networks

Year	Development
1943	McCulloch–Pitts neuron model
1958	Perception
1962	Adeline and Madeline
1974	Backpropogation algorithm outlined
1982	Kohonen self-organising network
1983	Reinforcement learning applied to control
1988	Radial bases functions
1991	Modular neural networks
1992	Soft computing
1997	Neural-fuzzy systems

The learning process may be supervised in which case the network is provided with sets of matching input and output pairs of training data and the learning rules chosen to reinforce those active inputs of any individual neuron which contributes to achieving the desired output while reducing the influence of those active inputs which contradict the desired output conditions. This general form of learning is known as Hebbian learning after the work of Donald Hebb who postulated that for two active neurons with outputs y_i and y_j then the change Δw_{ij} in the associated weight will be determined by the relationship:

$$\Delta w_{ij} = X \cdot y_i \cdot y_j \qquad (6.1)$$

where X is referred to as the learning rate.

With unsupervised learning, unlike supervised learning, there is no predetermined set of output conditions corresponding to a particular set of inputs. Instead the network autonomously develops its own set of output responses to the given input stimuli according to the learning algorithm used. Networks capable of unsupervised learning are often referred to as self-organising networks.

Artificial neural networks are therefore essentially pattern classifiers able to recognise a particular form of input and to generate an appropriate response, particularly in the presence of incomplete or noisy input data.

6.3 THE ADELINE

The adaptive linear element or Adeline was suggested by Widrow and Hoff in 1960 and has the configuration shown in Fig. 6.2. Each of the inputs $(x_1, ..., x_n)$ can take the value $+1$ or -1 and has associated with it a real valued weight $(w_1, ..., w_n)$ which can be either positive or negative. The x_0 input is held at $+1$ and acts as a reference level. The output (Sum) of the summing block of the Adeline is then given by the relationship of equation (6.2):

$$Sum = \Sigma\ w_i\ x_i + w_0 \tag{6.2}$$

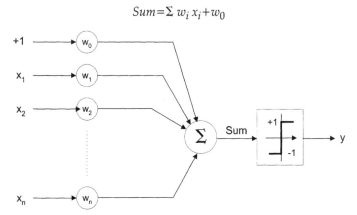

Figure 6.2 The ADELINE.

The final output (y) is then determined by the relationships:

$$y = 1 \text{ for } Sum > 0$$

and $y = 1$ for $Sum = < 0$

6.4 THE PERCEPTRON

The configuration of the basic perceptron was proposed by Rosenblatt in 1959 and consists of an input unit, an association unit and a response unit as shown in Fig. 6.3. Only the last of these, the response unit, has adjustable weights; hence this configuration is often referred to as a single-layered perceptron. Referring to Fig. 6.3, the output of this basic form of the perceptron is of the form of equation (6.3):

$$y = F\left(\sum_{i=0}^{n} w_i a_i\right) = F(k) \tag{6.3}$$

where $F(k) = 1$ for $k > 0$ and $F(k) = -1$ for $k = <0$.

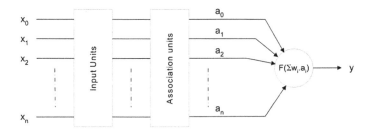

Figure 6.3 The PERCEPTRON.

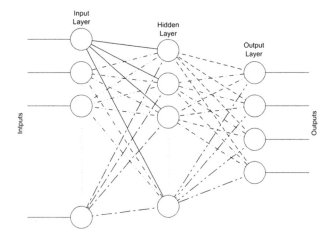

Figure 6.4 Three-layer perceptron.

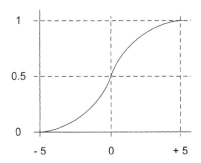

Figure 6.5 The sigmoid function.

6.4.1 Multi-layered Perceptron

Single layered perceptrons are limited to the solution of linearly separable functions. Multi-layered, feed-forward networks of perceptrons provide additional layers of weights which, when combined with more sophisticated learning rules, overcome this limitation. Consider the three-layer network shown in Fig. 6.4 consisting of three layers of neurons.

1. *Input layer*. This receives the input signal and acts to fan out the signal to the next layer of the network. No processing is carried out by this layer.
2. *Hidden layer*. This layer receives the outputs of the input layer and acts to process the signals by the adjustment of weights.
3. *Output layer*. This is the final layer and again acts to process the signal through the adjustment of weights. The outputs from this layer constitute the outputs from the network.

The introduction of a multi-layered structure, which may extend beyond the use of a single hidden layer, requires the use of the derivative of the error function as part of the process of setting weights and hence the hard limited output function of the basic perceptron cannot be used. Instead, the sigmoid function defined by equation (6.4) and having the form shown in Fig. 6.5 is often used.

$$y = \frac{1}{1 + e^{-\beta x}} \tag{6.4}$$

6.4.2 The Delta rule

In the course of the training process, the weights w_0 to w_n are individually adjusted after the presentation of each training set or *epoch* by an amount w_i such as to reduce the error between the desired output and the actual, weighted output. This condition forms the basis of the Delta rule, also known as the least mean squares rule or Widrow–Hoff rule, which attempts to minimise the squared error between the desired and actual outputs. Using the Delta rule, the error term (δ) is expressed as the difference between the desired output (D) and the weighted output of the neuron (Sum) such that:

$$\delta = D - Sum \tag{6.5}$$

where δ is the error between the desired output and the actual output when:

$$\Delta w_i \ \alpha \ x_i \ \delta \tag{6.6}$$

for $i = 0$ to n

The error function (ϵ) is then the mean square error over the full set (P) of training sets used in which case:

$$\epsilon = \frac{1}{P} \sum_{j=1}^{P} E_j = \frac{1}{P} \sum_{j=1}^{P} \delta_j^2 \tag{6.7}$$

where:

$$E_j = \delta_j^2 \tag{6.8}$$

and is the error associated with the jth training set or epoch.

Also:

$$\delta_j = (D_j - Sum_j) \tag{6.9}$$

and:

$$Sum_j = \sum_{i=1}^{n} (w_i x_i) \tag{6.10}$$

In which case:

$$\epsilon = \frac{1}{P} \sum_{j=1}^{P} \left(D_j = \sum_{i=1}^{n} \{w_i x_i\}_j \right)^2 \qquad (6.11)$$

Now let

$$\Delta w_i = -\kappa \times \frac{\partial \epsilon}{\partial w_i} \qquad (6.12)$$

where κ is a constant.

From equation (6.7):

$$\frac{\partial \epsilon}{\partial w_i} = \frac{1}{P} \sum_{j=1}^{P} \frac{\partial E_j}{\partial w_j} = \frac{1}{P} \sum_{j=1}^{P} \left[\frac{\partial E_j}{\partial \delta_j} \times \frac{\partial \delta_j}{\partial w_j} \right] \qquad (6.13)$$

Now, from equation (6.8):

$$\frac{\partial E_j}{\partial \delta_j} = 2\delta_j \qquad (6.14)$$

also

$$\frac{\partial \delta_j}{\partial w_i} = \frac{\partial \delta_j}{\partial Sum_j} \times \frac{\partial Sum_j}{\partial w_j} \qquad (6.15)$$

Putting

$$\frac{\partial \delta_j}{\partial Sum_j} = -1 \qquad (6.16)$$

and

$$\frac{\partial Sum_j}{\partial w_j} = x_{ij} \qquad (6.17)$$

Substituting in equation (6.11) gives:

$$\frac{\partial \epsilon}{\partial w_j} = -\frac{2}{P} \sum_{j=1}^{P} \delta_j x_{ij} \qquad (6.18)$$

Hence from equation (6.12):

$$\Delta w_j = \frac{2\kappa}{P} \sum_{j=1}^{P} \delta_j x_{ij} \qquad (6.19)$$

6.4.3 Backpropagation

The training of a multi-layer, feed-forward network of the form of Fig. 6.4 takes place as a result of supervised learning using the backpropagation algorithm. During training, the network is presented with a series of training sets or *epochs* each consisting of a matched input and output pattern. The weights of the hidden layer are then adjusted by the backpropagation of errors from the output layer so as to decrease the difference between the actual and desired output patterns. The learning rule is therefore a modification of the basic delta rule as given by equation (6.20) in which the constant κ is referred to as the *learning rate* or *learning coefficient*:

$$\Delta w_i = \kappa \delta x \qquad (6.20)$$

The training process generally involves a large number of epochs in order to establish the required distribution of weights. When trained, the network provides a means of classifying an arbitrary set of input patterns with no knowledge of the mathematical relationship between the input and output.

6.4.4 Convergence

The success of the training process is generally measured using the root mean square (RMS) error function of equation (6.21). As the training sequence progresses the RMS error will decrease with a value below 0.1 generally taken as indicating that the training is complete:

$$\epsilon_{\text{RMS}} = \sqrt{\frac{\sum_k \sum_j \left(d_{kj} - y_{kj}\right)^2}{n_k N_0}} \qquad (6.21)$$

where d_{kj} is the desired value for output neuron j after presentation of the kth training set, y_{kj} is the actual output produced by neuron j after presentation of the kth training set, n_k is the number of training sets, and N_0 is the number of output neurons.

6.4.5 Learning difficulties using backpropagation

Ideally, the result of the training should be convergence onto a global minimum. However, it is also possible for the system to encounter a local minimum, resulting in a non-ideal solution. A number of methods are however available to try to prevent this occurring. The most common of these are described in the following sections.

Addition of noise

The addition of a small amount of noise to the weights may well have the effect of disturbing the system from the local minimum. However, the amount of noise needed depends on the nature of the region surrounding the local minima, something which is generally unknown, and there is therefore an element of luck associated with the avoidance of local minima by this means.

Decreasing the learning rate

If the learning rate (κ) in equation (6.20) is set initially to a high value, the system will start by taking large steps through the solution domain. As learning proceeds, reducing the learning rate will result in smaller steps allowing the system to settle down with reduced risk of overshoot. This approach also offers the opportunity of avoiding local minima in the initial stages though at the expense of needing a longer time to converge.

Momentum

The inclusion of a momentum term in the form of a constant (v) operating on the change of weights (Δw_i) as in equation (6.22) can also be used to avoid local minima and aid convergence. The presence of the momentum term means that a large change in weights will produce a correspondingly large change in the subsequent weight adjustment, decreasing as the value of (Δw_i) reduces. This means that the network is less likely to get stuck in a local minimum early on in the learning process and will support convergence in regions of shallow gradient by forcing the network more rapidly 'downhill'. The presence of the momentum term is of particular value in enhancing convergence along shallow gradients.

$$\Delta w_{i+1} = \kappa x_i \delta + v \Delta w_i \tag{6.22}$$

Addition of extra nodes

The number of nodes in the hidden layer can influence the behaviour of the network both positively and negatively. Determining the effect of varying the number of nodes in the hidden layer on network performance can therefore often be a matter of trial and error.

6.4.6 Applications of multi-layer perceptron networks

Pattern classification

One of the best known applications of a multi-layer perceptron network to pattern classification is the NETtalk system developed by Sejnowski and Rosenberg in the late 1980s. NETtalk uses a neural network to isolate individual phonemes and generate speech and is capable of responding to words not included in its training data.

Other pattern classifiers include the analysis and classification of ultrasound data, for instance in relation to the detection and classification of defects in welds, the detection of cancers, the classification of radar images and sonar data and the analysis of noise signatures for applications such as the on-line condition monitoring of machine behaviour.

Image analysis

Multi-layer perceptron networks have been used for a variety of applications in image analysis including feature recognition, the isolation and classification of object types and feature recovery. For instance, neural networks can be trained to recognise specific objects within a crowded image field or in the presence of noise. In other applications, neural networks have been trained to recognise individuals in and from photographs and are being used in security applications for this purpose.

Control

An early application of neural networks to control was to provide a solution to the pole balancing problem of Fig. 6.6 in which the motion of the cart must be controlled to keep the pole upright. This used an adeline network which took the cart position $\{x(t)\}$ and velocity $\{\dot{x}(t)\}$ together with the pole angular position $\{\theta(t)\}$ and the angular velocity $\{\dot{\theta}(t)\}$ as inputs and generated as output the velocity of the cart $\{\dot{x}(t)\}$ required to keep the pole upright.

More recently, neural networks have been used for system identification in which the neural network is trained by one or other of the strategies of Fig. 6.7 to represent the transfer function of the plant, enabling it to be used in a model based controller. Robot controllers have also been developed in which a target point (P) for the robot is generated and fed into the network which then generates a position vector (ϕ). The manipulator then moves to this position and the new position of the end effector (P') determined. This new position is then input to the neural network to produce a position vector (ϕ). The neural network is then trained using the error signal ϵ_{robot} where:

$$\epsilon_{robot} = \phi - \phi' \tag{6.23}$$

This error signal is then used to adjust offsets for subsequent motion.

Financial modelling

Multi-layered perceptron networks have been used for a number of applications in financial modelling including market prediction, commodities markets and mortgage analysis. In the basic model, the training data for the neural net-

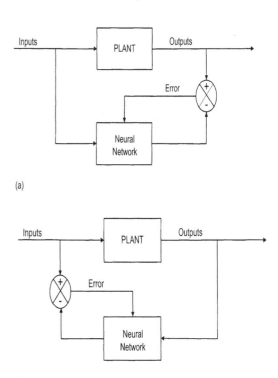

(a)

(b)

Figure 6.7 Neural networks in the control of plant.

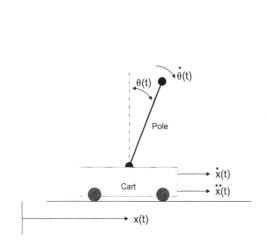

Figure 6.6 Pole balancing cart.

work would utilise historic data on share prices and other financial indicators together with the resulting market trends. Once trained, the network would then be used to predict market trends and variations.

6.5 KOHONEN NETWORK

The Kohonen network shown in Fig. 6.8 is a form of self-organising feature map consisting of an input layer of neurons together with a competitive layer which also forms the output of the network. The two layers are fully interconnected with each input neuron being connected to each of the neurons in the output layer.

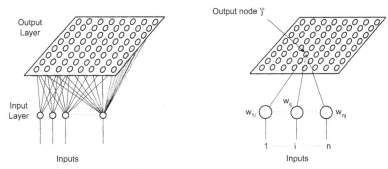

Figure 6.8 Kohonen networks.

Training of the Kohonen network is by means unsupervised learning. Following the presentation of an input vector (**V**) to the input layer, the neurons in the output layer 'compete' to identify a single output or solution neuron using a learning algorithm which generates weights such that the output nodes forming the competitive layer represent the probability density function of the input vectors. The input vector has the form given by equation (6.24) with the weights (**W**$_j$) associated with the jth node in the competitive or output layer then being defined by equation (6.23):

$$\mathbf{V} = [v_1, v_2, v_3, \ldots, v_n\}$$
(6.24)

$$\mathbf{W}_j = [w_{1j}, w_{2j}, w_{3j}, \ldots, w_{nj}]$$
(6.25)

where n is the number of nodes in the input layer.

The values of the weight vector **W**$_j$ would typically be set initially to a value of 0.5 ± a small random value. Following the presentation of a training vector at the input, the distance (d) between the input and output vectors is then calculated as:

$$d_j = \sqrt{\sum (v_i - w_{ij})^2}$$
(6.26)

where d_j is the distance between the input vector and output node j, v_i is the input to node i, and w_{ij} is the weight between input node i and output node j.

The output node with the shortest value for distance (d_j) is then declared the best match and 'wins' the competition. The weights of this node and all the other nodes in its defined neighbourhood are then adjusted according to the relationship:

$$\Delta w_{ij} = \alpha_{ij}(v_i - w_{ij}) \tag{6.27}$$

where α_{ij} is the learning rate.

For all nodes in the neighbourhood of the winning node:

$$w_{ij\text{NEW}} = w_{ij\text{OLD}} + \Delta w_{ij} \tag{6.28}$$

Referring to equation (6.26), the learning rate ij is a function of the distance of any individual node from the winning node with a typical relationship being the Gaussian form shown in Fig. 6.9. As training progresses both the learning rate α_{ij} and the size of the neighbourhood around any individual output node would be progressively reduced. Typically, the learning rate would be varied according to the relationship:

$$\alpha_{ij:n} = \alpha_{ij:0} \left(1 - \frac{n}{N}\right) \tag{6.29}$$

where $\alpha_{ij:0}$ is the initial value of the learning rate, $\alpha_{ij:n}$ is the value of the learning rate following the nth epoch of training, and N is the number of epochs in the training set, while the width (h) of the neighbourhood is varied according to:

$$h_n = h_0 \left(1 - \frac{n}{N}\right) \tag{6.30}$$

where h_n is the width of the neighbourhood after the nth epoch of training, and h_0 is the original width of the neighbourhood.

6.6 ADAPTIVE RESONANT THEOREM NETWORKS

A particular problem of the multi-layer perceptron networks is the inability to learn additional or new information without disturbing their current level of training. The adaptive resonance theory (ART) architecture is a self-organising network structure which allows the system to switch between a learning or plastic state in which the network parameters may be modified and a stable or fixed state for operation.

The basic configuration of an ART network is shown in Fig. 6.10 and is seen to consist of an input or comparison layer and output or recognition layer. Each neuron in the comparison layer is connected to all neurons in the recognition layer using the feedforward weight vectors ($\mathbf{w_f}$) while each neuron in the recognition layer is connected to those in the comparison layer using the feedback weight vectors ($\mathbf{w_b}$). In addition, there are also provided control signals S_1 and S_2 together with a reset signal which acts to compare the inputs with a vigilance pattern in order to determine if a new class pattern should be created for any given input pattern.

The operation of the ART network can be considered in terms of a number of discrete phases as described in the following sections.

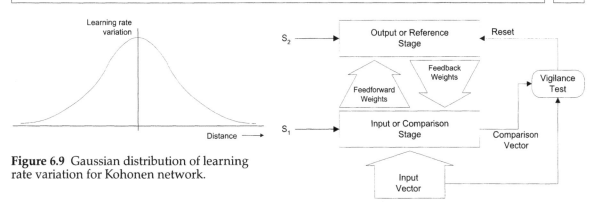

Figure 6.9 Gaussian distribution of learning rate variation for Kohonen network.

Figure 6.10 ART network structure.

6.6.1 Initialisation

During the initialisation phase the control signal S_1 acts to switch the associated nodes between their input and comparison modes and is set to 1 whenever a valid (non-zero) input is received but is forced to zero (0) when any node in the recognition layer is active. The control signal S_2 is used to enable or disable the nodes in the recognition layer and is set to 1 for any valid input pattern but becomes zero (0) if the reset signal indicates a failed vigilance test.

The weight vectors $\mathbf{w_f}$ and $\mathbf{w_b}$ are initialised according to equations (6.31) and (6.32):

$$\mathbf{w}_{f:ij} = \frac{1}{1+n} \tag{6.31}$$

where n is the number of nodes in the input layer, m is the number of nodes in the output layer, $\mathbf{w}_{f:ij}$ the forward weight between node i in the input layer and node j in the output layer, and $0 \le i \le n{-}1$ and $0 \le j \le m{-}1$. Also

$$\mathbf{w}_{b:ij} = 1 \tag{6.32}$$

where $\mathbf{w}_{b:ij}$ is the backward weight between node I in the input layer and node j in the output layer.

The vigilance threshold (v) will be set to a value between 0 and 1 ($0 < v < 1$).

6.6.2 Recognition

During the recognition phase each neuron in the comparison layer receives three inputs as follows:

- a component of the input pattern;
- a component of the feedback pattern;
- the control signal S_1.

A neuron in the comparison layer will output a 1 only if any two of the three inputs are high. This is the two-thirds rule suggested by Grossberg and Carpenter in their early work on the ART network.

The input vector (S_{in}) is then compared with the feedforward weight vector associated with each of the individual nodes in the output layer and the output node with the best match determined by computing the output node with the largest value for the scalar (dot) product of the input vector and the feedforward weight vector as in equation (6.33). Lateral inhibition between nodes in the output layer is then used to inhibit all other output nodes. The winning node then transfers its stored classification pattern in the form of the feedback weight vector $\mathbf{w}_b$ back to the comparison layer:

$$P_j = \sum_{i=0}^{n-1} \mathbf{w}_{f:ij} \mathbf{s}_{in} \geq \| \mathbf{w}_{f:ij} \cdot \mathbf{S}_{in} \| \tag{6.33}$$

for $0 \leq j \leq m-1$ where P_j is the result for node j in the output layer.

6.6.3 Comparison

During the comparison phase, the input layer is receiving both the input vector and that feedback from the winning node in the output layer together with the control signal $\mathbf{S}_1$ which will be set to zero as the output layer has an active node. Using the two-thirds rule, the input vector and the feedback vector will be ANDed together to produce a new vector, referred to as the comparison vector ($\mathbf{S}_{comp}$), at the output of the comparison layer. This vector is now passed to the reset circuit along with the input vector.

6.6.4 Reset

The reset circuit evaluates the dot product of the input and comparison vectors ($\mathbf{S}_{in}$ and $\mathbf{S}_{comp}$) to generate a count of the number of matching 1's in each pattern. This is then divided by the bit count of 1's in the input vector as in equation (6.34) to generate a value (τ) which is compared with the vigilance threshold (v):

$$\tau = \frac{\sum S_{in} S_{comp}}{\sum S_{in}} \tag{6.34}$$

If the computed value is greater than the vigilance threshold, then the classification is considered as complete and the membership is as indicated by the active node in the output layer. Weights are then adjusted for the best match according to the relationships:

$$\mathbf{w}_{f:ij}(x+1) = \mathbf{w}_{f:ij}(x)\mathbf{S}_{in} \tag{6.35}$$

and

$$\mathbf{w}_{b:ij}(x+1) = \frac{\mathbf{w}_{f:ij}(x)\mathbf{S}_{in}}{0.5 + \sum_{i=0}^{n-1} \mathbf{w}_{f:ij}\mathbf{S}_{in}} \tag{6.36}$$

If the vigilance threshold is not achieved it is assumed that a best match has not been achieved and the network enters its search phase.

6.6.5 Search

In the search phase the network attempts to find a node in the output layer which provides a better match than that previously obtained. The search phase begins by disabling the previously selected output node and setting its output to zero, forcing S_1 to zero and preventing that particular node from being involved with any further comparisons with the current input vector.

The network then repeats its comparison phase to select a new output node; this process is then repeated until an output node is found which passes the test for the vigilance threshold. If no satisfactory node is found the network then declares the input vector as a previously unknown class and assigns it to a previously unassigned node in the output layer.

6.6.6 Training

The ART network is very sensitive to variations in its parameters during training, particularly to the choice of the vigilance threshold with low values resulting in a low resolution process with few class types and a high value producing a sensitive network with fine resolution. The initialisation of the feedforward weight vectors to low values is also important to avoid any individual vector dominating the training process.

6.6.7 ART network summary

The version of the ART network described is that known as ART-1 and demonstrates the basis of the approach used by the theorem. The ART-2 network is similar to ART-1 but has the input layer divided into a number of functional layers to allow for more complex data matching and also incorporates developments in areas such as feature matching and noise suppression. The ART-3 network has the same basic topology as ART-2 but the defining equations are modified to more closely represent the behaviour of chemical neurotransmitters.

6.7 HOPFIELD NETWORK

A Hopfield network consists of a single layer of fully interconnected neurons as shown in Fig. 6.11. This topology together with the fact that each node communicates bi-directionally with every other nodes means that the network acts recursively to reach a stable condition. In which case the state of the network can be described by the state vector **U** such that:

$$\mathbf{U} = (u_1, u_2, u_3, ..., u_n) \tag{6.37}$$

in which u_1 to u_n are the states of the individual nodes.

Additionally, the connections between nodes I and j will have associated with them a weight w_{ij}. Typically, but not always, the weights in each direction will be the same such that:

$$w_{ij} = w_{ji} \tag{6.38}$$

In operation, each node of the Hopfield net will be updated in a random sequence with the process being repeated until there is no further change in the state of any of the neurons. During the updating process, the response of an individual

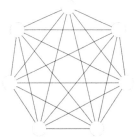

Figure 6.11 Hopfield network with seven neurons.

neuron is determined by the weighted sum of its inputs according to the relationship:

$$S_i(t+1) = \sum_{i \neq j} u_j(t) w_{ij} + \theta_i \tag{6.39}$$

The output u_j is then determined by reference to a threshold function ϕ such that:

$$u_i(t+1) = \begin{cases} +1 \text{ for } S_i(t+1) > \phi \\ -1 \text{ for } S_i(t+1) < \phi \\ u_i(t) \text{ otherwise} \end{cases} \tag{6.40}$$

Each state of the Hopfield network may be described by the energy function of equation (6.41):

$$E = -\frac{1}{2} \sum_{i \neq j} \sum ui_i u_j w_{ij} - \sum_i \theta_i u_i \tag{6.41}$$

Setting both the offset (Θ) and the threshold (ϕ) equal to zero then, for a change in node j, the associated energy is given by:

$$E_j = -\frac{1}{2} \sum_{i \neq j} \sum u_i u_j w_{ij} = -\frac{1}{2} u_j \sum_{i \neq j} u_i w_{ij} \tag{6.42}$$

If following the update procedure the output of node j is unchanged, there will be no change in the value of E_j. However, should the output of node j change then the difference in energy ΔE_j will be:

$$\Delta E_j = E_j(t+1) - E_j(t) = \frac{1}{2} \{u_j(t+1) - u_j(t)\} \sum u_i w_{ij} = \frac{1}{2} \Delta u_j \sum u_i w_{ij} \tag{6.43}$$

If u_j changes from -1 to 1 then:

$$\Delta u_j = 2 \tag{6.44}$$

Since for a change from -1 to 1 the weighted sum is greater than zero (0) when:

$$\sum u_i w_{ij} \geq 0 \tag{6.45}$$

in which case, from equation (6.43):

$$\Delta E_j \leq 0 \tag{6.46}$$

Similarly, if u_j changes from 1 to -1 then:

$$\Delta u_j = -2 \tag{6.47}$$

when for a change from 1 to -1 the weighted sum is less than zero (0) when:

$$\sum u_j w_{ij} < 0 \tag{6.48}$$

and

$$\Delta E_j > 0 \tag{6.49}$$

The network is therefore bound to converge with the energy function taking on successively lower values as the network approaches steady state.

Unfortunately, as the Hopfield network only seeks a minima, there is no way of knowing if the stable state corresponds to a local or a global minima. Indeed, there is no means other than restarting the network using a different set of conditions of disturbing it from a local minimum.

6.7.1 Application of Hopfield networks

The major application of Hopfield networks is as a form of associative memory. In this role, the network contains the same number of nodes as there are elements in each of the stored patterns. For the kth pattern in the set, the associated state vector $\mathbf{P_k}$ is defined by:

$$\mathbf{P_k} = (p_{k1}, p_{k2}, p_{k3}, ..., p_{kn}) \tag{6.50}$$

Assuming m patterns in total, the weights are determined by the relationship:

$$w_{ij} = \sum_{k=1}^{m} (2p_{ki}-1)(2p_{kj}-1) \tag{6.51}$$

The pattern for which a match is required is applied to the net which is then allowed to stabilise when the resultant state vector in the stable state represents the pattern as 'recognised' by the network.

A new pattern to be added to the set is 'learnt' by recalculating the weights to take account of the added pattern. It has been established that a network can typically store up to $0.15n$ states where n is the number of nodes in the network.

6.7.2 Boltzmann machine

As has been indicated, a Hopfield network may well converge on a minimum which may not be the global minimum. In order to overcome this tendency a controlled amount of noise, analogous to thermal noise, can be added to the system in order to attempt to disturb the solution from any local minima in which it is tending to settle. In the case of the Hopfield network this involves the use of a probabilistic update rule, the resulting network being referred to as a Boltzmann machine.

Referring to equation (6.52), it is seen that the energy associated with a particular node is:

$$\Delta E_j = -\frac{1}{2} \Delta u_j \sum u_j w_{ij} \tag{6.52}$$

The node will switch to a lower energy state according to the probabilistic update rule of equation (6.53):

$$P_j = \frac{1}{1+e^{-\Delta E_j/T}} \tag{6.53}$$

where T is the simulated temperature of the system.

If P_x is the probability of the network settling into a state of energy E_x the Boltzmann distribution has the form:

$$P_x = ke^{-E_x/T} \tag{6.54}$$

If P_y is similarly the probability of a state with energy E_y then:

$$\frac{P_x}{P_y} = \frac{e^{-E_x/T}}{e^{-E_y/T}} = e^{-(E_x-E_y)/T} \tag{6.55}$$

As the network settles to thermal equilibrium then if E_x is a lower energy state than E_y:

$$\left\{ \begin{array}{c} E_x < E_y \\ e^{-(E_x-E_y)/T} > 1 \\ \text{and} \\ P_x > P_y \end{array} \right\} \tag{6.56}$$

Thus, as the network approaches its thermal equilibrium state it is more probable that it will settle into a lower energy state.

If the value of 'temperature' (T) is high, then the system will rapidly reach equilibrium but there is no indication as to whether the solution reached is the best available. By gradually reducing the temperature as the system settles the probability of reaching an optimum solution is increased as the use of high 'temperatures' initially will act to cause the system to be dislodged from a local minima. As the temperature is reduced the probability of dislodgement is reduced but so is the probability of moving back up the energy states. This process of gradual temperature reduction is referred to as simulated annealing after the cooling process used in the heat treatment of metals.

6.8 IMPLEMENTATION OF NEURAL NETWORKS

The operation of neural networks can be simulated by software on a variety of platforms. In many applications such a software format may well provide the range of performance levels required, particularly as processor speeds increase.

Where there is a need to build the neural processor into a stand alone system then implementations based on neural processors or the use of digital signal processing (DSP) chips have been used. There has also been significant work to develop high speed artificial neurons for specialised applications such as image processing. These will typically be implemented using VLSI technology and offer a high degree of parallelism for high speed applications.

6.9 FUZZY SYSTEMS

Natural language communication between individuals contains a range of uncertainties and ambiguities such as 'quite tall', 'very tall', 'middle aged', 'quite compact', 'relatively few' and so forth which are resolved in conversation by an individual's own knowledge and understanding of the topic under discussion. Such natural language forms are consistent with the way in which humans perceive and respond to changes in their environment. Thus for instance, the concept of 'a comfortable temperature' will encompass a wide range of temperatures depending on an individual's personal preference. Similarly, a group of Formula 1 drivers are likely to have a different idea of what constitutes 'fast' than an average motorist.

Fuzzy systems therefore have their origin in the fact that in the real, analogue, world there are no absolutes. Rather, there is a gradual transition from one state or condition to another and this is reflected in the use of language to describe a particular condition. Even then there are shades of meaning. For instance, how does a clear day translate into a cloudy day and then into an overcast day. Fuzzy sets reflect the real world and our perceptions of and ways of describing that world. Thus phrases and statements such as:

'James is *tall*.'
'The weather is *cold*.'
'Turn the volume *down a little*!'
'The queue is moving *rather slowly*.'
'If the mixture is *slightly too thick*, then add a *little* water.'

all have meaning within the context in which they are made.

A *fuzzy proposition* will therefore contain words such as 'tall', 'average' or 'short' which define the nature of the fuzzy sets TALL, AVERAGE and SHORT respectively. Associated with the fuzzy proposition may also be a *fuzzy qualifier* such as 'very', 'quite' or 'extremely' which further define the proposition. Thus the fuzzy propositions:

'Robert is *quite* tall.'
'Robert is *very* tall.'
'Robert is *extremely* tall.'

all use a qualifier to convey additional meaning.

In addition to fuzzy qualifiers, *fuzzy quantifiers* or *hedges* such as 'most', 'normally' and 'few' will be used to further define propositions. Thus the statements:

'A person of 6 ft would *normally* be considered as tall.'
'*Most* projects suffer overruns.'
'There are a *few* instances of very low scores on this course.'
'The price would *normally* be slightly less than our competitors.'

all use fuzzy quantifiers.

The ideas behind what is termed *fuzzy logic* and its application to the control and operation of a wide range of complex engineering systems can be attributed to the work of Lotfi Zadeh in the mid-1960s and papers such as that entitled 'Fuzzy Sets' and published in the journal *Information and Control* in 1965. In this and other related papers of the same period, Zadeh sets out the mathematical background for fuzzy sets and their inter-relationships from which has developed a wide variety of applications as illustrated by Table 6.2.

Though Table 6.2 is far from comprehensive, it nevertheless serves to illustrate the scope for the application of fuzzy systems in the control and operation of engineering systems. The advantage of using a fuzzy system for these and other operations is that their operation reflects the way in which a human operator would control the system and alleviates the need for a detailed mathematical model. When combined with neural networks to produce fuzzy-neural, or neuro-fuzzy, controllers a learning capacity can be included enabling the system to adapt to unforeseen changes in system conditions when in operation.

6.10 FUZZY SETS

Conventionally, membership of a set is defined by the *characteristic function* for

Table 6.2 Applications of fuzzy systems

Application	Fuzzy operation
Engine management	Controls the operation of a car engine taking into account throttle setting, temperature, fuel/air mixture and so forth
Anti-lock brakes	Controls the application of the brakes according to vehicle speed, wheel speed, wheel spin and deceleration rate
Transmission control	Adjusts the frequency and stops of the lift according to passenger density and location
Blast furnace control	Controls the temperature, charge and discharge conditions of the blast furnace
Image/pattern recognition	Enables the discrimination between characters and shapes of similar format
Planning and scheduling	Adjusts schedules in response to current conditions, for instance by diverting buses in response to passenger demand
Air conditioning	Provides a stable environment taking into account factors such as external conditions, room occupancy and heat gains/losses
Washing machines	Adjusts wash cycle time in response to measured levels of dirt in water, size of load and fabric type
Video camera	Stabilises image for hand shake and controls focusing

that set such that the characteristic function has a value of 1 if the object is a member of the set and is zero otherwise. Thus:

$$\mu_A(x) = \begin{cases} 1 \text{ if } x \text{ is a member of set } A \\ 0 \text{ if } x \text{ is not a member of set } A \end{cases} \tag{6.57}$$

where the objects x are members of the *Universe* X which defines the total set space.

Sets to which the relationship of equation (6.57) applies are referred to as *crisp sets*.

In a *fuzzy set* the characteristic function is replaced by the *membership function* which can assume a value between 0, representing no membership, and 1, implying full membership. The actual value of the membership function in the range 0 to 1 is then referred to as the grade or level of membership.

In order to better understand this concept of crisp and fuzzy sets, and the relationship between them, consider the concepts of 'short', 'average height' and 'tall' as related to a group of people ranging in height from 1 m (≈ 5 ft 3 in) to 2 m (≈ 6 ft 7 in). Using crisp sets we may chose to define everyone less than 1.62 m (≈ 5 ft 4in) as 'short', everyone in the range 1.62 m to 1.78m (≈ 5 ft 10in) as of 'average height' and everyone taller than 1.78m as 'tall'. These definitions can then be represented graphically as in Fig. 6.12.

If instead a fuzzy representation was used the distribution of short, average height and tall individuals will blend into each other as illustrated by the relationships of Fig. 6.13 in which it is seen that there is a significant overlap between the sets defining 'short' and 'average height' and between those for 'average height' and 'tall'. Thus an individual 1.8 m (~ 5 ft 11 in) in height would have a membership function of 0.5 for the set of 'average height' and 0.4 for the set 'tall'.

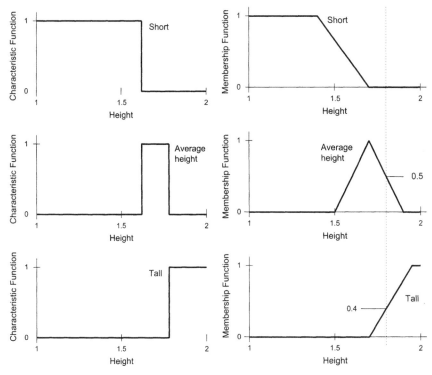

Figure 6.12 Crisp sets for groups short, average and tall.

Figure 6.13 Fuzzy representation of the sets for SHORT, AVERAGE HEIGHT and TALL.

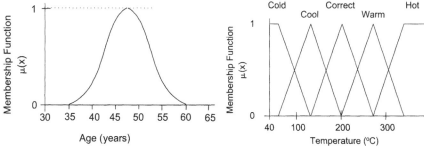

Figure 6.14 Fuzzy set for MIDDLE AGED.

Figure 6.15 Universe of discourse containing fuzzy sets for COLD, COOL, CORRECT, WARM and HOT.

6.10.1 Domains

The domain of a fuzzy set is the universe of values making up the set. The choice of domain is determined by the range of values that are to be represented by the fuzzy set in the context of the system for which it is being used. Consider the fuzzy set of Fig. 6.14 defining the condition 'middle aged' in the range from 35 years to 60 years. In this case, the domain is bounded by the limits of 35 years and 60 years even though there will be individuals with ages both above and below these limits since these are the points at which the value of the membership function reaches 0.

The fuzzy set MIDDLE AGED of Fig. 6.14 may be expressed in the form:

$$\mu_A[x] = [0/35, 0.23/40, 0.88/45, 0.88/50, 0.23/55, 0/60] \tag{6.58}$$

or

$$\mu_A[x] = 0/35 + 0.23/40 + 0.88/45 + 0.88/50 + 0.23/55 + 0/60 \tag{6.59}$$

in which the value of the membership function is followed by the associated temperature. Thus an age of 40 has associated with it a membership function of 0.23.

More generally, a finite fuzzy set containing n elements can be represented in terms of the series of fuzzy singletons (μ_i/x_i) making up the set such that:

$$\mu_A[x] = \mu_1/x_1 + \mu_2/x_2 + \ldots + \mu_n/x_n \tag{6.60}$$

6.10.2 Universe of discourse

The system space often contains a number of overlapping fuzzy sets as in Fig. 6.15 for a range of temperatures from 40°C to 400°C. This range from 40°C to 400°C is referred to as the *universe of discourse* and applies to the system or solution space with each of the individual fuzzy sets occupying that system space having its own domain. The total number of fuzzy sets associated with the universe of discourse is then referred to as the *term set*.

6.10.3 Alpha-cut

The alpha-cut (α-cut) or alpha-level of a fuzzy set is defined as a set containing all those domain values of a fuzzy set associated with a minimum value of α for the membership function. The α-cut may be either strong in which case:

$$\mu_A(x) > \alpha \tag{6.61}$$

or weak when:

$$\mu_A(x) \geq \alpha \tag{6.62}$$

The α-cut acts to restrict the size of the domain. Thus referring to Fig. 6.16 a value α = 0.3 will restrict the domain to the range from 41 years to 54 years.

6.10.4 Support set

Consider the situation of Fig. 6.17 in which the points at which the curve departs from and returns to zero do not coincide with the domain limits of 35 years and 60 years respectively. In this case, the range from 40 years to 55 years defines the support set.

6.10.5 Representation of fuzzy sets

The shape of the curve defining the fuzzy set is often of importance in defining the behaviour of the system with which the fuzzy set is associated. Typical shapes associated with fuzzy sets are shown in Fig. 6.18 and the choice of the representation used will in general be associated with the understanding of system behaviour.

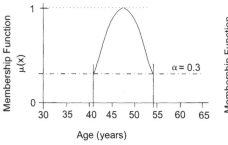

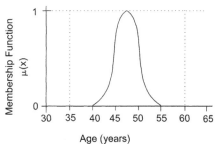

Figure 6.16 α-cut at 0.3 for fuzzy set MIDDLE AGED.

Figure 6.17 Support set for fuzzy set MIDDLE AGED for domain from 35 years to 60 years.

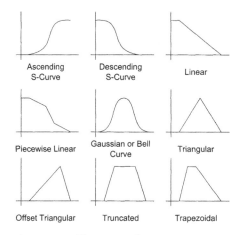

Figure 6.18 Fuzzy set shapes.

6.11 SET OPERATIONS

6.11.1 Crisp sets

For a crisp set the Boolean relationship of equation (6.57) can be written as:

$$\mu_A(x) = \begin{cases} 0 \text{ if } x \notin A \\ 1 \text{ if } x \in A \end{cases} \tag{6.63}$$

in which the logical operators $\in$ and $\notin$ are as defined in Table 6.3.

For such crisp sets there is no ambiguity about membership and the basic set operations of inclusion ($A \subset B$), complement ($\sim A$), intersection ($A \cap B$) and union ($A \cup B$) result in a new series of sets as illustrated by Fig. 6.19. A further operation, the exclusive-OR ($A \oplus B$) results in a set which contains all those elements which are in either A or B only.

Table 6.3 Logical operators

Symbol	Meaning
$\in$	Member of set
$\notin$	Not a member
$\cap$	Set AND or INTERSECTION
$\cup$	Set OR or UNION
$\subset$	INCLUSION
$\supset$	IMPLICATION
$\sim$	Set NOT, COMPLEMENT or INVERSION
$\varnothing$	NULL or EMPTY set
$\mu.[x]$	FUZZY membership function
$\{x\}$	CRISP or BOOLEAN membership function
$\wedge$	Logical AND
$\vee$	Logical OR
$\oplus$	exclusive-OR

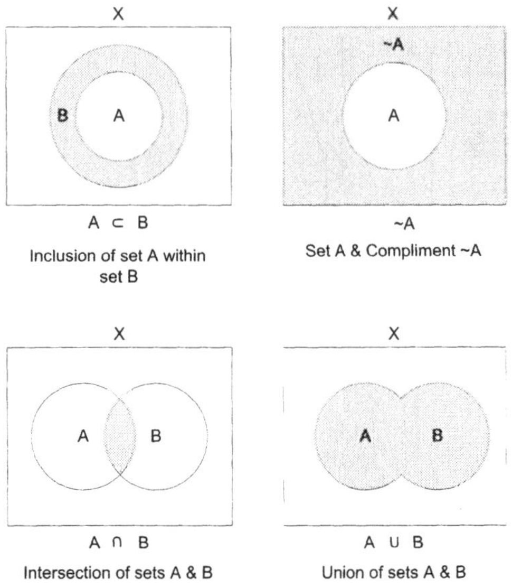

Figure 6.19 Crisp set operations.

6.11.2 Fuzzy sets

The nature of fuzzy sets means that though the same operations of intersection, union and complement exist, they have a different interpretation to those for crisp sets. For fuzzy sets:

Intersection	$A \cap B = \min(\mu_A[x], \mu_B[y])$	(6.64)
Union	$A \cup B = \min(\mu_A[x], \mu_B[y])$	(6.65)
Complement	$\sim A = 1 - \mu_A[x]$	(6.66)

These relationships are illustrated in Fig. 6.20.

6.12 FUZZY REASONING AND CONTROL

Fuzzy reasoning or *fuzzy inference* forms the basis of fuzzy expert systems. In a similar way to a conventional expert system, the fuzzy expert system will be structured around a knowledge base structured around a set of fuzzy rules. These will be structured in the same:

'IF .... THEN ....'

form as in a conventional rule base but will contain both fuzzy quantifiers and fuzzy qualifiers.

Consider the pole balancing or inverted pendulum problem referred to earlier and shown again in Fig. 6.21 in which the requirement is to control the motion ($\dot{x}(t)$ and $\ddot{x}(t)$) of the cart to maintain the pole upright. The information available for control purposes is the angle of the pole ($\theta(t)$) and the angular velocity ($\dot{\theta}(t)$). Working from this information and defining the possible range of values for $\theta(t)$ and $\dot{\theta}(t)$ set out in Table 6.4:

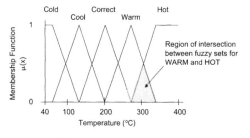

Figure 6.20a Fuzzy set operations.

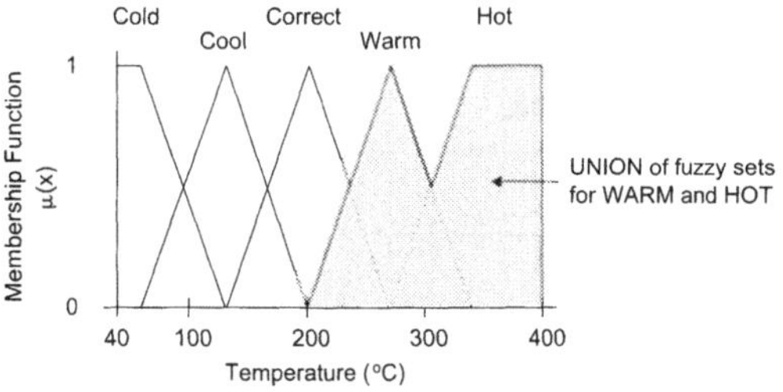

Figure 6.20b Fuzzy set operations.

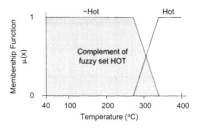

Figure 6.20c Fuzzy set operations.

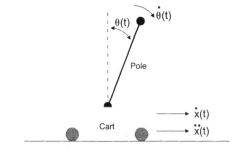

Figure 6.21 Pole balancing cart.

Table 6.4 Fuzzy sets for inverted pendulum

Set	Descriptor
Negative large	NL
Negative medium	NM
Negative small	NS
Zero	Z
Positive small	PS
Positive medium	PM
Positive large	PL

A set of rules governing the operation of the controller can now be drawn up along the lines of:

IF the angle ($\theta(t)$) of the pendulum is negative medium AND the angular velocity ($\dot{\theta}(t)$) is near zero
THEN set the motor speed (ω) to positive medium

IF the angle ($\theta(t)$) of the pendulum is positive small AND the angular velocity ($\dot{\theta}(t)$) is near zero
THEN set the motor speed (ω) to negative small

and

IF the angle ($\theta(t)$) of the pendulum is near zero AND the angular velocity ($\dot{\theta}(t)$) is near zero
THEN set motor speed (ω) to zero

For the system as defined with 7 fuzzy sets used to describe each of $\theta(t)$ and $\dot{\theta}(t)$, the number of possible fuzzy rules would be $7 \times 7 = 49$. In practice, the

$\theta(t)$

	NL	NM	NS	Z	PS	PM	PL
NL				NL			
NM				NM			
NS				NS	PS		
Z	NL	NM	NS	Z	PS	PM	PL
PS			NS	PS			
PM				PM			
PL				PL			

$\dot{\theta}(t)$

Figure 6.22 Fuzzy associative memory map for the pole balancing problem with 15 active rules.

number of active rules used would be determined by the performance requirements. These active rules can displayed in the form of a *fuzzy associative memory* (FAM) map as shown in Fig. 6.22 for a system with 15 rules in which the outputs are given in the shaded boxes. The fuzzy sets themselves are shown graphically in Fig. 6.23.

Similar matrices for other groups of rules could be generated for the pole balancing problem and Fig. 6.24 shows a FAM with 11 rules in which the conditions positive large and negative large have been ignored for both $\theta(t)$ and $\dot{\theta}(t)$.

In practice, the system would be configured as in Fig. 6.25 in which the information about the system conditions by the sensors is fed to all the rules

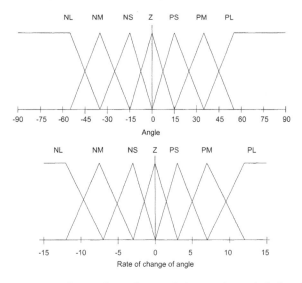

Figure 6.23 Fuzzy sets for angle and rate of change for pole balancing problem.

$\theta(t)$

	NM	NS	Z	PS	PM
NM			NM		
NS			NS	Z	
Z	NM	NS	Z	PS	PM
PS		Z	PS		
PM			PM		

$\dot{\theta}(t)$

Figure 6.24 Fuzzy associative memory map for the pole balancing problem with 11 active rules.

making up the fuzzy rule base or FAM simultaneously. Because of the fuzzy nature of the system and the fact that the individual fuzzy sets will overlap more than one of the rules will normally be activated to produce an output. These outputs are then combined and fed to the *defuzzifier* which generates a value which is fed to the actuators to control the system.

With reference to the pole balancing problem and the associated FAM of Fig. 6.22 consider the situation of Fig. 6.26 in which $\theta(t)$ has a value of $+10°$ and $\dot{\theta}(t)$

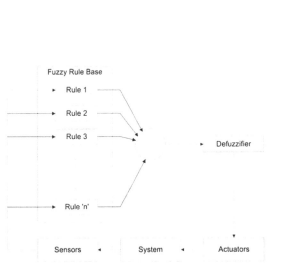

Figure 6.25 Basic structure of a fuzzy controller.

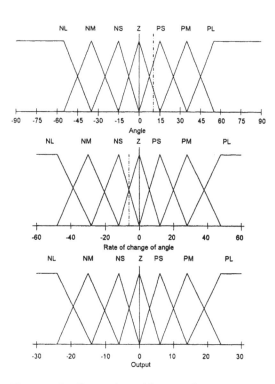

Figure 6.26 Operation of fuzzy rules.

a value of $-6°/s$. Reference to Fig. 6.22 will show that the following rules will be fired:

  IF the angle $(\theta(t))$ of the pendulum is near zero AND the angular velocity $(\dot\theta(t))$ is near zero
    THEN set the motor speed (ω) to zero

IF the angle $(\theta(t))$of the pendulum is near zero AND the angular velocity $(\dot\theta(t))$ is negative small
    THEN set the motor speed (ω) to positive small

  IF the angle $(\theta(t))$ of the pendulum is positive small AND the angular velocity $(\dot\theta(t))$ is near zero
    THEN set the motor speed (ω) to negative small

  IF the angle $(\theta(t))$ of the pendulum is positive small AND the angular velocity $(\dot\theta(t))$ is negative small
    THEN set the motor speed (ω) to negative small

or

IF $\theta(t) = Z$ AND $\dot\theta(t) = Z$ THEN $\omega = Z$
IF $\theta(t) = Z$ AND $\dot\theta(t) = NS$ THEN $\omega = PS$
IF $\theta(t) = PS$ AND $\dot\theta(t) = NS$ THEN $\omega = NS$
IF $\theta(t) = PS$ AND $\dot\theta(t) = NS$ THEN $\omega = NS$

Referring to Fig. 6.27 it can be seen that for the first of these active rules the value of the membership function for $\theta(t)$ is 0.33 and for $\dot\theta(t)$ it is 0.5, similar relationships can be found for the other rules shown by Figs 6.28, 6.29 and 6.30.

In order to generate the output level associated with each rule the *minimum* of the membership functions for $\theta(t)$ and $\dot\theta(t)$ generated by that rule. Thus for the rule:

IF $\theta(t) = Z$ AND $\dot\theta(t) = Z$ THEN $\omega = Z$

the output value is 0.33. This can be expressed by the relationship of equation (6.67) and is shown in Fig. 6.27:

$$\omega_z = \min \mu_z^\theta(+10), \mu_z^{\dot\theta}(-6) = \min (0.33, 0.5) = 0.33 \tag{6.67}$$

For the remaining rules the relationships and resulting output values are obtained from equations (6.68), (6.69) and (6.70) respectively as follows:

$$\omega_{PS} = \min \mu_z^\theta (+10), \mu^{\dot\theta} (-6) = \min (0.33, 0.5) = 0.33 \tag{6.68}$$

$$\omega_{NS} = \min \mu^\theta (+10), \mu_z^{\dot\theta}(-6) = \min (0.68, 0.5) = 0.5 \tag{6.69}$$

$$\omega_{NS} = \min \mu^\theta (+10), \mu^{\dot\theta} (-6) = \min (0.68, 0.5) = 0.5 \tag{6.70}$$

Having obtained the values for the membership functions for motor speed they can be combined as shown in Fig. 6.31 to produce a fuzzy output which is passed to the defuzzifier to produce an output value which will be supplied to the motor driver.

6.12.1 Defuzzification

The process of defuzzification is used to convert the fuzzy outputs from the fuzzy rule base into a single, crisp value which can be supplied as the control signal to the actuators. A number of techniques are available for this purpose

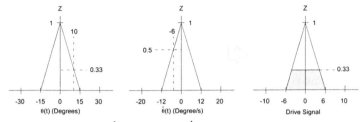

Figure 6.27 Fuzzy rule IF $\dot{\theta}(t) = Z$ AND $\dot{\theta}(t) = Z$ THEN $\omega = Z$.

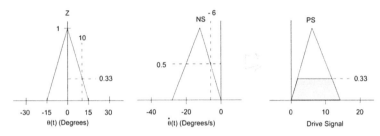

Figure 6.28 Fuzzy rule IF $\dot{\theta}(t) = Z$ AND $\dot{\theta}(t) = NS$ THEN $\omega = PS$.

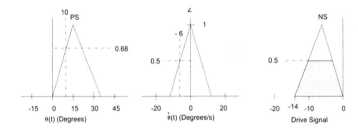

Figure 6.29 Fuzzy rule IF $\dot{\theta}(t) = PS$ AND $\dot{\theta}(t) = Z$ THEN $\omega = NS$.

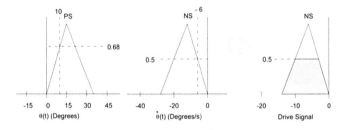

Figure 6.30 Fuzzy rule IF $\dot{\theta}(t) = PS$ AND $\dot{\theta}(t) = NS$ THEN $\omega = NS$.

of which the most commonly used are the *mean of maxima, centre of maximums* and *centre of gravity* methods.

Mean of maxima

Using the mean of maxima method the limits of the region associated with the maximum value of the fuzzy output are first identified and the mean value of these limits is then used as the crisp output value. Referring to Fig. 6.32 it can be seen for the example under discussion that the maximum region covers output values in the range from -3 (S_{max}) to -10 (S_{min}). The crisp output value is then:

$$\text{Mean of maxima output} = \frac{S_{max} + S_{min}}{2} = \frac{-3 + (-10)}{2} = -6.5 \tag{6.71}$$

Centre of maximums

The centre of maxima method is often used where the fuzzy output has a number of levels. Using this method the highest and next highest plateau's are identified and the midpoint between the centres of these two plateau is used as the output. Referring to Fig. 6.33 it can be seen that these centres lie at values of -6.5 and 4.67 respectively, in which case the centre of maximums method generates an output value of:

$$\text{Centre of maximums output} = \frac{4.67 + (-6.5)}{2} = -0.92 \tag{6.72}$$

Centre of gravity

Using the centre of gravity or *centroid* method the first moment of area is used to locate the position of the centre of gravity of the fuzzy output along the *x*-axis. This can be expressed as:

$$\text{Centre of gravity output} = \frac{\Sigma(\text{Moments of individual areas})}{\text{Total area under curve}} \tag{6.73}$$

Applying this relationship to the fuzzy output of Fig. 6.31 gives the value of -1.05 shown on Fig. 6.34.

6.13 DEVELOPING A FUZZY SYSTEM

Fuzzy controllers have been shown to be particularly suited to applications where the process to be controlled is ill-defined, is not well understood or changes with time but for which a clear statement of the required performance indicators exists and where human operators have developed a control strategy based on their understanding of system behaviour. Examples of this type of system include boiler control, furnace control, vehicle speed control, train operation and scheduling, lift control, speech and pattern recognition, planning and forecasting and many others.

The process of developing a fuzzy control system follows very closely the procedures used in the creation of a conventional expert system. In each case, a knowledge engineering process is used to establish the criteria for the operation of the system and to define the fuzzy rules to be used and the shape of the associated fuzzy sets. This requires that the knowledge engineer brings together the views and opinions of the experts consulted, including in particular the resolution of any conflicts that arise. It is also important at this stage to identify the performance criteria for the system as these will determine the

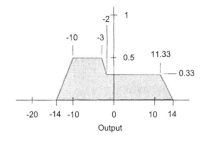

Figure 6.31 Fuzzy output produced by active rules.

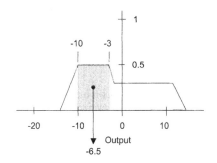

Figure 6.32 Defuzzification using mean of maxima method.

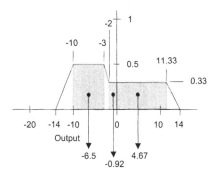

Figure 6.33 Defuzzification using centre of maximums method.

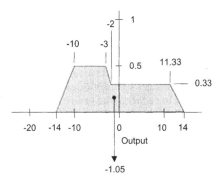

Figure 6.34 Defuzzification using centre of gravity method.

parameters to be measured and will influence the shape of the fuzzy sets used and their domains as will the choice of defuzzification method used.

The size of the rule base is a particular concern with complex systems and for this reason it is necessary to seek to break down the rule base into a series of related modules, each of which is a more manageable size. Using this modular strategy it is possible to restrict the number of rules while still retaining the overall performance.

Once the fuzzy controller is brought into service it will then be necessary to refine the system to achieve the required levels of performance. This is done in the first instance using a simulation of the controlled system to support revising and refining the rule base, changing the shape functions for the associated fuzzy sets and the method of defuzzification until the desired performance levels are achieved. Once a sufficient degree of confidence in the achievable performance had been achieved, the controller would be applied to the real system.

6.14 FUZZY-NEURAL SYSTEMS

As can be seen from Table 6.5, fuzzy systems and neural networks are to a significant degree complementary and the combination of a neural network within a fuzzy system provides a learning mechanism which can be used to autonomously adjust the rule base in response to changes in system conditions and behaviour. The resulting self-organising and self-learning systems can then be applied to the control of systems even where there is significant uncertainty in the understanding of the way in which the system is likely to behave or where it has proved impossible to generate a comprehensive rule base.

Table 6.5 Comparison between fuzzy and neural systems

Advantages	
Fuzzy systems	*Neural networks*
Knowledge representation is structured around linguistic terms in a form understandable to people.	Learn from example and do not require prior knowledge in their construction.
Prior knowledge can be used in structuring thefuzzy rule base enabling the incorporation of human expertise at an early stage in the design of the system. As the system knowledge is encoded in a series of structured rules it is easy to make changes to individual rules without a need to redefine the entire system. Can handle incomplete, uncertain and imprecise data.	Multiple parallel processing can result in high speeds.

Disadvantages	
Fuzzy systems	*Neural networks*
Choice of membership function and defuzzification processes can have a significant impact on performance.	The operation of the network is not transparent to the user so the process of reaching a conclusion is not apparent. The operation of the network means that the recovery of knowledge in the form of rules is in general impossible. Retraining to a new task can result in a loss of already learned knowledge. Unable to handle incomplete, uncertain and imprecise data.

Software | 7

7.1 INTRODUCTION

Software systems are probably the most complex artefacts that the human race builds, and software is a product that must be engineered like any other. Software engineering techniques have been developed to make this possible. They are applicable wherever more than one person is required to produce software, or wherever the user is distinct from the producer. Thus large multi-national projects such as Eurofighter Typhoon and Airbus require considerable investment in process, methods and tools to ensure an effective collaboration.

The opposite to good software engineering practice is often referred to as 'code and fix' or 'hacking'. Table 7.1 illustrates the differences between the two approaches.

Table 7.1 Features of good and bad software practice

Software engineering	Hacking
Professional	Amateur
Rigorous specification	Immediate coding
Extensive testing and errors traced	Debug to get working
Views of user respected	Cure faults during use by 'patching'
Contact agreement	Not costed
Coherent design	Badly structured program
Clear coding	Difficult to follow – few comments
Support documents	No documentation
Little maintenance required	Expensive to change and maintain
Good user interface	User expected to adapt to program

There are four basic requirements from software engineering. It must provide software developers with:

1. a structure to co-ordinate and manage software development and hence minimise the risk and complexity of building software systems;
2. methods to enable developers and customers to explore, and hence agree, the nature of the software system as early as possible. This provides the opportunity for a firm contract between the developer and the customer with agreed costs and time-scales;
3. techniques to minimise the effects of change both during development and operation;

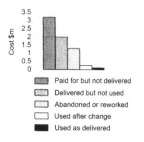

Figure 7.1 Defence
software – problems with
real-time systems
software.

4. the means to reduce errors at all stages of the software process and hence reduce the maintenance costs.

The importance of software engineering can be appreciated by considering Fig. 7.1 which shows the results of some defence system software projects.

7.2 SOFTWARE DEVELOPMENT PROCESS MODELS

Following a series of expensive disasters in the late 1960s involving huge over-spending on software projects, it was realised that an informal 'code-and-fix' approach to software development is not adequate for projects as soon as more than one party is involved. The first response to this crisis was the development of the well-known *waterfall model* of Fig. 7.2.

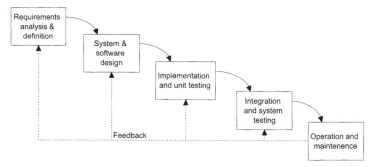

Figure 7.2 The waterfall model with feedback.

This is a document-led approach that lays great stress on the completion of one phase before embarking on the next. This strategy enabled some degree of control to be imposed on the software development and procurement processes, and it is probably still the most popular model for government procurement. In practical terms some element of feedback and rework is invariably required. However, this is a major contributor to cost overspends and so there has been increased emphasis on 'early testing' of each completed stage in an attempt to eliminate errors before proceeding to the next stage.

It has been found that the waterfall model is not well suited to many types of system and it has many detractors. Problems arise particularly with highly interactive end-user applications and AI type programs, where it is too difficult to define detailed requirements at the start of the project. The result can be extensive and elaborate requirements documents that do not really represent the client's needs, followed by the development of large quantities of unusable code.

The relative development costs for the different stages of a range of software projects are shown in Table 7.2 from which two obvious conclusions can be drawn:

1. The cost of actually coding the computer programme is generally a relatively small component of the total cost.
2. Maintenance costs can dominate to the extent that the purpose of software engineering can be re-defined as 'to minimise maintenance costs'.

Table 7.2 Relative costs in the software development process

System type	% Costs			
	Requirements/design	Implementation	Testing	Maintenance
Command and control	46	20	34	
Space	34	20	46	Typically add
Operating systems	33	17	50	200% to 400%
Scientific	44	26	30	for long-life
Business	44	28	28	software

Other development models emerged and these, together with the waterfall model, are listed, with their main characteristics, in Table 7.3.

The last of these, the spiral model, is worth further explanation. The model was proposed to provide more flexibility, particularly for those projects where the waterfall model is inadequate. The model is presented diagrammatically in Fig. 7.2 from which it is seen that the process starts at the centre of the spiral and goes through successive rounds, eventually emerging with an implemented system. It makes the assumption that each round of the process involves risks and that prototypes and simulations may be useful for minimising these risks. Thus the regions shown between dashed lines in Fig. 7.3 represent options that may be taken up as a result of the preceding risk analysis.

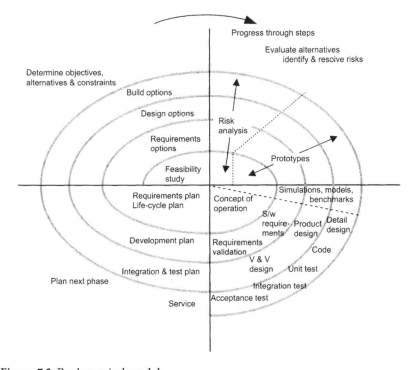

Figure 7.3 Boehm spiral model.

For complex systems, each system module will have its own spiral, which are nested within a spiral for the integration of the whole system. Indeed complex modules may be planned as a series of sequential spirals with the output from

Table 7.3 Software development models

Development model	Characteristics
Waterfall	Distinct phases e.g. requirements specification, design, implementation and unit test, integration and system test, installation and acceptance test. Document led, ideally with testing and 'signing off' before commencing the next stage
Evolutionary programming	Rapid development of a working system which is then tuned to work properly and added to in increments. Difficult to distinguish from code-and-fix but used where early requirements are difficult to specify such as in AI programs
Prototyping	A rapid prototype is produced for the purposes of refining requirements, but is then discarded and the final system properly designed and implemented from scratch
Incremental	Uses the linear aspects of the waterfall model but applies the steps of analysis, design, coding and testing to segments of the whole problem. The software may therefore be delivered in stages. Later segments may be influenced by operational experience gained from those delivered early
Reusable components	Systems are build by combining previous program modules. In essence this is akin to a super-high-level language. Most appropriate at present to specialist domains. Some adjustment may be necessary to initial requirements in order to capitalise on existing components
Formal transformation	A formal specification (i.e. logically consistent and complete) is automatically verified and then converted to executable code. Only currently realisable for small, simple systems
Spiral model	A complex process model that attempts to incorporate any of the above models when appropriate. Uses risk analysis to determine the most suitable route.

one forming the input to the next. This is how iterative modifications are fed back from, for instance, the design stage to revised requirements. The hub of the spiral, sometimes referred to as round 0, consists of a feasibility study to identify options. The risks associated with each of these options is considered, along with constraints, and this starts off the spiral process.

The radial distance from the centre of the spiral represents accumulating project costs. The different process models listed in Table 7.3 should not be thought of as being mutually exclusive. It is common to use a combination of processes. Indeed, the spiral model can be thought of as a meta-model that incorporates the others when advantageous.

7.3 SOFTWARE METHODS

Software development process models are concerned with *what* is done, i.e. with identifying the main steps in the development process. Methods, on the other hand, describe *how* the development proceeds, usually by identifying a series of useful techniques and tools. There are many competing methods; however, two dominant method paradigms have evolved:

1. *Process-oriented methods.* These represent software programs essentially as a series of processes or functions that transform input data into output data and are regarded as the traditional or conventional approach. Thus an automatic washing machine program might take data from, say, the user and an internal clock, and use a process called 'control_wash_programme' to convert it into output data for driving the motor and the pump. A commonly adopted functional method is structured analysis (SA), developed by DeMarco, which can be followed by structured design (SD), developed by Yourdon and Constantine. SA/SD is often collectively referred to as the 'Yourdon approach'.
2. *Object-oriented (O-O) methods.* These represent software programs as a collection of independent entities or objects that offer services and communicate via messages. The objects often map onto real world objects. Thus an automatic washing machine program might contain an object 'wash_controller' that can send messages to the objects 'motor' or 'display'. A widely used object-oriented method is the object modelling technique (OMT), developed by Rumbaugh *et al*.

7.4 REQUIREMENTS

A key feature of all the process models, with the possible exception of evolutionary programming, is the formulation of a detailed requirements specification, against which the end result can be compared. As explained in Chapter 4:

1. requirements state 'what is required' not 'how' it will be achieved;
2. requirements should be stated in terms that are precise enough to allow the final program to be tested against them. This means that vague terms such as 'faster than average' or 'high-quality interface' should be avoided.

Chapter 4 indicated the importance of requirements interpretation in ensuring that the right product is produced. This chapter shows how those ideas relate to software.

7.4.1 Requirements capture

Requirements capture, or requirements gathering, is the process of determining exactly what the client wants and the precise nature of the problem. The first stage is to establish a development team and a programme of regular meetings where requirements are reviewed and evolved. The team should include representatives from both the developer and the client and possibly including marketing and manufacturing representatives where appropriate. Some developments take many years and so the developer must be flexible and prepared to make changes, particularly to functional requirements, as hardware changes.

Two available techniques are:

- facilitated application specification techniques (FAST);
- viewpoints analysis (discussed in Chapter 4).

Fast

Using a car security system as an example, the FAST approach goes through the following steps:

1. After a few preliminary meetings the text of a joint 'product request' is jointly drafted. This is merely one or two pages long and summarises the project in simple text.
2. Before the next meeting each team member makes five lists:
    (a) entities (objects) that are part of the system environment (e.g. window, door, radio, intruder)
    (b) entities produced by the system (e.g. alarm, immobilisation, information)
    (c) entities used by the system (e.g. pressure detector, motion detector, remote control, intruder event)
    (d) services that react with the entities (e.g. monitoring, priming, cancelling)
    (e) constraints and performance criteria, i.e. non-functional requirements (e.g. cost below £300, user friendly, easy to install, low power consumption)
3. Each individual's lists are considered and merged without debate to form a master list.
4. The master list is then refined by debate until a consensus is reached.
5. Small sub-teams are formed to expand specific list items into mini-specs. Thus remote control may be expanded to include:
    (a) contained in key fob
    (b) less than 50 grams
    (c) easy battery change
    (d) damp protection
    (e) low power consumption
    (f) etc.
6. The whole team then considers each mini-spec. Issues for later resolution are recorded.
7. Each team member produces a list of validation criteria for each mini-spec. (e.g. the remote control power consumption should be low enough to provide a minimum of a thousand operations on one battery)
8. A consensus list of validation criteria is produced.
9. One or more team members is nominated to write a draft requirements definition document.
10. This is circulated for comment and then agreed at a final meeting.

7.4.2 Requirements analysis

The purpose of requirements analysis is to arrive at a thorough, complete and unambiguous description of the problem, which can then be used as the basis for program design. In structured analysis this is achieved by constructing three models of the required system. These models are regarded as being *orthogonal* and hence providing largely independent descriptions of the system.

Data entity modelling

Data entity modelling sets out to define and understand the variables that are input, transformed and created by the program. It is usually regarded as the least important model in structured analysis, particularly for small mechatronic devices, where relatively simple data items are being manipulated. However, data entity modelling can be a valuable precursor to 'data design', where the trend is strongly away from individual global variables towards user-defined abstract data-types hidden within program modules. There are also initiatives such as STEP from ISO which aim to define standards for data definition and transfer with the aim of enabling the easy transfer of data between different users and different computer packages. Thus the manufacturer of a particular temperature sensor would pass a data model of the sensor to a customer which contains all the attributes of the sensor such as its dimensions, range and signal characteristics, together with a CAD representation. The customer can then combine the sensor model with those of other components to build a complete system.

Data entities may be represented as rectangles, as in Fig. 7.4, where the object name is shown above the line and the attributes below it. On the left in Fig. 7.4 is a data model which may be used to describe any type of tilt sensor while on the right of the figure is an 'instance' of the entity in which the attributes have been filled in to depict a specific sensor. The attributes listed should represent the data that will enable the interfacing software to be produced.

The principal tool used for showing the interaction between data models is the *entity-relationship diagram*. Many conventions exist for drawing these and here the *object modelling technique* (OMT) convention is used.

Attributes may be omitted for clarity in entity relationship diagrams. Referring to Fig. 7.5 which shows an association between a robot navigation system object and the tilt sensor object, the line connecting the two data entities indicates that there is an association between them. The nature of the association is written above the line. Symbols on the end of the line indicates the number of entities involved in the association, referred to as *cardinality*. The solid circle means 'many' (zero or more), a plain line indicates 'exactly one' and an empty circle means 'optional' (zero or one). Alternatively, if the precise number is known it can be written adjacent to the end of the line. Thus Fig. 7.5 can be interpreted as:

'A navigation system reads zero or more tilt sensors.'

Further important relationships are *generalisation* and *aggregation*. Generalisation indicates that other entities are a more specialised form of a general entity. This condition is shown in Fig. 7.6 in which the line symbol (Δ) indicates that the lower three entities are 'forms of' the upper entity.

Aggregation indicates that one or more entities are 'part of' another entity. This is shown in Fig. 7.7 in which the line symbol ($\Diamond$) indicates that a navigation system can be formed from a satellite GPS receiver, a flux-gate compass and several tilt sensors.

Functional modelling

Chapter 4 contained a description of functional modelling using the DeMarco structured analysis method as applied to complete systems. Indeed, and particularly in a mechatronic system, the decision as to whether a function will be implemented in hardware or software is often unclear at the requirements stage. It is therefore in most cases sensible to leave the definitive allocation of functions to particular technologies to as late as possible in the design stage. Remember, requirements are concerned with 'what' is to be done not 'how'.

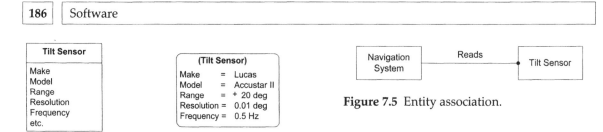

Figure 7.4 Data entities.

Figure 7.5 Entity association.

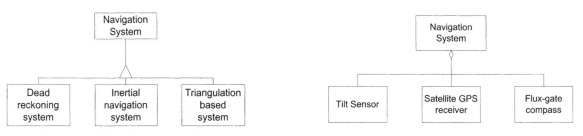

Figure 7.6 Generalisation.

Figure 7.7 Aggregation.

However, if the primary aim is to develop software the following additional points should be considered:

- The dataflows shown on the DFDs should be true data that is processed by software. Thus with reference to the example in Chapter 4 of a washing machine, the software DFD should not contain items such as 'mains_power' or 'waste_water'.
- Initially ignore trivial dataflows such as requests for data or error messages.
- The external dataflows entering a child diagram must exactly balance those shown on the parent process. However, the data dictionary may be required to check this. For example, a parent process on the washing machine may have 'user_input' as a data flow, whereas the child diagram may expand this by having inputs to two child processes; 'wash_program' and 'wash_temperature'. In this case the data dictionary will contain the information:

user_input = wash_program + wash_temperature

- Try to separate major data inputs on the left of the diagram from outputs on the right of the diagram. This helps with later structuring of the program during the design stage.
- Different lettering can be used to indicate a primitive process.

These points are demonstrated in the following much simplified example of a bank automatic cash till or auto-teller in which only the basic function of withdrawing cash is considered. It should also be remembered that there is no single 'correct' answer with this type of analysis.

Referring to the context diagram of Fig. 7.8, it can be seen that, on the whole, inputs have been separated from outputs, the account database being an exception as it has both an input and an output function. Thus, 'cash' is shown being output to the 'hole in wall' rather that back to the 'customer'. Unlike 'waste_water' in the washing machine example, 'cash' is actually a real data value as it contains information on the actual amount of money.

The lines shown dotted on the level 0 diagram of Fig. 7.9 indicate an extension to the basic DeMarco method proposed both by Ward and Mellor and

Hartley and Pirbhai to indicate control data. Control data is often a simple binary signal from one process to another to indicate that a process is complete or cleared to proceed. Note also that 'messages' has been expanded to 'card_message' + 'transaction_message'. This expansion must be recorded in the associated data dictionary.

In the level 1 diagram of Fig. 7.10 for the 'Handle card' process, upper case has been used to indicate that 'card reader' is a primitive process. This must be described by a process specification. The expansion of the level 1 process 'card validate' to level 2 is shown in Fig. 7.11. The further expansion of the resulting level 2 process 'Check PIN' to a level 3 diagram is shown in Fig. 7.12 at which point only the primitive processes 'REQUEST PIN', 'READ PIN' and 'VALIDATE PIN' remain, permitting no further expansion.

An extract from a data dictionary is shown below with entries listed in alphabetical order for ease of reference:

account_details = account_PIN + balance

card = card_details *coded on magnetic strip*

card_details = bank_details + date + account number

card_message = bank_message + date_message + PIN_message

date_message = "card expired"

messages = card message + transaction message

pin_message = [PIN_request, PIN_invalid]

PIN_number = 4{real numbers}4 *input by user*

request_amount = cash withdrawal amount in £s
etc.

Example process specifications are:

Process 1.1 CARD READER: translate magnetic strip into the following
data:–
bank_details
date
account–number
magnetic_PIN

Process 1.2.3.2 READ PIN: $i = 4$
For each Ni while $i > 0$
Do the following
get Ni
store as PIN_digits.

Process 1.2.3.3 VALIDATE
PIN: If PIN_digits =
account_PIN
Do the following
Card_ok true
etc.

Dynamic modelling

It is often useful to represent the dynamic behaviour of the system. This is something which none of the methods considered thus far can accommodate.

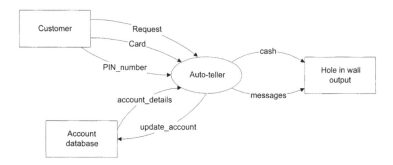

Figure 7.8 Context diagram of auto-teller.

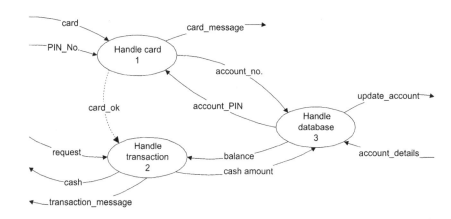

Figure 7.9 Level 0 diagram.

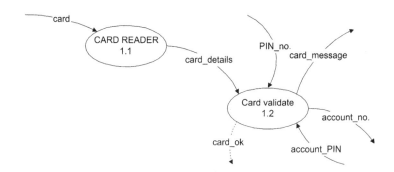

Figure 7.10 Level 1 diagram – 'Handle card'.

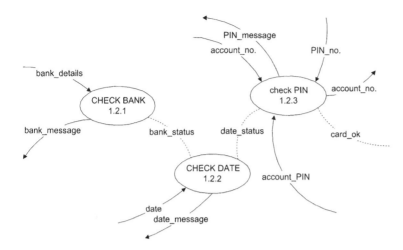

Figure 7.11 Level 2 diagram – 'Card validate'.

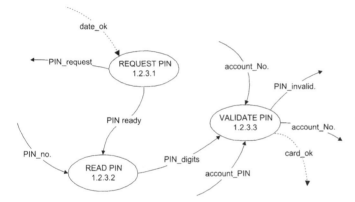

Figure 7.12 Level 3 diagram – 'check PIN'.

The dataflow diagrams of structured analysis show the flow of information between processes and some control information, but they do not give a clear picture of the behaviour of the system with time. One way of representing time based behaviour is with the aid of state transition diagrams.

The state transition diagram is particularly appropriate for those real-time systems which are also *finite state machines* in which different functions are performed at different times, depending upon signals from a time clock or some other external event. Many mechatronic systems fall into this category.

Clearly the automatic washing machine and the bank cash dispenser come under the definition of a finite state machine and Fig. 7.13 shows a finite state diagram for the bank cash till. Note that the *states* shown in the boxes of Fig. 7.13 are all 'doing' states. The arrows joining the boxes indicate the *events* or *stimulus* that trigger a transition from one state to another. Clearly, each state must generally have at least one arrow entering it and one arrow leaving it.

7.4.3 The software requirements document

The software requirements document is the basis of a contract between the

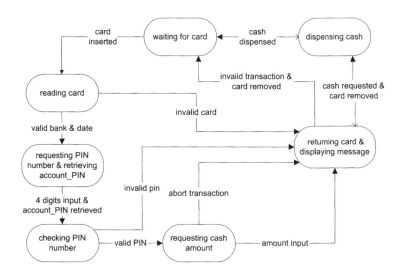

Figure 7.13 High-level state transition diagram of auto-teller.

system procurer (client) and the system producer (software engineer). Considerable work is required to produce the software requirements document for a complex system and hence it may be necessary to let a separate contract for its production. A software requirements document typically consists of the following parts:

1. An introduction describing the need for the system, what it does and how it compliments the business objectives of the client. The *system evolution* should be described, where it came from and what is likely to happen to it under future development.
2. The *system model* shows the relationship between the system components and how they fit into the working environment. A simple abstract dataflow diagram should be included.
3. The *functional requirements definition* which describes in natural language together with tables and diagrams as appropriate what the system does for the user. This is suitable for use by management.
4. The *non-functional requirements definition* which defines the constraints placed on the designer in the form of standards, hardware, data bases and so forth.
5. A glossary of technical terms containing detailed information which is more specialist in nature, but which are required for a formal contract.
6. Appendices, including
    (a) *Appendix 1:* The *functional requirements specification*. This is an amplified and more formal form of the functional requirements definition and is primarily for the use of the software developers. It follows from a detailed requirements analysis that looks at the system from three viewpoints:
       ● data modelling;
       ● functional modelling;
       ● dynamic modelling.
    (b) *Appendix 2:* The *non-functional requirements specification*. This gives information on items such as specific response times, memory requirements and so on.
    (c) *Appendix 3:* The *hardware requirements* definition. This should contain both minimal and optimal configurations for the hardware.

(d) Appendix 4: The *data requirements* of the system. These need to be made as clear as possible as they could influence the working practices of the user organisation.

7.5 DESIGN

Having thoroughly analysed the software requirements the next step in the waterfall model is to design the hardware system and the data and program structures. Here, the concern is only with the software, and the basic aim is to produce *easily maintainable* code. The principal method used is decomposition into modules. A good modular structure has low coupling and high cohesion.

Coupling is a measure of the interdependence of modules. With high coupling (undesirable) it is likely that changes to one module will affect the proper functioning of many others. High coupling results from:

1. the use of global variables, keep data private if possible;
2. the exchange of control information as opposed to simply data;
3. lots of communication between modules;
4. early 'binding' of variables, for instance through pretending that a variable is a constant. Consider for example a program to calculate the amount of income tax payable. The current tax rate could be included in the program in any of the following ways, which demonstrate progressively later binding:
   (a) coded separately into each module. Each module would then have to be changed every time the tax rate changes.
   (b) declared in a single module which then passes the value to the other modules as required.
   (c) read-in at 'link time' when the various modules of the program are linked together for compilation.
   (d) read-in at run time from a file containing the current tax rate.
   (e) entered manually at run time following a screen prompt.
   For this example perhaps (e) would be the most appropriate choice.
5. branches to other modules at the same level;
6. the use of 'pathological' connections such as GOTOs, which generally create unmanageable code.

Cohesion can be seen as the inverse of coupling and is a measure of the degree of functional relatedness of a module. If a module has good cohesion, splitting it results in increased coupling. If it is not possible to think of a good name for a module then it probably lacks cohesion. Table 7.4 sets out different levels of cohesion.

Table 7.4 Levels of cohesion

Level of cohesion	Description
Coincidental	Parts not related but bundled together
Logical	Components perform similar functions such as input or error handling
Temporal	Activities occur at the same time, e.g. start-up procedure
Procedural	Make up a control sequence
Communicational	All parts use the same input data
Sequential	The output from one element forms the input to another
Functional	All parts are necessary to perform a single function

7.5.1 Structured design

Structured design carries on from structured analysis and the dataflow diagram (DFD). It is a top-down approach in that it uses a hierarchy of modules each with a single entry and single exit point together with a further graphical tool in the form of the *structure chart*, the purpose of which is to show the organisation of the software into modules and sub-modules.

The first step in converting a DFD to a structure chart is to distinguish between input and output processes and central transformations. The aim is to find the input and output processes that are furthest from the terminators. What is then left in the middle are the central transformations. A process which checks the format of data, or adds information, is still classed as input. A process which sorts data or filters it is classed as a central transformation.

Figure 7.14 shows a specimen dataflow diagram with simple input and output streams. The dotted lines indicate the partition between input, output and central transformations. Referring to Fig. 7.15, the resulting structure chart starts with a single rectangle representing the whole system at the top. Working outwards from the central transformation, process number 3 in Fig. 7.14, it can be seen that there is a need to receive a value of C and to output values for D and E. A module is therefore required to perform each of these actions and these, together with the central transformation module, are placed on the next level down. The chart is then extended by considering the dataflows further away from the central transformation and adding extra lower levels.

The arrows with open circles alongside the connecting lines are used to indicate the direction of dataflow. It should be noted that data always flows vertically between modules on different levels and never between adjacent modules on the same level. This is the essence of the top-down structure.

Unfortunately, for real programs the situation is not usually so clear. Structure charts work most satisfactorily with conventional data-processing programs which have clear input, processing and output streams. Many mechatronic systems do not really fit this model and often contain many parallel 'sense-process-act' loops which may mean that the separation of input from output is not desirable. Nevertheless, structure charts may still be helpful in imposing a good top-down structure on a program.

The following shows how the auto-teller DFDs could be converted into a structure chart. The first step is to divide the level 0 diagram, shown originally in Fig. 7.9, into input, output and central transformations. It could be argued that 'Handle card' is principally concerned with data input, from the magnetic strip. 'Handle transaction' and 'Handle database' could be considered to be primarily central transformations. This leaves no output processes!

The outputs which do exist (messages, cash and updating the database account) are handled as sub-processes of the others. An alternative would be to create a separate 'Handle output' process but this would reduce cohesion and increase coupling by splitting functionally related activities such as 'Handle database'. The resulting partitioning is shown in Fig. 7.16.

The structure chart derived from the DFDs is shown in Fig. 7.17. Only those modules concerned with the 'Handle card' process are shown. An extra refinement is the addition of arrows with solid circles. These represent control flow or *flags*.

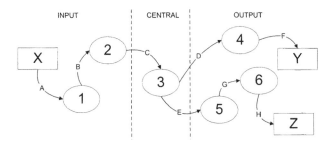

Figure 7.14 A data diagram with simple input and output streams.

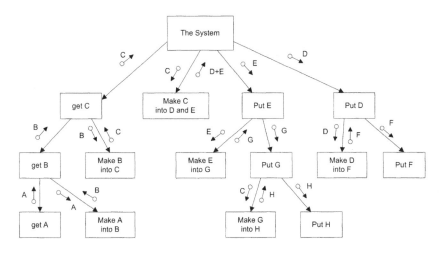

Figure 7.15 Structure chart.

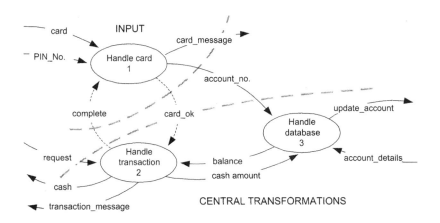

Figure 7.16 Level 0 diagram with partition.

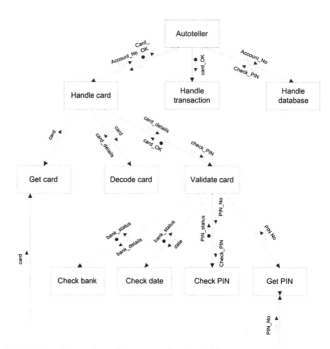

Figure 7.17 Structure chart for part of auto-teller.

7.6 OBJECT-ORIENTED METHODS

Object-oriented (O-O) development can still be related to the waterfall model in that the first three stages are analysis, design and implementation. As suggested in Table 7.3, with object-oriented methods there is often a tendency to develop software incrementally. A complication is that the stages can be mixed with process-oriented development. Thus a process based analysis can lead to an object-oriented design which in turn can be programmed in a non-object-oriented language. However, a consistent approach throughout is clearly preferable and one such approach is the *object-modelling technique* or OMT.

An object-oriented system is viewed as a collection of interacting objects that communicate by passing messages rather than a collection of processes that communicate by passing data. Objects should be based on real world entities, and because of this, it can be argued that object-oriented approaches have a better *mapping* to real word problems, hence they are more easily understood. The concepts of low coupling and high cohesion are taken to extremes, hence, in theory, producing easily maintained and extendible systems with good prospects for re-use. Some experts argue that object-oriented methods have superseded process-based methods but others argue that some problems are much more suited to a process-based approach and that the two methods can be complementary.

7.6.1 Objects

An *object* is simply any useful, well-delineated entity which has meaning for the system being considered. Objects are very similar to the data entities discussed earlier, apart from one important enhancement. In addition to having attributes, objects can also provide *operations*. An operation is a function or process that can be performed by the object. This means, for example, that a sensor object could calibrate or zero itself and then send a message to a 'sensor manager' object to say that it is ready.

There are four key concepts of the object-oriented approach: *abstraction*, *encapsulation*, *inheritance* and *polymorphism*. Each of these will be considered briefly.

Abstraction

Those aspects of a problem that are important are isolated from those aspects which are unimportant. All abstractions are incomplete and many different abstractions of the same thing may be made for different purposes. A 'good' model of something is one which captures the crucial aspects of a problem and omits the others. Thus in the above example, a tilt sensor will also have attributes of size and colour (Fig.7.18), but these aspects may be irrelevant to the particular problem under consideration so they are omitted (Fig. 7.18). A good abstraction will mean that the class can be re-used throughout the whole development lifecycle of the software.

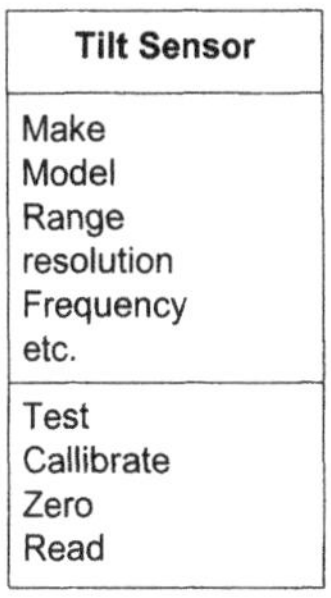

Tilt Sensor
Make
Model
Range
resolution
Frequency
etc.
Test
Callibrate
Zero
Read

Figure 7.18 Object class.

Encapsulation

Also referred to as *information hiding*, this is a term that has been widely used in programming for many years. Generally it refers to the practice of separating the *external* characteristics of an object, which can be accessed by other objects, from the *internal* implementation details of the object itself, which are hidden from other objects. It is therefore a separation between the interface to an object and its operations and means that objects have very low coupling and consequently that an internal operation can be changed without having knock-on effects to the rest of the system.

Inheritance

Recognising that certain object classes are special cases of other, more generalised, classes reduces the number of completely distinct cases that must be understood and analysed. This can greatly simplify the modelling of a problem and assist in its solution. In software systems it is also the basis for code re-use.

Figure 7.19 presents an object diagram which shows that both a tilt sensor and a rotary potentiometer are 'forms of' the sensor class, which in turn is a 'form of' the electronic equipment class. A tilt sensor class may be termed a sub-class of the sensor class. A sensor class may therefore be termed a 'superclass' of the tilt sensor class.

Notice how the attributes and operations have been distributed. The electronic equipment class has attributes such as 'Make' and 'Model' which are common to *all* electronic equipment. The sensor class *inherits* those attributes from the super-class and then adds further attributes which are specific to all sensors, and so on. Figure 7.20 shows an example of a tilt sensor instance which shows attributes corresponding to all its inherited attributes in addition to its own.

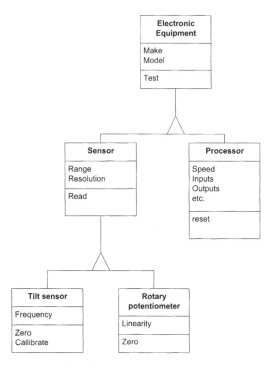

Figure 7.19 Inheritance of attributes and operations.

Polymorphism

It can be seen from Fig. 7.19 that because of inheritance, all electronic components should be able to carry out an operation that consists of some kind of test routine on themselves. Clearly the most appropriate test for a sensor will be very different from that for a processor. The specific routine for each different class is known as a *method*. Thus operations are implemented by methods.

Polymorphism is the name given to the ability of classes to interpret a demand for a common operation in such a way that it is appropriate. This is a powerful concept as it enables another object, for instance a 'sensor manager' class, to demand a test from any sensor without having any knowledge concerning the particular method that will be employed. This would apply even if some new type of sensor was to be added to the system in the future, provided it came with its own test method.

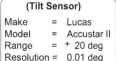

Figure 7.20 Object instance.

7.6.2 The object-modelling technique approach

The object-modelling technique (OMT) is one of the most popular methods for developing O-O systems and has some similarities to the structured analysis/ structured design (SA/SD) approach discussed earlier. Both approaches base the requirements analysis on three models of the system:

1. the data entity model of SA/SD becomes the *object model* in OMT. The main difference is the addition of operations to the OMT data objects.
2. the *dynamic* or *behavioural* model is described using state transition diagrams in both approaches. OMT supplements this with event trace diagrams.

3. functional models in the form of dataflow diagrams are used by both methods. In OMT, the processes of a DFD correspond to the actions identified on the state transition diagram and are eventually implemented as object operations.

Both approaches also advocate the maintenance of a comprehensive data dictionary throughout the development process. Where they differ is in the emphasis given to the various models. In SA/SD the functional model is the most important and the data entity model the least important. In OMT, the object model is the most important and the functional model the least.

Event trace diagrams

An *event trace diagram* should be drawn for each working scenario of the software. The aim is to capture all the input and output events for the object classes involved. Each object involved in the transaction is represented by a vertical line with the class name at the top, arrows then indicate the direction and nature of the interaction. Figure 7.21 shows an event trace diagram for the autoteller which depicts a transaction with no problems. Other diagrams would need to be produced for problematic scenarios such as a card being out of date or a PIN number typed in wrongly.

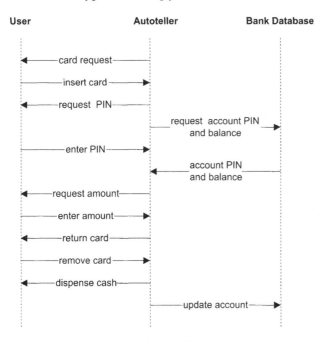

Figure 7.21 Event trace for successful autoteller transaction.

OMT application

The application of OMT is not easy and requires experience for effective implementation. The following steps briefly describe the process:

1. Analysis stage:
   (a) Object model
      - start with an initial high-level description of the problem written in text;
      - identify the principal object classes. A good starting point is to extract the nouns from the problem description;
      - add associations between classes to build up the object model;
      - add attributes to the object classes, individually at first;
      - simplify the attribute lists by exploiting inheritance;
      - complement the object model diagram with a data dictionary that describes each object class.
   (b) Dynamic model
      - prepare event scenarios that describe what actually happens;
      - prepare an event trace for each scenario;
      - produce a state transition diagram for each object class that has dynamic behaviour.
   (c) Functional model
      - identify inputs and outputs;
      - use levelled dataflow diagrams to show functional relationships;
      - describe functions in a process specification.
   (d) Refinement
      - extract important operations from the functional model and the state transition diagrams and add to the object model;
      - check the models and test them using typical scenarios;
      - develop detailed scenarios such as error conditions and add to the models.
2. Design stage:
   (a) System design
      - partition into subsystems taking into account coupling and cohesion;
      - identify concurrency in the form of parallel operations;
      - allocate subsystems to processors;
      - develop database strategy;
      - investigate interfaces to external resources such as other computers;
      - establish a means of implementing software control.
   (b) Object design
      - start with the object model from the analysis stage;
      - ensure that all processes from the DFDS and events from the state transition diagrams are represented by an operation in the object model;
      - design algorithms for all the operations;
      - optimise objects attributes and associations to improve efficiency, improve access paths to data and avoid re-computation of complex expressions.

7.7 TESTING

All software contains errors. It is a measure of the complexity of software production that even the most rigorously produced software for safety-critical applications is likely to contain several errors for each hundred lines of code. Fortunately these errors are often benign; however disasters involving loss of life have been caused by software. The design and implementation of an appropriate test regime in order to reduce the risk of failure is therefore an important part of software development. A successful test is one which exposes errors. The testing of software has two components: *verification* and *validation*, sometimes abbreviated as V & V. These attempt to answer the following questions:

1. Verification:
     Is the product correct?
     Is it free from errors and does it meet the specification?
2. Validation:
     Is it the correct product?
     Does it do what the customer wants?
     Is the specification correct?

7.7.1 Testing strategies

Verification and validation (V & V) testing should not in general be carried out by the software developer but by an independent tester. The degree of independence, for instance whether or not it occurs in a different company, depends upon the safety integrity level of the application. It is important to recognise that testing is not an activity which begins when the development is complete. Testing activities should take place throughout the whole of the development life-cycle.

There exists a conventional 'V' model for testing as shown in Fig. 7.22. This implies that testing starts after coding. The problem with this is that the early requirements stage would not be verified until the final acceptance testing. Figure 7.23 suggests that the most expensive errors to correct are those that are made early and corrected late (see also Chapters 1 and 4). Even with simple coding errors it has been estimated that carrying out fixes in the field can cost up to 25 000 US dollars per fix. Because the cost of correcting software errors rises exponentially during the development process the emphasis should be on the avoidance of errors and the *early* detection of errors that do occur by an appropriate testing regime.

A refinement is the 'V' model with 'early test case preparation' shown in Fig. 7.24, which requires the designer to design the test necessary to prove compliance with the specification. The testing is still done independently. However, because the early detection of errors can result in significant savings, an even better model involves 'front-loaded testing'. This is referred to as the 'W' model and is shown in Fig. 7.25.

However, the question arises as to how, for example, requirements are verified when the software has not yet been designed and coded. This is done by a combination of:

1. requirements review meetings with the client;
2. rapid prototypes to explore user interface issues and basic functionality;
3. specification of tests to be carried out later;
4. if *formal methods* are used to define the requirements in terms of logic, then computer tools can be used to check their consistency.

7.7.2 Testing techniques

Most verification is carried out using 'white box' techniques. These assume that the tester has knowledge concerning the internal working of the program.

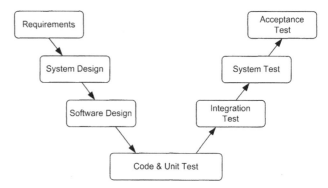

Figure 7.22 The 'V' testing model.

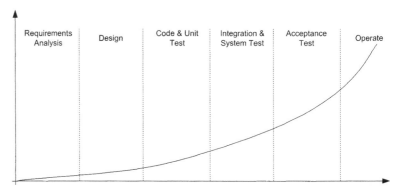

Figure 7.23 Increasing cost of errors.

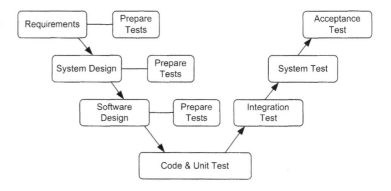

Figure 7.24 The 'V' model with early test preparation.

Thus a test will be designed to exercise each function or object within each module. Validation is carried out using black box techniques where no knowledge of the program structure is assumed and the program is exercised over the complete range of possible inputs and the results compared to the customers expectations. There are two forms of white box verification:

- static verification using inspection, analysis, and formal verification tools;
- dynamic testing by running the implemented code.

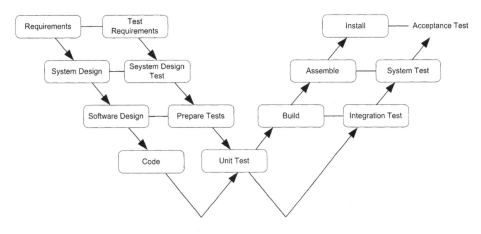

Figure 7.25 The 'W' model with early testing.

Static verification

Static verification can be by manual inspection of the code by a third party, or automatic static analysis is now possible with software tools. These provide information such as that shown in Table 7.5

Table 7.5 Static analysis outputs

Control flow analysis	Checks for loops with multiple exits or entries Unreachable code
Data use analysis	Variables not initialised or declared Variable written twice Variables not used Parameter type mismatches – array bound violations
Interface analysis	– consistency of routine and procedure declarations – procedure never used
Information flow analysis*	identifies all input variables on which output values depend
Path analysis*	– identifies all paths through the program

* Can generate significant volumes of output

Dynamic testing

Clearly, modules are first tested individually. However, they must then be combined for integration testing. There are three basic techniques used:

- top-down;
- bottom-up;
- thread testing

Top-down testing
Consider the typical module structure shown in Fig. 7.26. Top-down testing starts with the module at the top of the tree (M 1) and this is tested with *stubs* to simulate lower-level modules. The stubs can be much simplified programs or data that will exercise the module over its full range. This is shown as 'Test 1' in Fig. 7.27. Changes should be incremental, so the next step is to add module

M 2.1 with its associated stubs, so that the integration of M 1 and M 2.1 can be tested. M 2.2 would then be added and so on until all modules have been tested.

Bottom-up testing
Test drivers are written to exercise lower level components as shown in Fig. 7.28. After individual tests on the bottom-level modules they are integrated by the test driver. One module is then added at a time as the tester works up the tree until the system is complete.

Bottom-up testing is not generally recommended as testing cannot begin until the lowest levels have been implemented. Also, the high-level functioning of the program cannot be seen until the end of testing. On the other hand test drivers are often easier to produce than stubs.

In practice, software developers may want to re-use lower-level modules from a previous contract and so a combination of top-down and bottom-up may be employed.

Thread testing
This technique can be used for the difficult task of testing real time systems. Individual processes are tested first using either a top-down or bottom-up strategy. Principal *threads* within the program structure are then tested with a full range of stimuli. Multiple threads are then added and the 'switching' tested.

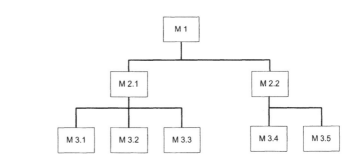

Figure 7.26 Typical program module structure.

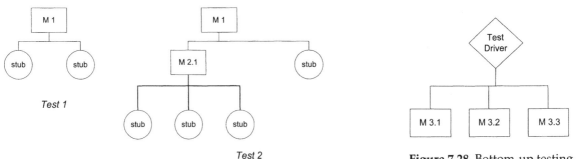

Test 1

Test 2

Figure 7.27 Top-down testing.

Figure 7.28 Bottom-up testing.

7.8 CASE STUDY: THE ROLE OF SOFTWARE IN AUTOMATED AND ROBOTIC EXCAVATION

It must be clear that the purpose of a research project is different from that of commercial product development and that consequently the development processes will differ. The crucial difference is that a commercial product has a well defined end result, which in turn generally permits the formulation of specific requirements, whereas a research project, by definition, is more open-ended. In order to give the project focus a set of top-level requirements were established. These are based on those developed in Chapter 4 and are given in Table 7.6.

Table 7.6 Top-level requirements for LUCIE excavator

---

1. The project will concentrate on the task of trenching, and be able to produce a good quality smooth-bottomed trench
   (a) It should adapt to different soil types without human intervention
   (b) The required accuracy of the finished trench is ±25 mm
   (c) It should cope with obstructions, such as boulders in the trench
   (d) The required cycle time should be similar to a human i.e. 12–15 s
2. The system must eventually accept simple trench data from a CAD system and auto-program itself to carry out the task
3. The excavator must move itself around a construction site in order to:
   (a) Complete a particular tenching task
   (b) Move between different tasks
4. The system must be capable of incremental development
5. The system will be a self-contained with no cables to external computers
6. The system should operate as safely as an equivalent manual excavator

---

7.8.1 Process issues

A major difficulty with this class of intelligent robot is the difficulty of obtaining a complete and unambiguous definition of the system requirements before moving to the design stage.

As was described in the case study section at the end of the chapter on artificial intelligence (Chapter 5), it is clear from the observation of skilled human operators that they constantly vary their strategy and tactics to maintain efficiency as ground conditions vary. In doing so, they make extensive use of many sensory feedback loops, for instance by listening to the sound of the engine, to optimise the path of the excavator bucket.

In Chapter 5, the knowledge engineering procedures used to learn how human operators dig trenches was described. Even if the human approach was fully understood, it would still leave a significant problem in knowing how closely it should be emulated or mimicked by an automated system. Furthermore, experimentation requires at least a hardware prototype system, not just the software, and the development of a one-fifth scale model in a form which allowed for tactical experimentation as well as providing support for software development is also described in Chapter 5. In addition, a rapid prototyping system based on the use of a model of the excavator operating in virtual space was also developed to provide additional support for software development, and in particular for the design of the operator interface. This prototyping system is described in the case study section of Chapter 8 on human–machine interfaces.

It is recognised that in future, computer simulation is likely to play an increasing role in robot development. However, in the case of the LUCIE system the problem of realistically modelling soil and ground conditions meant that such an approach was inappropriate and the whole system development process described here adopted instead.

Complex autonomous robots inevitably contain many sub-systems, some of which may be bought off-the-shelf, particularly at the prototype stage. In the case of LUCIE , examples of this are a satellite positioning system, a low-level controller for the hydraulic valves and a CAD/CAM interface for converting site plans and digging instructions. The software for each of these sub-systems is likely to be written in a different computer language and a vital aspect of the development process is therefore the early identification of a lucid high-level architecture which emphasises high modularity.

A modern excavator has a very high power-to-weight ratio, about ten times that of a conventional industrial robot. In fact, an excavator arm can apply a force which is sufficient to turn itself over. This, combined with mobility, means that safety is an important issue and this is discussed more fully in Chapter 9.

The problems of system development can therefore be summarised as follows:

1. It is not possible to write the complete system requirements in advance and hence they must be further refined throughout the development process.
2. A stage of knowledge acquisition is required together with some means of representing the knowledge and incorporating it into the control system.
3. The undefined nature of the task means that there is a need to be able to tune the system during operation. This points to the necessity for either a 'learning' or a 'training' facility.
4. A working hardware prototype is required for the refinement of the requirements and operation.
5. A flexible development strategy is required without the need for the expensive redeveloping of a large amount of real-time software.
6. Large systems will usually consist of many off-the-shelf components and so a top-level architecture which stresses high modularity is vital for the flexible integration of sub-systems.
7. It must be possible to validate the safety of the complete system.

Because of the particular problems mentioned above it is inevitable that the development process will contain more stages than in many projects. The following steps are discussed in the context of the spiral model which provides the necessary flexibility:

1. *Problem definition and scope.* This is to define the principle objectives of the robotic system.
2. *Knowledge acquisition.* This stage may involve observation and interrogation of human operators so that the robot's operational process can be understood and the functionality and performance of the system defined.
3. *Preliminary system requirements definition.* A high-level description of what is required from the robotic system is written.
4. *Global system decomposition.* The overall system is broken down into manageable modules which can be developed relatively independently. Such a rational decomposition is one of the most vital steps in system development. The risks associated with developing each module should be considered, and as a result of this, the most appropriate process model(s) adopted.

In the context of the spiral model, these first four stages can be considered to be part of the round 0 for the 'whole system' spiral.

5. *Detailed requirements specification.* This is undertakes for those sub-systems which can be adequately defined. If modules can be tightly specified at an early stage, then this should be carried out using a conventional specification technique such as DeMarco's structured dataflow diagrams and state transition diagrams for the control requirements. The adoption of a waterfall type approach is then appropriate for further development.
6. *Design and implementation of a rapid feasibility prototype.* This would be followed by experimental testing and refinement of requirements. A rapid prototype of both the hardware and software is required for further refinement of the requirements, particularly for those modules not included at stage 5. For large robots it may be advantageous to also build a software or hardware scale model. In general the rapid prototype will use off-the-shelf hardware components, which will probably bare little resemblance to a finished production system, and will be far from optimum in terms of cost, robustness, compactness and performance. In terms of the spiral model this stage will occupy a complete spiral for each module and form the input to other spirals at stage 7. The modules will be integrated into a rapid prototype by completing the 'whole system' spiral.
7. *Design and implementation of a development prototype.* To be followed by experimental tuning and production of a detailed requirements specification. This stage will build upon the experience gained at stage 6 with the aim of building a prototype which has the full functionality of the final system, but which also has additional capabilities for tuning and adjustments so that the detailed requirements can be finalised. It would be expected that many of the hardware and software components would be re-usable in the production version.
8. *Design and implementation of the production version.* This is intended to meet the detailed requirements defined in stage 7 and is likely to contain more highly optimised but less adaptable software, but supported by a professional user interface.

What is apparent from the above is that the waterfall model's clear chronological distinction between requirements specification and system design has been lost. In reality, this is the case for most complex systems.

With simpler robots it should be possible to merge steps 6 and 7. The success of the design of the high level system architecture at stage 6 can be measured by how similar it is to the final product at the end of stage 8. For one-off or low volume robots the development would cease at stage 7.

8 User interfaces

8.1 INTRODUCTION

The human–machine interface (HMI) is the means by which users communicate their intentions to the mechatronic system and select the parameters that govern system operation. The design and operation of this interface may well therefore be a major factor in determining the success or otherwise of a mechatronic system.

The design of a human–machine interface is a complex process which must take account of the nature and form of the information that the interface is intended to both receive and transmit, the appropriate human factors or ergonomics, the environment in which it is intended to be used and the skills and knowledge of its users. Referring to Fig. 8.1, Rasmussen has suggested that human–machine interface design should be considered in terms of three different level of behaviour, skill-based, rule-based and knowledge-based, as follows.

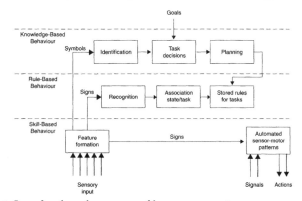

Figure 8.1 Levels of performance of human operators
(Courtesy of J. Rasmussen, 1983, Skills, Rules and Knowledge: Signals, Signs and Symbols and other Distinctions in Human Performance Models, IEEE Transactions on Systems, Man and Cybernetics, SMC-13, No 3, May/June, pp 257–266 © 1983 IEEE)

8.1.1 Skill-based behaviour

Skill-based behaviour is concerned with the carrying out of activities, such as driving a car, riding a bicycle or swimming which form smooth, continuous and highly integrated patterns of activity and behaviour with little evidence of conscious control. When operating at this level the sensed information is perceived as a series of signals in time and space which have no direct significance other than their relationships in space-time but which act as cues or signs. Thus, a car driver will make an assessment of the severity of a bend from prior indicators leading into the bend and then continuously adjust this assessment as a result of other visual signals as the bend is reached and negotiated.

8.1.2 Rule-based behaviour

Rule-based behaviour implies the availability of a set of stored rules or procedures which define the sequence of actions to be taken in response to a given set of conditions. These rules and procedures may have been derived empirically, communicated from and by other sources or are the result of a conscious planning process on the part of an individual. During the execution of the rule-based behaviour information is largely perceived as signs such as the position of a pointer or the colour of a light. The individual is then required to interpret the rules in response to the information available and then to act accordingly.

8.1.3 Knowledge-based behaviour

Knowledge-based behaviour is essentially goal oriented and is concerned with the task identification, decision making and planning activities associated with establishing the requirements for achieving the desired goal. As such, the output of knowledge-based behavioural activities will often form a major input into rule-based behavioural activities. At this level, much of the information is symbolic in form and relates to the general level of activity without the detail required at the level of rule-based behaviour.

In designing the human–machine interface it is therefore essential that the role of the interface is placed into the appropriate behavioural context and recognised that a failure to do so could have disastrous, or near disastrous consequences.

For instance, at the Three Mile Island nuclear plant in Harrisburg, Pennsylvania, USA, the operators based their actions on their interpretation of what were in fact a series of implied rather than direct measurements. Hence, the confirmation of a signal having been sent to open or close a valve was, in the absence of any direct measure and feedback of valve position, interpreted as meaning that the valve was in the state corresponding to that of the transmitted signal rather than its actual position. Thus, a valve which stuck open would, following the transmission of the signal to close, be perceived by the operators as closed!

Table 8.1 sets out some of the major factors that influence the design of an effective human–machine interface. From these it can be seen that factors such as the social, cultural and work environments in which the interface is to be used play a significant part in influencing interface design and it must therefore be borne in mind that the design of the human–machine interface is as much about human behaviour, communications, interaction and personal skills as it is about hardware and software.

8.2 THE DESIGN OF THE HUMAN–MACHINE INTERFACE

As with any design process, it is essential to ensure that that which is being designed is what is actually required and not simply the designer's perception

Table 8.1 Factors influencing the design of the human-machine interface

Factor	Impact
User	Motivation
	Job satisfaction and enjoyment
	Personality
	Skills and experience
	Level of training
Organisational factors	Working practices, organisation and politics
	Job specification and role
	Training and skills development
Environmental factors	Working environment including noise, heating, lighting, ventilation, space allocation and so forth
Human factors (ergonomics)	Seating and posture
	Organisation and layout of equipment
Health and safety	Stress
	Muscular-skeletal disorders including repetitive strain injury
	Eyesight
	Exposure
Interface	Input and output devices used
	Speech and dialogue
	Natural language systems
	Displays and the use of colour, icons, graphics and multi-media formats
Task structure	Levels of complexity, novelty, difficulty, repetitiveness, etc.
	Level of involvement of user in task definition
	User skills and competencies
System	Hardware and software
	Applications
	Environment
Motivation	Increase productivity, improve quality, decrease costs, reduce error rates, reduced labour force, removal of workers from hazardous environments, support for innovation and new product development, etc.
Constraints	Costs
	Timescale
	Available staff skills and access to training
	Building structure
Machine oriented factors	Dialogue structure and techniques
	Graphics
	Virtual reality systems
	Language
	Input and output devices

of what is required. In the case of a human–machine interface this means ensuring that the design process is user-centred and that it concentrates on listening to users and identifying user needs. It must also be a highly iterative process involving testing with the prospective users to ensure that their needs are being met and to refine these where there was uncertainty in the initial needs analysis. Only when this is done will the interface be capable of meeting the demands placed upon it.

8.2.1 The user

In designing a human–machine interface, it is not often that there will be an identifiable single user for that interface. Rather, as for instance would be the case of a photocopier or automatic teller machine (ATM), there is going to be a large number of different users, each of whom might have the same final objective but who have different levels of skill and confidence as well as differing physical attributes.

In addition to the primary user group for any interface, there may well be other users who require access not only to the primary functions but also to secondary or hidden functions not available to the primary user group. Such users would include maintenance and service personnel, the persons who refill the machine and those who clean it.

Each of these users has their own particular human–machine interface requirements which must be considered along with those of the primary user group. Buur and Windum have suggested that the user perception of a particular interface should be considered in relation to five categories:

1. *Technical perception*. This relates to the nature and function of the system and the degree of control over its operation that is exercised by the user.
2. *Ergonomic perception*. This considers the organisation and layout of the interface to ensure that the appropriate human factors are taken into account.
3. *Psychological perception*. This relates to the way in which the interface is perceived by the users and the degree of comfort they have in using it.
4. *Pedagogical perception*. This is concerned with the speed and ease with which a user can learn to use the interface and the way in which individual functions are identified and the meanings of the various symbols and icons used.
5. *Social perception*. This is concerned with the way in which the interface integrates with and fits into the associated social context.

Each of these different perspectives of the human–machine interface is in turn linked to aspects of user psychology and cognition and to the nature of memory.

Cognition and memory

Cognitive science is concerned with the understanding of how humans learn and, as a consequence, how they acquire knowledge and with the relationships between areas such as working (or short term) memory and long term memory and the way in which the stored information is accessed and processed. Figure 8.2 suggests how the human processing functions are shared between various functional areas and memory in order to describe the ways in which humans responded to various forms of stimuli.

In the context of this and other models of human information processing, the working or short-term memory represents that part of the memory in which current information is held while it is acted upon. Working memory is therefore primarily associated with activities such as the receipt of input data and the preparation of output data together with a range of sorting, storing, planning and retrieval activities. The amount of information that can be stored in the working memory is relatively small and the capacity to hold that information is limited in time. In particular, working memory is characterised by:

- operation on a 'last in, first out' basis;
- the loss of recently learned material as a result of distraction;

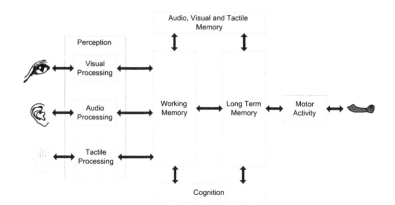

Figure 8.2 Human information processing operations.

- inhibition of recall as the result of the presence of other conflicting and irrelevant inputs;
- there is the possibility of confusion between inputs of similar content and context which can inhibit recall;
- the immediate memory of complex data such as images tends to be poor.

The implications for the design of the human–machine interface are:

- a need to minimise the levels of possible distraction for tasks involving working memory;
- the structuring of the information into chunks in such a way that it does not overload the working memory either in terms of quantity or retention time;
- the need to present images together with an associated descriptive text.

On the other hand, long-term memory represents the main information store of any individual human. Recall of information from the long-term memory can be rapid, particularly when the information required is frequently used, while in other instances there may be a substantial delay during which period attention may well have been diverted to other activities, implying some form of background processing of data.

In order to achieve the observed levels of performance it must be assumed that the information stored in long-term memory must be highly organised. A semantic model assumes some form of structure in which nodes of similar objects or classes of objects are grouped together which are then linked together while an alternative approach treats knowledge as a series of linked schemata based on prior experience. Forms of knowledge representation include:

1. *Analogy*. This takes the form of a series of images representing an object or class of objects such as a chair or a table.
2. *Propositional*. The propositional representation of knowledge is structured around statements of the form 'the cake is in the larder'.
3. *Distributed*. Distributed knowledge may be considered as a series of linked nodes in which the links define the nature of the stored knowledge.
4. *Connectionist*. A connectionist network brings together analogy and propositional forms of representation as a series of interconnected nodes in which knowledge is represented by the activation states of the nodes and their links.

Human factors

Human factors engineering or ergonomics is concerned with the relationship of humans to their working environment and is concerned with factors such as body size, strength, ageing and sensory behaviour.

Body size

In designing the human–machine interface, it must be remembered that there is no such thing as a normal or average person and that the system will require to be used by individuals of different and differing stature. In designing the interface, consideration must therefore be given to features such as the position and location of the controls in relation to the expected position of the user, including variations in the position of the user.

In dealing with body size, anthropomorphic data is generally classified as either static, derived from static subjects in fixed postures, or dynamic, relating to the swept volumes associated with different types and classes of movement together with the effects of clothing. The designer must therefore consider:

- the range of potential users taking into account age range, sex, physique and so forth;
- national characteristics. For instance, users in one country may prefer push buttons for setting and operating a washing machine while those in another may prefer a rotary knob implying a different facia design but the same underlying system;
- identify the purpose and function of all controls associated with the interface;
- consider the effects of the chosen design in relation to very large or very small individuals taking into account reach, posture and variations in body position, accessibility and so forth;
- the ability of the user to control both their position and the position of the controls.

Strength

In designing the interface, care must be taken to ensure that its operation and use does not place any undue strain on the user and that the energy expenditure required in using the interface matches the pattern of use.

Age

Age and ageing affects the capability of individuals to carry out tasks in a variety of ways as illustrated by Table 8.2. Ageing factors to be taken into account when designing the interface are:

1. The faster, more difficult and more complex a motor task is then the more likely it is to be affected by ageing.
2. Older people can maintain levels of performance in areas with which they are familiar. However, they may well be reluctant to accept changes in technology as they are then unable to use their greater experience to compensate for physical differences.
3. Older people should not be expected to acquire new working habits which are in contradiction to their previous habits.

Sensory behaviour

Humans receive information about their environment by means of a number of sensory sub-systems responding to a range of physical stimuli. The primary

Table 8.2 Effects of ageing on human performance

	General effect	Specific instance	Consequence
Sensory performance	Decreased efficiency	From age 20 to age 60 performance will tend to reduce by around 50 to 60 percent of maximum	Vision requires improved lighting conditions
			For hearing generally no consequences for the majority of work environments
			Reduced versatility
Physical capacity	Relatively small changes until around age 60 after which the rate of decrease increases	From age 20 to age 60 strength reduces to around 75 to 80 percent of maximum	Increased emphasis on postural aspects
			Speed of heavy work reduced
Compensatory factors	Increased experience		Greater rage of situations already encountered
			Increased reliability

sensory systems, and those most generally associated with the design and operation of a human–machine interface, are vision and hearing. Other sensory inputs include touch, taste, smell, temperature, pressure, orientation and rotation and an individual's ability to be aware of the position of their limbs (kinaesthetics). Table 8.3 shows the response of certain of these human sensory systems to stimuli.

In order to achieve an effective human– machine interface, the intended users must be the focus or centre of the design process with the emphasis placed on satisfying user needs. The adoption of a user centred approach to the design of the interface means that the interface designer must be aware not only of the product but also of its intended users and their requirements. The interface designer must therefore:

Table 8.3 Human sensory performance levels

Sensation	Sensory organ	Stimulation	Lowest detectable signal	Largest tolerable or practical signal	Frequency – Sensitivity	
					Lower limit	Upper limit
Vision	Eye	Electromagnetic radiation in the visible spectrum	10^{-6} ml	10 ml	300 μm	1500 μm
Hearing	Ear	Amplitude and pressure variations in surrounding media	< 100 N/m^2	2 000 N/m^2	20 Hz	20 000 Hz
Touch	Skin	Deformation of surface	Around 4×10^{-9} to 10^{-7} joules at the fingertips			
Vibration	No specific organ	Amplitude and frequency	25×10^{-5} mm at fingertips	Around 40 dB above threshold		Around 10 000 Hz at high intensity

- focus on user needs and tasks from the beginning of the design process and including the development of user guides;
- ensure that user's cognitive, social and other characteristics are understood and accommodated within the design process;
- measure user reactions through the extensive use of prototypes, manuals, interfaces and operating systems together with associated simulations;
- adopt an iterative approach to the design, incorporating feedback from trials with users.

Like any design process, the design of the user interface must comply with the typical design model shown in Fig. 8.3 linking the hardware requirements of the interface to the underlying software and thence to the system itself.

8.2.2 Input devices

There are currently available a large number of possible devices for the provision of user input and the choice of the correct input device is therefore an major consideration in the design of the user interface. To be effective the input device chosen must:

- be appropriate for the purpose for which it is intended;
- be suited to the environment in which the interface is to be used;
- provide the user with feedback in response to the actions taken;
- match the human factors needs of the intended users.

In practice, no single input device is likely to match the needs of all possible users of the system and the requirement is normally therefore that of choosing that which is most suited to the needs of the largest number of potential users. This in turn will require the interface designer to make a series of judgements and trade-offs, supported throughout by user feedback, as to the most appropriate form of interface.

Common input devices are described in the following sections.

Keyboards

Keyboards are discrete entry devices consisting of a series of buttons or switches which the user can select either individually or in combination to perform a particular action. A large number of different types of keyboard are available ranging from the conventional QWERTY keyboard and numeric keypads to more specialised forms such as that found on automated teller machines or the ECLIPSE keypad of Fig. 8.4 and intended for use by the physically disadvantaged.

The major problems associated with the use of keyboards is the level of skill required to use them effectively. Thus while a simple keypad may be the best solution for many applications, as the complexity of the input requirement increases, so does the level of user skill required to make effective use of the keyboard.

Soft keys

In some circumstances, the user may be presented with a limited number of keys whose function changes with user input. Thus the operations associated with the buttons arranged alongside the display of an automated teller

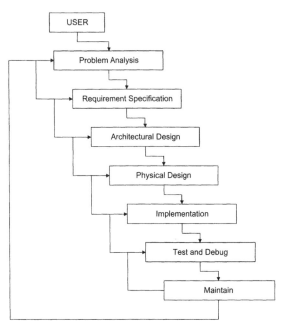

Figure 8.4 ECLIPSE keyboard
(*Courtesy of MARDIS Ltd*)

Figure 8.3 Human–machine interface design process.

machine will change in relation to the requirements of the user to access different types and levels of information.

Touch screens

A touch screen allows the user to make an input simply by touching the active part of the display in response to a prompt by the system. Touch screen displays can be made highly intuitive and are therefore easy to learn and to use and provide a direct and immediate interaction with the system. They do not, however, support fine control and manipulation of on screen objects and can be tiring to use for certain positions and orientations of the screen.

Mice, tracker balls, etc.

The mouse, along with tracker balls, touch pads and joysticks has become for many users the primary means of issuing commands to a computer-based system using a point and click approach based on screen icons. Other functions enable the user to drag and position objects on screen and to control precise motions such as those required when drawing.

Three-dimensional systems

Three-dimensional (3D) tracking systems enable the user to manipulate screen objects in three dimensions by means of associated hand movements. A typical 3D wand would use a magnetic or ultrasonic reference to enable motions in X, Y and Z directions to be achieved along with rotation about each of these axes.

Speech

Speech is the principal means of communication between individuals. Speaker dependent systems which can be trained to recognise and respond to an individual are becoming increasingly available. However, even when combined with context searching and editing such systems are still prone to error, particularly when the voice of the individual to which they have been trained changes, for instance as a result of a cold. The achievement of speaker independent systems which do not require extensive training therefore remains a major objective.

It is however certain that speech will assume an increasing importance in the design of human–machine interfaces at all levels in coming generations of mechatronic systems.

Special needs

In addition to the conventional input devices referred to, a number of other input devices have been developed to meet the special needs of the physically disadvantaged. These include:

- head mounted pointers;
- monitoring of eye movements either by the reflection of light of the eyeball or by direct measurement of the activity of the eye muscles;
- mouth operated systems.

8.2.3 Output devices

The role of the output device is to provide the user with information and feedback as to the current systems status and to prompt the user as to the actions to be taken. Some output systems are described in the following sections.

Vision

Of the human senses, vision is that which is primarily used to gain information as to the local environment and hence the visual aspect is often the most important consideration in designing the user interface. This means making appropriate use of features such as colour, shape and position to communicate with the user which, in turn, has led to the widespread adoption of symbols such as those shown in Fig. 8.5 for the communication of information.

Figure 8.5 Graphics symbols.

In computing, the development of graphical user interfaces (GUIs) with multiple windows is now commonplace in most computer systems enabling the user to hold in place many different sources of information simultaneously. The addition of images, both still and moving, together with sound has led to the development of a wide range of multi-media systems.

However, and despite the ability of human vision to discriminate between large numbers of different shades and colours it is also capable of being fooled, as for instance in the case of the classic optical illusion of Fig. 8.6. This can be a particular problem where multi-function displays such as those found in aircraft cockpits are used as the user may tend to 'see' what they are expecting to be displayed rather than what they have selected to be displayed. Thus, a user under conditions of stress may think they have requested engine hydraulic pressure to be displayed but have actually selected engine temperatures. They may well, however, continue to interpret the displayed information as

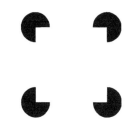

Figure 8.6 Optical illusion.

hydraulic data. This may be contrasted with an 'old fashioned' cockpit with individual displays for each item which meant that the data was received by looking in a particular direction.

In a similar context, the choice between analogue displays and digital displays needs to be carefully considered. Analogue displays are generally better at conveying a general impression of the magnitude and trend of a measured parameter, as for instance in the case of a car speedometer, while digital displays are generally much more effective in displaying precise data in a numeric form.

Speech and sound

Though speech as an input medium is still developing, synthesised speech is now widely used for output in an increasing number of applications including aircraft, vehicles and domestic systems. A particularly effective application is in association with systems for the disabled such as the ECLIPSE keypad referred to earlier (Fig. 8.4).

Multi-media systems integrate speech with visual images in a variety of formats. Thus the user of a computer based reference such as an atlas or encyclopaedia may be able to not only view a map of a country or region but to see related images and to hear the natural sounds or ethnic music associated with that region.

8.2.4 Virtual reality

Virtual reality (VR) has gained prominence in areas such as aircraft cockpit simulators and interactive games. In its total immersion form, the user is equipped with a stereo vision system, usually helmet mounted, stereo sound and possibly some form of tactile sensor in the form of a glove. Equipped in this way, the wearer can experience a presence within a computer generated world and can interact with and manipulate that world, even to the extent to being able to pick up and move objects. Such systems are, and are likely to remain, for the immediate future at least, relatively expensive and therefore a specialised form of user interface.

At perhaps a more mundane level, desk top reality systems create high resolution images on a computer screen, enabling the user to fly through the virtual world by means of a 3D wand or joystick such as those referred to earlier. Desk top reality systems have found extensive applications in areas such as architecture and planning and in the visualisation of sequences of activities.

An area of increasing interest is the creation of virtual environments to support interactive group working, particularly where the group members may be physically separated and using the computer as a means of communication as well as for the transfer of information. Computer supported co-operative working (CSCW) implies the possible existence of a number of groups of individuals separated by both distance and time who wish to share information and exchange views on common projects and objectives. The resulting categories of interaction involved in CSCW systems may then be classified as in Table 8.4.

The creation of a virtual environment within a CSCW system enables participants to view each other as if they were sitting round a single table with access to shared files. Information is then passed between groups by means of techniques such as the electronic whiteboard by which information generated

Table 8.4 Classification of CSCW interactions

	Same time	*Different times*
Same place	Face-to-face meetings (Classrooms, video conferencing)	Asynchronous interaction
Different places	Synchronously distributed (Shared editors and windows)	Asynchronous distributed (email, bulletin boards)

at one location is automatically made available to all other participating locations.

Other applications of virtual environments include the use of mobile television systems which will enable groups of individuals to view and respond to an observed scene. Thus, a surgeon can communicate with colleagues during an operation or a civil engineer can interact with a design team with, in each case the remote group being able to see the same images as those present.

8.3 TELE-OPERATION

Remote or tele-operated systems present a particular problem for the designer of the human–machine interface in that they must not only provide information on the operation of the system itself but also of its environment. In addition, they must also take account of the fact that as the distance between the operator and the remote system increases so does the transmission delay increase. Thus, the control of a system in Australia from the UK via a satellite link will result in a delay of around 0.75 s from the transmission of the instruction to the operator observing the response. While for many applications the communications distances and the delay are much less, depending on the application delays of 0.1 s can in some instances be disturbing and cause problems for the operator.

Various strategies have and are being evaluated for the control of tele-operated system, many of which revolve around the use of a remote stereo vision system slaved to the head motions of the operator to provide as near a conventional view as possible. However, on the Pathfinder mission to Mars the transmission delays are such as to prohibit direct communication with and on-line control of remote systems. The route for the Sojourner vehicle of Fig. 8.7 is therefore planned by the controller on Earth using stereo images provided by the television cameras on the lander to establish a series of waypoints. These waypoints are then transmitted to the Sojourner vehicle which then proceeds between waypoints using its own sensors to detect and respond to any obstacles not identified by the operator on Earth.

Nevertheless, and despite the use of complex remote environments, tele-operated systems such as those used for operations in areas of high risk to humans, such as sub-sea and nuclear environments, tend to show a productivity significantly less than those achieved by similar systems operating under direct operator control. This is in part due to the problems of communicating the true conditions to the remote operator but is also a result of the way in which the machine intelligence is integrated with that of the operator. Developments in machine intelligence as well as in operator interface design are likely to provide opportunities for significant improvements in performance in future.

8.4 SUMMARY

The correct design and functioning of the human–machine interface is a major factor in determining the success or otherwise of any mechatronic product and must therefore be afforded an appropriate level of effort in the design process. The advent of computer based tools for the rapid prototyping and visualisation of the interface structures and the opportunity presented to obtain effective user feedback are therefore likely to have a major role in the design of the interface while techniques such as virtual reality will also have an increasing role to play in this process.

In designing the interface, it is necessary to consider all aspects of user requirements, particularly where the user group extends over a wide range of individuals with different skills, understanding and abilities, something which is especially true in the case of the physically disadvantaged. Where possible, the interface should avoid the need for specialist knowledge on the part of the user and should be organised to present the information required in as clear, structured and organised a manner as possible using appropriate techniques and methods including vision and speech.

8.5 CASE STUDY: PROTOTYPING AND USER INTERFACE DESIGN

In a conventional excavator the operator is required to directly control the individual degrees-of-freedom of the excavator arm by means of a pair of joysticks, each of which controls two degrees-of-freedom. This means that even for an expert operator capable of controlling three, and occasionally four, degrees-of-freedom simultaneously, the achievement of actions such as true straight-line motion is difficult and it is even more difficult for an average or novice operator capable of controlling only two degrees-of-freedom simultaneously under normal conditions.

The addition of extra degrees-of-freedom to the excavator arm as suggested by Fig. 8.8 will place additional demands on the operator. The introduction of computer based control of the actuators in response to operator commands allows for the introduction of a much more instinctive form of operator control focusing on the motion of the tip of the bucket or other end effector and allows the joint motions to be determined by the machine itself.

In many applications, it is essential that the excavation is carried out to a defined profile, failures to achieve which can result in significant penalties, particularly where a need to backfill results. With a conventional excavator, this requires the operator to work in co-operation with other individuals who will measure the trench profile and feed instructions back to the operator. Another area of difficulty is that the bucket may not be directly visible to the operator, for instance as a result of the shape and depth of the trench or because the bucket is operating underwater. In both cases, it is difficult for the operator to ensure that the bucket is following a track which will fill the bucket as quickly and effectively as possible. The ability of the on-board computer to record the current and previous tracks of the bucket and to hold information as to the desired profile of the excavation therefore offers possibilities of enhanced operational efficiency.

For a system such as an excavator, the development costs are high and it is not generally possible to produce multiple prototypes and, in particular, the opportunities to evaluate different forms of operator interface are limited. The use of a rapid prototyping approach based on the use of desk-top reality

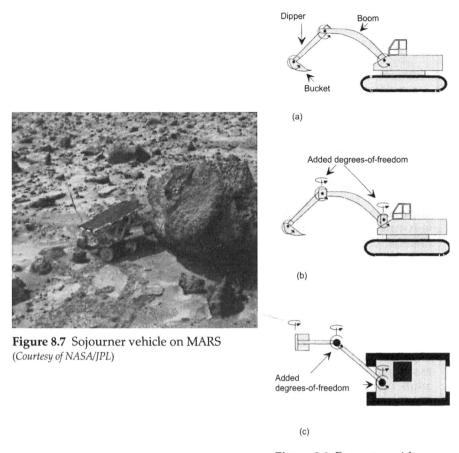

Figure 8.7 Sojourner vehicle on MARS
(*Courtesy of NASA/JPL*)

Figure 8.8 Excavator with added degrees-of-freedom.

affords a means of developing and evaluating in a relatively low cost environment different operator strategies as well as different machine configurations.

8.5.1 Rapid prototyping system

The basic configuration of the rapid prototyping environment used is shown in Fig. 8.9 in which case the development procedure is as follows:

1. *Stage 1*. The machine kinematics are modelled and displayed on the machine emulator using a desktop reality environment. The operator then controls this model using the joysticks or other control device under test and receives information, including the simulated operator displays on the system emulator. The configurations of the system emulator and the machine emulator are such that the information passed between them is identical in content and form to that used on the real machine. Thus the machine emulator receives the actuator commands from the system emulator and returns the joint sensor values and other data.
2. *Stage 2*. Once the software and kinematics have been tested in Stage 1, the machine emulator can be replaced by the real machine for evaluation

purposes. At this stage, the system software is still running on the system emulator which also provides the operator displays.

3. *Stage 3*. Once the software has been proven on the real machine it can be transferred to the target processor which can then replace the system emulator to control the simulated machine on the machine emulator. The target processor will also be responsible for driving the operator displays.

4. *Stage 4*. The complete system is installed on the real machine.

The prototyping approach as outlined enables the rapid reconfiguration of both the system software and the operator interface as well as allowing different kinematic architectures to be readily evaluated without the need to build either a model or a full scale prototype. By transferring the model into the virtual world of the machine emulator using a standard drawing format, design changes can be rapidly evaluated against a range of tasks and activities. In addition, the system can also be used for operator training and task planning in complex environments and to provide visualisation of the operation of the machine against a range of applications.

8.5.2 Operator interfaces

Chest packs

Visual feedback based on observation of the position of the bucket is a major source of information for excavator operators. Thus in situations where they are unable to see the bucket they are deprived of one of their major control inputs and performance generally deteriorates. In order to overcome this situation, operators have on occasion been provided with a set of joysticks mounted in a harness and worn on the chest to enable them to leave the cab and to position themselves where they can observe the motion of the bucket. Experience has shown that after an appropriate learning period the performance of the operators using the chest pack is generally equivalent to that which they achieved from the cab for this type of operation.

Bucket track displays

It is however not always possible for the operator to observe the path of the bucket either from the cab or elsewhere, as for instance when profiling a canal where the bucket will be submerged over most of its path. In this case an operator display of the form shown in Fig. 8.10 has been used to show the current position of the bucket and excavator arm as well as the area excavated. Knowing the width of the bucket the operator can be provided with an indication of the bucket contents and when it is full.

A refinement of the basic display is to include the desired profile of the excavation in the form of limits beyond which the machine is not normally allowed to operate. This profile could be entered into the system in a variety of ways, for instance directly by the operator or remotely by reference to an external system. In the latter case, the use of external references such as could be provided by a global positioning satellite (GPS) system would enable the profile to be automatically adjusted as the work proceeds.

Multi-degree-of-freedom 'joysticks'

The replacement of the conventional joysticks by a multi-degree-of-freedom

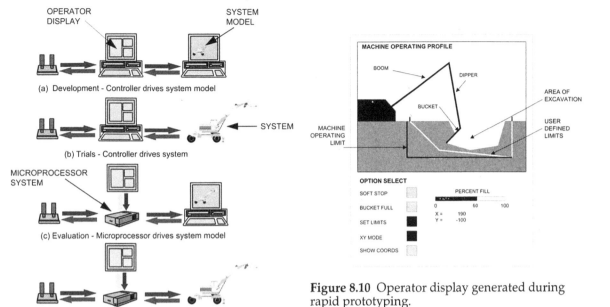

(a) Development - Controller drives system model

(b) Trials - Controller drives system

(c) Evaluation - Microprocessor drives system model

(d) Completed system

Figure 8.9 Rapid prototyping environment.

Figure 8.10 Operator display generated during rapid prototyping.

'joystick' mimicking the motions of the bucket enables the operator to concentrate on directing the bucket as required without the need to be concerned with individual joint motions. A number of possibilities exist for this purpose including the use of scaled, kinematically equivalent arms to force-ball type controllers in which the output is directly proportional to the force applied and platform mounted joysticks. In each case, the intent is that an action by the operator is directly translated into a motion of the bucket.

A problem observed with this type of user input is the degree of crosstalk between the individual degrees-of-freedom that results from the nature and construction of the multi-degree-of-freedom input device used. In practice however, the majority of the problems that resulted from this effect are minimal and could generally be overcome by imposing constraints on the allowable motion. Thus, when the bucket is in the trench a constraint is applied such that the motion could only be in the plane of the arm to ensure operation within the line of the trench only.

The use of a multi-degree-of-freedom input device provides a much more natural form of control than is used at present and when used with switchable constraints enables precise linear and other motions to be achieved. In addition, as the operator is only responsible for the motion of the bucket or end effector it supports the introduction of additional degrees-of-freedom into the machine kinematics.

Soft stops

Hydraulic excavators are an often abused item of construction plant with operators having a tendency to simply open the valves to the maximum and letting the machine run into its end stops under full hydraulic flow, often resulting in severe shock loading. The introduction of computer based control means that the system can autonomously detect the approach of the end stop and take action to decelerate smoothly to that point, even though the operator continues to hold the demand at maximum. The result is less wear on the machine and a smoother ride for the operator.

System safety 9

9.1 INTRODUCTION

This chapter is concerned with the important issue of ensuring that mechatronic systems operate safely. A system which can pose a serious danger to humans or the environment is known as a *safety critical system*. There is no official designation as to when a system is safety critical or not, although some clients may specify that a new system is to be regarded as such when they procure it from a developer. This is important because the development process for safety related systems is much more rigorous, and hence expensive, than normal systems. Safety critical systems can be loosely divided into two categories, *primary* and *secondary*.

9.1.1 Primary safety critical systems

These are systems that can directly cause significant danger if they contain a fault or malfunction. They are often involved in controlling systems that contain large amounts of energy and are often real-time systems. Examples include systems in nuclear installations, chemical plants, guided missiles, medical lasers, oil and gas platforms and most fast moving vehicles. Thus an aircraft flight control system, a railway signalling system or a car ABS braking system would be classed as primary safety critical systems.

9.1.2 Secondary safety critical systems

These are systems that do not directly control large amounts of energy, but can nevertheless indirectly cause disasters. Examples include structural design software for aircraft wings, medical record systems, medical expert systems and machines for assembling aircraft wiring looms.

9.2 BACKGROUND TO SAFETY

The procedures for designing safety critical systems are contained in the international standard *IEC 1508 Functional Safety: Safety Related Systems*, though at the time of writing this is still only in draft form. The standard is in seven parts and forms a generic framework on which application specific standards can be based. It covers electrical, electronic and programmable electronic systems and hence covers the spectrum of mechatronic devices and forms the basis of this chapter and will henceforth be referred to as the 'Standard'. The UK military standard *DEF STAN 00-56 Safety Management Requirements for Defence Systems* is generally compatible with IEC 61508 although there are some differences. Other national standards exist throughout the world mainly for military procurement.

In order to fully understand the safety process it is important to be clear about the meaning of certain key words and to be consistent in their use.

These terms and their relationships are represented diagrammatically in Fig. 9.1. This shows a system operating normally which develops a *fault* which in turn creates a *hazard*. At this stage, appropriate action can return the system either to its original operating state or some other safe state. Inappropriate action leads to an *accident*, which constitutes a failure of the safety system.

Accident	The actual event which causes harm.
Hazard	A situation, such as a fault, which has the potential to cause harm.
Risk	A combination of the probability of a hazard[†] occurring and the severity of the resulting accident.
Level of safety	The level at which the risk is at an acceptable level

† To be fully consistent this should be the probability of an accident occurring. This in turn depends upon both the probability of the hazard occurring *and* the probability of the hazard causing an accident. The draft standard appears confused here.

As expected, *absolute safety* is not a valid concept in the design of real systems. The aim is to achieve a *level of safety* which is defined relative to an acceptable degree of risk. If the existing risk of causing an accident is too high, it must be reduced. Several options, deployed either singly or in combination, are available to achieve the desired level of safety.

1. The control system could be manufactured such that the risk of the hazard occurring is acceptably low. This may for instance involve adopting a dual controller architecture. The controller itself thus becomes 'safety related', and must be subject to the discipline of the Standard.
2. A protection system could be provided so that in the event of the hazard occurring, the protection system has an acceptable probability of intervening to return the system to a safe state before an accident occurs.
3. External factors such as a safety barrier could be invoked. This could reduce the probability of the accident occurring and/or reduce the severity of the accident to acceptable levels.

9.2.1 The 'ALARP' principle

A model which is influential in forming the modern approach to safety is the 'as low as reasonably practical' or 'ALARP' model developed by the UK Health and Safety Executive (UKHSE). Figure 9.2 shows a graphical representation of this model which divides the level of risk into three categories:

1. An intolerable region in which the level of risk must be reduced or, if this is not possible, the system must be abandoned.
2. An acceptable region where most people would be prepared to accept the risks; usually in return for some perceived benefit.

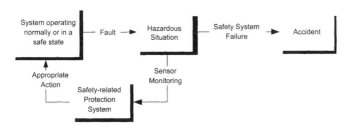

Figure 9.1 Safety relationships.

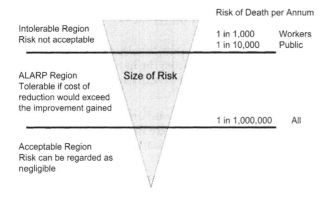

Figure 9.2 The ALARP model.

3. An ALARP region where every reasonable attempt should be made to ameliorate the risk, but where this process stops when the costs outweigh the benefits.

The numerical values associated with the boundaries are those currently suggested by the UKHSE. However, they point out that these can only be fixed in relation to current practice and cultural values. The figures are based on a 'notional person' who is assumed to be exposed to the hazard for the whole year. It is presumed that the risk values in the ALARP model refer to *total* risks. Thus a worker using a particular piece of machinery should be considered to be exposed to a combination of all the possible risks from that piece of machinery, whereas a member of the public living close to a factory should be considered to be exposed only to those external risks that might arise from the operation of the factory.

The ALARP model can be used in either a quantitative mode or a qualitative mode.

9.2.2 Safety versus availability

Yet another perspective which throws some light on safety issues is that illustrated by Fig. 9.3. This shows how tensions exists between safety, system availability and cost. This is particularly relevant to mobile robots as it would be easy to make a robot which was safe but hardly ever available for work because of stringent restrictions on separating the robot from other activity. Furthermore it may only be possible to make the robot available **and** safe by adding expensive sensing and processing.

Figure 9.3 Safety tensions.

9.3 THE IEC 1508 APPROACH

At the heart of the IEC 1508 approach lies the concept of the *overall safety lifecycle*. This is set out diagrammatically in Fig. 9.4 and will be considered in some detail. Emphasis will be placed on the key procedures rather than aspects such as management, configuration control, validation or documentation.

In particular, Boxes 3, 4, 5 and 9 of Fig. 9.4 will be expanded upon.

9.3.1 Hazard and risk analysis, Box 3, Fig. 9.4

There are three principal objectives associated with this activity:

1. identification of the hazards under all reasonably foreseeable circumstances;
2. identification of the chain of events that lead to the hazards and the final consequences that could result from the hazards (the accident);
3. determination of the risk associated with each hazard (namely its frequency and severity).

The frequency or probability of a hazard arising may be expressed quantitatively or qualitatively. The analysis may necessitate several iterations as the design evolves and usually requires at least a *preliminary hazard analysis* (PHA)

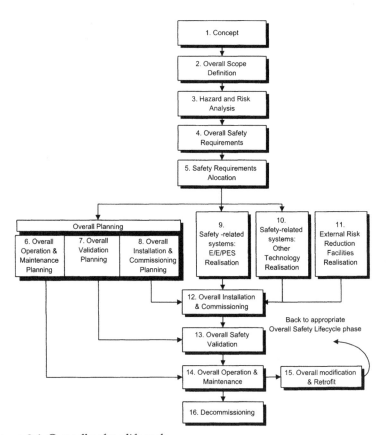

Figure 9.4 Overall safety lifecycle.

to identify the main areas of concern followed by a more rigorous *detailed hazard and risk analysis*. The opportunity should always be sought to remove hazards entirely by modifying the system or the way in which it is used. The starting point for the analysis should be the *equipment under control* (EUC) without any specific safety systems in place.

Hazard analysis

The Standard does not specify any methods or techniques for carrying out the hazard and risk analysis. Three of the more commonly used techniques are:

- hazard and operability studies (HAZOP);
- failure mode effect and criticality analysis (FMECA);
- fault tree analysis (FTA) combined with event tree analysis (ETA).

Each of these techniques will be briefly reviewed. A common feature of all the techniques is that they are a group activity that should be carried out by an experienced team of engineers, software and safety experts. It should however be noted that any single method is usually insufficient in itself for application to mechatronic systems and it is generally necessary to examine combinations of methods.

Hazard and operability analysis
HAZOP analysis was developed for chemical plants which consist of a network of pipes and the basic principles of the technique are published by the Chemical Industries Association. It has since been adapted for use in programmable systems and *Defence Standard 00-58* explains how the approach can be generalised so that it is applicable to networks of wires and dataflows.

A HAZOP study usually involves of a team of specialists who systematically question every aspect of every part of a system and its operation using a set of key 'guide-words' such as:

NO or NOT
MORE or LESS
AS WELL AS
etc.

to establish how deviations from the planned operation may cause hazardous situations. Thus the flow of data from a safety sensor will be subjected to the key words and the effect on the system considered of any errors or omissions in that dataflow.

The study may result in several different theoretical deviations from normal operation for each aspect or component studied. Each is then considered in turn to establish how it is caused and what consequence it produces. Some of the causes may be unrealistic and some consequences may be rejected as trivial or meaningless. However, some of the deviations with realistic causes and subsequent realistic consequences will be potential hazards. These are noted and examined at a later stage to establish how they may be reduced or preferably eliminated.

The use of the HAZOP approach will generate many hypothetical situations in a mechanistic manner and the success or failure of the study will depend upon three main factors:

1. the accuracy of the data, schematics and engineering drawings upon which

the study is based;
2. the technical expertise of the team members;
3. the team retaining a sense of proportion in their examination of the seriousness of hazards identified.

Failure mode effect and criticality analysis

This is an established technique found in many engineering applications. Expertise of individuals is again employed when carrying out failure mode effect and criticality analysis (FMECA). It is a bottom-up approach in which inductive reasoning is employed to identify levels of criticality and to investigate methods of reducing these problems. Using FMECA, the objective is to determine the features of a product design, its production and distribution which are critical to various modes of failure. The elements of FMECA are employed during design to perform the following tasks (see also Chapter 4):

1. identify individual product or system components;
2. list all possible failure modes of each identified system or component;
3. determine the probable effect that each mode of failure would have on the overall function of the product or system;
4. identify all the possible causes of each of the determined failure modes;
5. assess the failure modes on a numeric scale, e.g. 1 to 10, to determine, using experience, reliability data and judgement, values for
   (a) P, the probability of each failure mode occurring ($1 =$ low, $10 =$ high)
   (b) S, the criticality or seriousness of the failure ($1 =$ low, $10 =$ high)
   (c) D, the difficulty of detecting the onset of failure ($1 =$ easy, $10 =$ very difficult)
6. Calculate the criticality index rating (C) by determining the product of the 3 categories above ($C = PSD$) and tabulate all of the findings.
7. Annotate briefly the action required to rectify or reduce C.

After this stage has been completed for all foreseen possibilities, the FMECA results can be ranked to establish areas of high criticality which are 'must improve' areas down to those risks which can be tolerated. The problem of human involvement in the assessment will dictate the acceptability of the results of FMECA studies and the technique only identifies accidents that arise from failures, not incorrect requirements specifications. However, as was suggested in Chapter 4, a failure mode effect analysis (FMEA) or FMECA strategy can be used effectively at the functional level during the design process to identify areas of possible risk to the design team.

Fault tree analysis and event tree analysis

Fault tree analysis (FTA) utilises a top-down or deductive reasoning approach to establish how a chain of events can be traced from a top level event. Each possible accident is analysed to discover what failure, event, or combination of these would cause the top level event. These events or actions are then linked by a tree structure to the top level event using logical AND/OR statements to establish relationships. The OR function is used to indicate that either one event OR another may cause the event above while the use of the AND function implies that all the ANDed events must be present for the link to be established.

Fault tree analysis provides the engineer with a means of systematically describing logical sequences of events leading to the occurrence of a critical top level event and of estimating the corresponding mathematical probabilities associated with that event. It can be used in either a quantitative or a qualitative mode and it is widely recognised as an ideal tool for the reliability analysis of complex systems.

Figure 9.5 shows a fragment of a fault tree for the accident 'Robot arm strikes human'. This exposes the following issues if the aim is to follow a *quantitative* route:

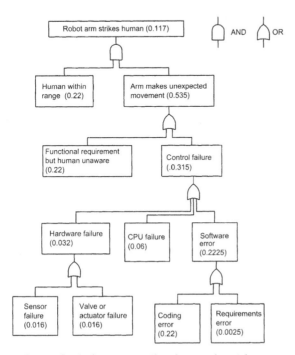

Figure 9.5 Fault tree for 'robot arm strikes human' accident.

1. Every bottom level event must have a probability associated with it. This then enables the probability of occurrence of the top level event to be calculated. From the axioms of probability, the probability of a high-level event occurring can be determined as follows:

For ANDed events:

$$P\,(A \cap B) = P(A) \times P(B) \tag{9.1}$$

for ORed events

$$P\,(A \cup B) = P(A) + P(B) \tag{9.2}$$

where $P(A)$ and $P(B)$ are the probabilities of the individual low-level events occurring. The above assumes the events are independent. As probabilities are always less than one, it follows that ANDed events are less likely to occur than ORed events. Thus one tactic in developing safety systems is to create ANDed events by introducing redundancy. The *general addition rule* is:

$$P\,(A \cup B) = P(A) + P(B) - P(A \cap B) \tag{9.3}$$

This applies when the events A and B are *not* independent. If a string of

ORed events ever sum to more than one, it indicates that they are not independent.

2. Even when trying to operate a strict quantitative approach, it will be necessary to make subjective judgements concerning some events. Table 9.1 enables some degree of conversion between descriptive probabilities and their numerical equivalents. These figures are an example based on UKHSE data for nuclear power stations.

3. The event probabilities must clearly have consistent units. The example events shown on the ALARP diagram of Fig. 9.2 were shown in terms of 'deaths per annum' and it is proposed to adopt this as the standard.

To convert the event frequencies shown in Table 9.1 to an annual basis it is necessary to make assumptions concerning the use of the system (hours/annum) and to use the probability *complement rule*. This ensures that, in all cases, probability <1. Thus if an event has a 50% probability of occurring ten times a year, the probability of it occurring at least once a year is high but still less than one. Thus the probability of a 'probable' event from Table 9.1 occurring in a year of 2500 working hours/year is given by:

$$P_{probable} = 1 - (1 - 100 \times 10^{6})^{2500} = 0.22 \tag{9.4}$$

for less frequent events the probability approximates to:

$$P(A) = P_{hour} \times 2500 \tag{9.5}$$

Table 9.1 Probability ranges

Accident frequency	Description (during life of system)	Numerical equivalent
Frequent	Likely to be continually experienced	$10\ 000 \times 10^{-6}$ / operating hour
Probable	Likely to occur often	100×10^{-6} / operating hour
Occasional	Likely to occur several times	1×10^{-6} / operating hour
Remote	Likely to occur some time	0.01×10^{-6} / operating hour
Improbable	Unlikely but may exceptionally occur	0.0001×10^{-6} / operating hour
Incredible	Extremely unlikely that the event will occur	$0.000\ 001\ 10^{-6}$ / operating hour

Thus all the frequencies can be expressed as yearly probabilities as shown in Table 9.2.

Table 9.2 Annual probability ranges

Accident frequency	Equivalent annual probability (use 2500 hour/year)
Frequent	0.99
Probable	0.22
Occasional	0.0025
Remote	0.000 025
Improbable	0.000 000 25
Incredible	0.000 000 0025

4. The particular hazardous events included in Fig. 9.5 are chosen to illustrate the difficulties in determining meaningful quantitative data. The hazards fall into four main categories:

(a) Events based on human actions such as 'human within range'. The probability of such events occurring is clearly related to training and operating procedures. These issues thus become an integral part of the safety case, and are referred to as external risk reduction facilities in IEC 61508. Where the safety of a system is dependent upon actual human action or intervention, Table 9.3 based on *Defence Standard 00-55* provides some sobering examples of the influence of human error on failures per event.

Table 9.3 Human error probabilities

General omission error, when there is no warning	0.01
Errors of omission when the actions are embedded in a well-rehearsed procedure	0.003
Simple arithmetic error with self checking	0.03
General error of supervision	0.1
Handover/changeover error	0.1
Decision error under high stress	0.2–0.3
Failure to act correctly in reasonable time after onset of high stress risis	0.3–1.0

(b) Hazards resulting from component failures such as 'sensor failure', 'CPU failure' or 'coding error' are collectively known as *random errors*. Figures for hardware failures should be based upon data for a whole module, if purchased as a unit, rather than individual components. Thus if a system includes a PC computer board, reliability data, based on previous experience, should be obtained for the complete product and not the individual components that make up the board.

This clearly provides an incentive to build safety related systems from well tested modules rather than by fabrication from basic components. Work is currently underway to establish data-banks of reliability data; however, in the absence of more specific data the values shown in Table 9.4 can generally be assumed. While these values may be adequate for crude comparisons between different types of architecture, they are clearly not ideal for real systems as, for instance, the reliability of sensors is highly dependent upon the type used.

Table 9.4 Random failure rates

Item	Typical failure rate ($\times 10^{-6}$ per hour)	Typical % safe failures
Power supply	5	90
Input circuit	0.2	50
Main processor	50	50
Output circuit	0.2	50
Sensors	13	50
Actuator	13	50

(c) Hazards resulting from intrinsic design errors or software faults such as 'software requirements error' are known as *systematic errors*. They are the most difficult type of error to identify and eliminate and, unlike random errors, it does not generally help to introduce redundancy into the system. IEC 61508 implies that probabilities cannot be associated with systematic errors and that they must be handled by increasing diversity and the rigour of the design process as the degree of risk increases.

The estimation of software error hazards also presents a difficult problem. It is accepted that all software contains errors; however, the impact of such errors is significantly reduced if the software is rigorously engineered and tested. It will be shown how the degree of risk associated with the system influences the software development process.

Another example of a design error is a *performance error*. An example of this type of error would be a perception system which becomes overloaded and unable to respond in time. Performance errors are particularly important for robots that operate in unstructured environments as their sensing needs are more difficult to define It is appropriate to identify such issues at the design stage in order to highlight the performance requirements necessary to eliminate the hazard from the fault-tree. Compliance should then be ensured by rigorous testing.

(d) Operational hazards such as 'arm making unexpected movement but human unaware' are not due to any errors or faults but simply reflect the fact that operations carried out by the system are sometimes risky. If this risk is unacceptably high, a safety related system must be introduced to reduce it.

The figures shown in brackets for the bottom level events in the sample fault tree of Fig. 9.5 are based on estimates of probabilities from either Table 9.1 or Table 9.4 (modified for per annum). Using the probability AND and OR rules given above, this results in a probability of occurrence for the top level event 'robot arm strikes human' of 0.117.

The major failing of FTA is that the initial identification of accidents is not covered. A technique that can help with this is the *event tree*. This is the inverse of the fault tree. The top event is a fault or hazardous event and the tree shows all the possible accidents that can result from the event. The combination of fault trees and event trees is essentially similar to the technique known as *cause consequence diagrams*.

Risk analysis

Having established either quantitatively or qualitatively the probability of an accident occurring, the next step is to combine this with the severity of the accident to arrive at the degree of risk. IEC 1508 leaves the classification of accident severity to specific application sectors. An example of severity categories is shown in Table 9.5.

Under a full *probabilistic risk assessment* (PRA), the loss is evaluated in the form of a complex risk vector which may take into account:

- loss of human life;
- reduction in life expectancy;
- days of illness;
- financial loss;
- environmental pollution.

Such an approach is only economically justified for major aspects of major projects, and even then the need to make subjective judgements concerning some parameters leads to large variances.

The accident severitys of Table 9.5 are combined with the probability ranges

Table 9.5 Accident severity categories

Category	Definition
Catastrophic	Multiple deaths
Critical	A single death and/or multiple severe injuries or severe occupational illnesses
Marginal	A single severe injury or occupational illness and/or multiple minor injuries or minor occupational illness
Negligible	At most a single minor injury or minor occupational illness

given in Table 9.1 to produce a risk classification scheme appropriate to the application. An example of this is shown in Table 9.6.

Table 9.6 Risk classification of accidents

Accident frequency	Consequence catastrophe	Critical	Marginal	Negligible
Frequent	I	I	I	II
Probable	I	I	II	III
Occasional	I	II	III	III
Remote	II	III	III	IV
Improbable	III	III	IV	IV
Incredible	IV	IV	IV	IV

The allocation of a particular accident to one of the classes, I to IV, provides a direct link to the ALARP approach and indicates what further action is required. This is summarised in Table 9.7.

Table 9.7 Interpretation of risk

Risk class	Interpretation
Class I	Intolerable risk
Class II	Undesirable risk, and only tolerable if risk reduction is impracticable or if the costs are grossly disproportionate to the improvement gained
Class III	Tolerable risk if the cost of risk reduction would exceed the improvement gained
Class IV	Negligible risk – no action required

In the case of the robot example considered above, a collision between a robot arm and a human would be likely to cause 'a single death'. Consequently the severity of the accident could be considered to be 'critical'. When combined with a frequency of 'probable,' Table 9.6 would indicate a 'Class I' risk which is 'intolerable'.

The 'hazard and risk analysis' element of the safety lifecycle of Fig. 9.4 concludes with the completion of the first stages of a document known as the *hazard and risk management description*. This contains a description of each identified accident and the components that contribute to it, together with the position in the risk class table of each accident. All assumptions made should be justified.

9.3.2 Overall safety requirements, Box 4, Fig. 9.4

The purpose of this activity is to specify an acceptable risk level, or level of safety, for each of the previously identified accidents and hence to define the required amount of risk reduction. The safety functions required to effect the risk reduction must then be identified. IEC 1508 also requires the degree of dependability, the *safety integrity level* or SIL, for each safety function to be specified. However, as argued below, it is contended that this is inappropriate at this stage and should be left for Box 5: Safety Requirements Allocation.

Although not explicitly stated in the draft standard, the hazard and risk management description document should be extended to include:

- the level of safety of each accident;
- the measures taken to reduce or remove hazards and risks.

The level of safety is again related to the particular application domain, and may even vary for different personnel within an industry. For example, it may be argued that military personnel in action could be subjected to higher target risks than civilian personnel at the base. Ideally, all risks would be reduced to Class IV where they can be regarded as insignificant. However the ALARP principle indicates that this is often not economically viable. A more realistic target is to aim for Class III, although even Class II may be acceptable in certain circumstances.

Table 9.8 shows the risk classification table with annual probabilities added and the robot collision example highlighted in the box. The arrows indicate two possibilities for risk reduction.

Table 9.8 Determination of the level of safety

Accident frequency	Annual probability	Consequence			
		Catastrophe	Critical	Marginal	Negligible
Frequent	0.99	I	I	I	II
Probable	0.22	I	I	II	III
Occasional	0.0025	I	II	III	IV
Remote	0.000 025	II	III	III	IV
Improbable	0.000 000 25	III	III	IV	IV
Incredible	0.000 000 0025	IV	IV	IV	IV

Either the severity of the accident could be reduced to 'marginal' and its frequency reduced to 'occasional' or the severity could remain constant and its frequency reduced to 'remote'. Either way the risk is reduced to Class III. The latter strategy will be considered.

When carrying out a quantitative analysis the level of safety takes the form of a *tolerable risk frequency*. It can be seen from Table 9.8 that this should be in the range 0.000025 to 0.000 00025 if a Class III risk is to be achieved, 0.000 025 being the maximum permissible. The difference between the initial risk frequency without safety measures, F_{np}, and the tolerable risk frequency, F_t, becomes the necessary risk reduction, ΔR. Thus for the situation of the example:

$$\Delta R = \frac{F_t}{F_{np}} = \frac{0.000\ 025}{0.117} = 0.000\ 21 \tag{9.6}$$

In practice, this risk reduction may be achieved by a combination of safety functions. In the example considered this could be by means of:

1. personnel training to reduce the risk of humans getting within range of the robot arm;
2. the introduction of one or more independent protection systems that contains a collision avoidance function.

The Standard argues that each function should be assigned an SIL. An SIL is a categorisation of a safety-related system according to its required degree of dependability and varies from an SIL of 1 (low) to an SIL of 4 (high) as is shown in Table 9.9. Such a categorisation has no meaning for external risk reduction facilities such as personnel training. Furthermore, it is not until the responsibility for risk reduction has been formally divided up between the safety systems that an individual SIL can be assigned. This activity is therefore best left for the 'safety requirements allocation' stage.

9.3.3 Safety requirements allocation, Box 5, Fig. 9.4

The responsibility for reducing the risk of an accident must be allocated to specific safety related systems, and in the case of *programmable electronic systems* (PESs), a SIL must be assigned. Although not stated in the Standard, this implies revisiting the hazard analysis techniques used previously to ensure that the necessary risk reduction has been achieved. Figure 9.6 shows the previous fault-tree changed in two ways:

1. It is assumed that training of personnel will make humans ten times less likely to stand within range and be ten times more aware of unexpected movements.
2. An independent collision prevention system has been added.

Without the collision prevention system the probability of the accident taking place is now 0.0074 against a target of 0.000 025. The failure protection system must produce further risk reduction, and because the system is ANDed its failure probability must be better than 0.000 025/0.0074 = 0.0034.

The bigger the risk reduction required, the more dependable the safety protective measures must be. In other words, they must have a higher safety integrity. The SIL for the safety protection measures is then obtained from Table 9.9.

In the example, a probability of failure of 0.0034 indicates an SIL of 2. The value of the SIL influences the architecture of the system, the degree of redundancy, and the rigour with which the hardware and software are developed. It also influences the requirements for on-line diagnostic testing and periodic proof testing of the system components. Other important points are:

1. If the safety argument assumes that the main control system for the equipment has a failure rate better than 10^{-1}, then the control system itself must be classed as a safety related system and hence subjected to the process described in the Standard. In the example the fault-tree indicates an assumed failure rate of 0.315 for the control system which means that it does not need to be designated a safety related system.
2. If risk reduction is to be allocated between more than one safety related PES, then each system must be truly independent. This means that they must be functionally and technologically diverse as well as being physically separated.

Table 9.9 Safety integrity levels

Demand mode – Probability of failure to perform its design function on demand _or_ _Continuous mode_ – Probability of a dangerous failure per year	_Safety integrity level (SIL)_
$\geq 10^{-5}$ to $<10^{-4}$	4
$\geq 10^{-4}$ to $<10^{-3}$	3
$\geq 10^{-3}$ to $<10^{-2}$	2
$\geq 10^{-2}$ to $<10^{-1}$	1

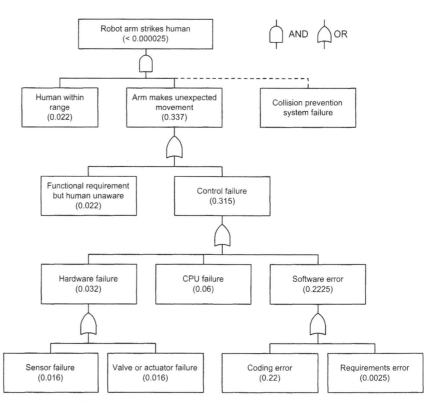

Figure 9.6 Fault tree analysis for safety requirements allocation.

3. It is possible that a particular safety-related PES will play a risk reduction role for different accidents with varying risk probabilities. In this case the highest required SIL is to be adopted throughout.

9.3.4 Realisation of PES safety related systems, Box 9, Fig. 9.4

The core of this stage of the process is the design and development of the safety related system to carry out the various safety functions that have been allocated to it. Firstly, a detailed requirements specification must be produced for each safety function that has been allocated to the system. It may also be appropriate at this stage, to further allocate sub-functions between hardware and software.

It is important to be clear about whether the system is a *protection system* or a *control system*, i.e. whether or not the system is independent of the main machinery controller. The principles involved are the same but different rules apply. A protection system will be assumed for the remainder of this example. The aim is to achieve adequate safety integrity for both the hardware and the software. In each case both random errors and systematic errors must be considered.

Hardware safety integrity (random errors)

The components that are used for building the system can be classified as either *Type A* or *Type B*:

- Type A components are well tried and tested. The failure modes must be well defined and all components must be fully tested. There must be good failure data based on at least 100 000 hours operation over at least two years with ten systems in different applications.
- Type B components are those that do not comply with all of the above. Microprocessors are considered to be type B components.

The principal means of checking hardware components are by means of off-line *proof checks*, in the form of periodic system checks, for instance once a month, and by on-line *diagnostic checks* which can be of various degrees of coverage.

Table 9.10 gives the abbreviated hardware requirements for different SILs.

The Standard also provides examples of system architectures which can be adopted without further checks. Alternatively, the designer can adopt a *bespoke architecture*, but this must be justified quantitatively and the Standard provides no guidance for how this is achieved. For example, with an SIL of 2, the Standard provides fifteen possible architectures ranging from a single processor system with high diagnostic coverage and 6 month proof checks, to a triple processor system with no diagnostics and 1 month proof checks. Different levels of diagnostic coverage may be required for the processor, the sensors and the final elements. Tables are provided to indicate what diagnostic coverage can be expected from different types of technology. For example, 'redundant actuators with cross-monitoring' can be considered to provide 'high' coverage.

Hardware safety integrity (systematic errors)

These are controlled by means of a series of recommended measures such as hardware diversity and on-line failure detection to identify inappropriate behaviour of the system. Higher SILs again require more diagnostic coverage.

Software

Both coding errors and systematic errors are controlled by the adoption of good software engineering practices – again more rigour required for higher SILs. For each SIL requirements are made concerning:

- the method of requirements specification;
- design techniques and methodologies;
- testing and verification techniques;
- programming style;
- language selection.

Table 9.10 Hardware requirements for different SILs

SIL	Type A components	Type B components
1	Proof checks	Proof checks plus • either medium diagnostic coverage • or system should continue to function with a single fault
2	Proof checks plus • either medium diagnostic coverage • or system should continue to function with a single fault	Proof checks plus • either high diagnostic coverage and system should continue to function with single fault and not likely to fail if a second fault occurs while the first is being repaired • or if not high diagnostic coverage the system should continue to function with two faults
3	Proof checks plus • either high diagnostic coverage and system should continue to function with a single fault • or if not high diagnostic coverage the system should continue to function with two faults	Proof checks plus • high diagnostic coverage with automatic detection • and system should continue to function with a single fault and not fail if a second fault occurs while the first is being repaired
4	Proof checks plus • high diagnostic coverage with automatic detection and system should continue to function with a single fault and not fail if a second fault occurs while the first is being repaired • quantitative hardware analysis carried out	Proof checks plus • high diagnostic coverage with automatic detection and • system should continue to function with two faults • quantitative hardware analysis carried out

For example, Table 9.11 shows an extract from the language selection requirements where: HR = Highly recommended; R = Recommended; NR = Not recommended; – = Neutral.

Table 9.11 Computer language selection

Language	SIL1	SIL2	SIL3	SIL4
Ada or Pascal	HR	HR	R	R
Ada or Pascal subset	HR	HR	HR	HR
Fortran 77	R	R	R	R
Fortran 77 subset	HR	HR	HR	HR
'C'	R	–	NR	NR
'C' subset	HR	R	–	–
Assembler	R	R	–	–
Basic	–	NR	NR	NR

9.3.5 IEC 1508 and mass-produced systems

There is no guidance provided within the Standard as to how systems which are to be mass-produced are to be treated. Clearly the safety criteria must be more strict for high-volume systems than for one-offs. Several approaches are possible:

1. When considering the *frequency* of an accident in order to evaluate the risk,

the whole population of systems could be considered, rather than an individual system.

2. The *severity* of the accident could be changed from, say, a single death to multiple deaths to reflect the greater number of systems in use.
3. The risk classification table could be adjusted to suit the particular application and reflect higher usage of the system. In other words the 'intolerable' risk class I would populate more of the boxes in the table.

Any of these approaches would be acceptable provided it is clearly explained and justified in the safety case.

9.4 CASE STUDY: THE ROLE OF SAFETY IN AUTOMATED AND ROBOTIC EXCAVATION

9.4.1 The approach

At the heart of the approach is the concept of a *safety manager*. This is conceived as an independent and distinct entity whose job it is to monitor the environment and to give permission for all behaviour which could have a safety critical component. This is a *behaviourist approach* in that it is concerned with achieving safe behaviour, but is not concerned with the processes that determine functional behaviour. In terms of IEC 1508, this is thus a *protection system* with a *non-safety-related* robot control system.

9.4.2 The process model

The approach generally followed the 'safety life-cycle model' shown in Fig. 9.4. Five key documents that provide the data for the safety analysis were adopted. This is shown in Figure 9.7 with the five key documents located at the top.

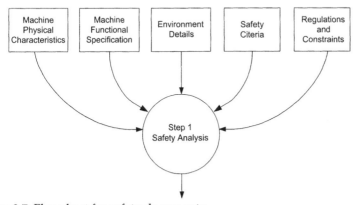

Figure 9.7 Flowchart for safety documents.

The *machine functional specification* and the *environment details* could be subject to iterative revision once the safety analysis has been performed as they may not be fully achievable within prevailing cost and technological constraints. The definition of the *safety criteria* is particularly difficult for systems of this class. From an engineering point of view, the robots will often be performing tasks previously carried out by humans, so a reasonable target would be to set the tolerable risk level to match previous safety performance. However, an engineering perspective on safety is often too narrow and does not take into account the social factors which influence acceptable levels of safety.

In essence, people, and their governments who draw up regulations, expect a higher level of safety and reliability from automated systems than they do from systems which are human-controlled. While this is in some respects, illogical it has much to commend it. Individual human failure normally results in a single incident, automated system faults may result in incidents in many different systems. Therefore, quite reasonably, automated systems must be designed and built to a higher standard of safety than is currently the case for equivalent human-operated systems.

A further complication arises from the actual application area of the robot. Such devices can greatly assist the formation of a safety argument for a process by removing workers from hazardous environments, but, on the other hand, the devices themselves constitute hazards. Consequently, it is clearly easier to justify, on a cost/benefit basis, the use of the robot in a hazardous environment such as nuclear decommissioning, than it would be on a general construction site.

Defining hazards

The next stage of the safety analysis involves using the information in the five key documents to carry out a hazard analysis. Initially hazards were defined in the most generic way to avoid making initial assumptions about system and environment or any interactions between the two, and to make sure no hazardous situations were omitted from the analysis before starting. The hazards were eventually classified as follows:

1. collision with an object on the surface;
2. collision with an underground object;
3. toppling of the robot.

Hazard analysis and problems with existing techniques

The purpose of hazard analysis is to work back from a set of hazard definitions to establish possible causes of hazards; and then with reference to the functional description, to establish situations in which they might arise. Detailed proposals can then be made concerning means of avoidance of the hazard, or failing that, detection and avoidance of a subsequent accident. Initially attempts were made to construct a fault tree. However on trying to apply FTA to the toppling hazard, the technique was initially found to be inadequate. Problems of three types were experienced:

1. It was hard to decide whether each level of deconstruction should represent a temporal change, a movement back in time, or a more detailed description of the event above, no movement in time.
2. It was impossible to represent the physical reasons why toppling occurred as discrete events, but these reasons had to be understood in order to know how the toppling could be caused.
3. The tree became nothing more than a disjointed representation of what was already known, and therefore not an analysis tool at all.

Of course, FTA is a well-established and widely-used technique. The problems were probably due to the two ways in which the analysis was unusual. Firstly, all fault trees are so named because they were devised to establish sets of faults which in combination could give rise to a particular hazard. Toppling can occur without a fault occurring simply as an unfortunate combination of circumstances in an unstructured environment. Secondly, most of the systems on which FTA has previously been used seem to have been control systems which work in structured environments. Unstructured environments give rise to very complex situations which are impossible to represent as conjunctions and disjunctions of discrete events. This limitation also applies to event tree analysis, and all other techniques which represent situations as combinations of discrete events.

The next stage was to express in mathematical terms why toppling occurs. It became apparent that understanding the physics of the problem gave much deeper insights into the causes of the hazard and how it can be avoided. For an object to topple, its centre of gravity must act outside the area of its contact with the ground in at least one vertical plane. In the case of the LUCIE excavator, it was determined that it was necessary to consider:

1. the slope of the ground;
2. the position of the excavator itself – particularly the distribution of mass in the plane of the boom arm, and the width of the base in that plane;
3. any external rotational force (e.g. wind pressure) which might be applied to the excavator.

With these factors in mind, the description of LUCIE's basic functionality was re-visited, in order to establish:

1. at which points during operation, if any, there is danger of toppling;
2. how the hazard might best be avoided, or, failing that, detected, and an accident avoided.

It was found that when investigating each of the three hazard classes mentioned earlier, that when trying to come up with a scenario in which available sensors could be deployed in order to avoid the occurrence of an accident, there was a trade-off to be managed. This was founded on the fact that there are two ways of ensuring safe operation:

1. add, within realistic cost and tech- nology constraints, extra functionality to handle a dangerous situation which may occur;
2. place limitations on the environment, or nature of the engineered system's interface to that environment, in order to ensure that a dangerous situation does not occur, external factors in IEC 1508 terms.

In general, placing restrictions on the environment was made a last resort, in order to have the greatest possible functionality while still maintaining safety, namely choose method 1 over method 2 wherever possible.

Once preliminary system design ideas had been formulated it was possible to re-visit the fault trees. The problem was discretised by requiring the sensors to measure certain factors within the environment, such as inclination of the vehicle, and to set self-imposed safety limits. Eight fault trees were produced covering every relevant combination of:

- collision with static and moving objects;

- toppling;
- travelling and digging.

A small part of the fault tree for toppling while digging is shown in Fig. 9.8. Collision with underground objects has not been considered yet as adequate sensing has not been developed.

Although the fault trees were carried out in a qualitative manner they proved valuable for the following reasons:

1. Several key hazards were seen to recur in all the fault trees. These clearly needed to be the focus of the protection system requirements.
2. The fault trees formed the basis of a qualitative risk analysis which identified a collision with a person whilst slewing as the highest level risk.

9.4.3 Safety system architecture

Hardware

The principal sensors used to achieve safety were:

- a two axis tilt sensor;
- a Leuze RotoScan RS 3 optical laser distance sensor, which can detected obstacles up to a distance of 15 metres. This can provide both a simple indication that an object has entered one of two pre-defined protection zones, and an RS232 serial string containing accurate positioning and range data to all surrounding objects. This is updated at 10 Hz. The sensor scans through 180° which means that at least two are required for full coverage (although in this project only one was used for cost reasons).

The sensors were connected to an independent 'safety manager' processor, which communicated with other sensors and processors via a CANbus as shown in Fig. 9.9

Software

The results of the safety analysis caused a revision in initial ideas about the concept of a safety manager. It was originally thought that the safety requirements specification would be concerned mainly with the functionality of the safety manager itself; and indeed the hazard analysis report contains many proposals about what the safety manager needs to be aware of in order to detect hazards, and actions it should take in the event of a hazard to avoid an accident. However, in implementing these proposals the safety manager depends very much on:

- the information to which it has access;
- the amount of control it has over the actions of other modules in the system.

It can only make decisions on the presence of a hazard if provided with enough information from the rest of the system, and can only carry out corrective manoeuvres if it can give commands to the boom and track controllers.

For these reasons, the first question to addressed after completing the hazard analysis was not the details of the safety manager's functionality but the manner and substance of communications between it and other modules. The CANbus system uses a broadcasting paradigm, which means all transmitted messages are sent to all nodes on the bus, which decide locally whether or not

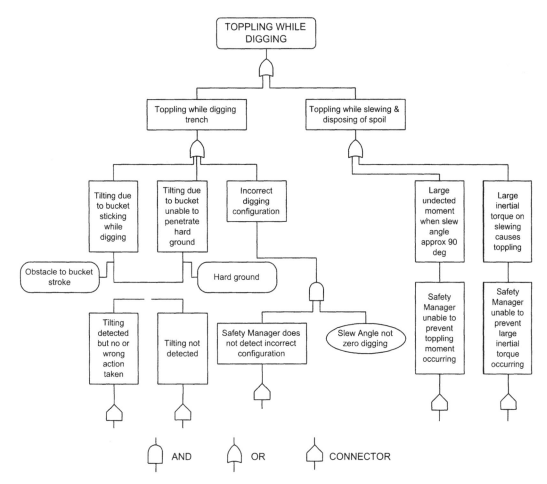

Figure 9.8 Fragment of fault tree analysis for toppling while digging.
(After C. Pace, MSc Dissertation, Lancaster University)

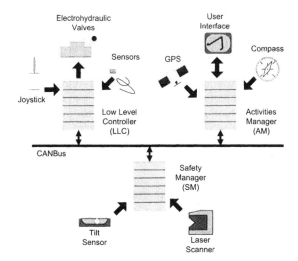

Figure 9.9 Hardware architecture.

GPS antenna on roof

Rotoscan laser sensor

Figure 9.10 The LUCIE excavator under autonomous operation.

the message is relevant to them. Thus the safety manager has access to all communication between all modules; and the manner of all communication, not just that specifically meant for the safety manager, has a bearing on safety manager functionality. A safe communication protocol was needed, or at least a communication philosophy, for the whole system. Factors influencing this philosophy include:

• The hazard analysis, the results of which indicate the type of information the safety manager requires about the actual position of the excavator and state of the immediate environment, and the intended imminent actions of other modules.
• The type of information which needs to be passed between modules.
• The fact that the safety manager must maintain safety in the event of component failure, and therefore needs to be aware of the status of all other modules in the system.

In short, the results of the hazard analysis had to form part of the input to the entire design process, not just the software design of the safety manager, if its proposals were to be successfully implemented. In the light of the importance of the CANbus, some of the initial hopes about the potential of the safety manager had to be moderated. Specifically, it is clear that safety manager *cannot* be the only safety-critical part of the system. It depends wholly on the information it is provided with on the CANbus. Therefore at least this bus, and the sensor modules which place the information on the bus in the first place, must also be safety-critical. The functionality of the safety manager eventually consisted of:

1. Checking the health of the other processors and, by means of high-level diagnostics, the health of the sensors. If high-level diagnostics were not possible, as was the case with the tilt sensor, then redundancy is required.
2. Checking the local environment for intruding objects.
3. Checking a data-base of pre-defined hazards and using global positioning data to see if they are in the locality.
4. Receiving proposed action commands from the activities manager.
5. Consulting a rule-base of safety rules to see if the proposed actions are safe.
6. Adding a token to the command and passing it to the low-level controller for implementation. The low-level controller can only act on a command if it has the token added.
7. In the event of a proposed action being unsafe, the safety manager may either stop the machine using the independent hardware link provided or permit the action at a slower speed. Thus, as a person walks towards the machine the safety manager initially monitors the approach but allows normal activity. As the person gets closer, the arm movements are slowed down. At a particular point, the movement is stopped. In order to facilitate this action , complex 'safe-state zones' were developed.

9.4.4 Prototype development and site trials

A complete prototype system was developed and implemented on the LUCIE excavator. A simplified safety manager has been produced in 'C', mounted on an independent onboard PC which is complete with a CANbus interface. The 'collision with an object on the surface' hazard has been taken as an example. As a person approaches the excavator, they are detected by the laser scanning sensor, and the safety manager firstly intervenes to slow down the movement of the arm, and, if the person continues to advance, the machine is stopped. Figure 9.10 shows LUCIE on site and excavating under autonomous control.

10 Manufacturing technology

10.1 INTRODUCTION

The impact of the mechatronic approach in manufacturing has been very significant in a number of areas from conceptual design of products and systems, as discussed in Chapter 3, through to shipping and after sales support. Within the manufacturing environment, the influence of mechatronics can be seen in manufacturing techniques based upon highly integrated CAD–CAM links and their extension to computer integrated manufacturing (CIM) to cover all aspects of machine performance, materials handling, assembly warehousing and despatch. This chapter will examine some of the areas where mechatronics is having the greatest impact and potential future effect, beginning with an appraisal of the scope of manufacturing in relation to different types of industry and organisation. These may be classified under four or five categories which can have quite different manufacturing cultures as exemplified by the implicit 'motto' which they may have.

Manufacturing environments often have an implicit culture which determines how the business behaves. For example, the traditional job-shop has multi-skilled staff who tend to pride themselves on their capabilities. Hence individualism prevails and managing people against time deadlines can be difficult. The prevailing culture of an organisation may be encapsulated in the motto which sums up the 'mood' and this may apply at all levels. Thus the manufacture of metals-based and general piece parts products, and much else, may be divided into broad classes of operational layout or configuration, determined by:

1. the nature of the product, especially its form or shape;
2. the complexity and degree of variability of the product specification;
3. the numbers or volume manufactured of the product.

The broad classes of manufacturing organisation under which may be considered the primary issues of operational characteristics and layout, the quantitative tools and mechatronic techniques used and the special challenges they represent are:

- job shop: very low production volume, high variety;
- batch or flow manufacture: intermediate volume, moderate variety;
- manual mass production: high volume, primarily manual assembly;
- automated mass production: high volume, use of automated handling and robots;
- continuous or process: manufacture of continuous media, liquids, powders, textiles.

In considering each of these categories, it will be found that different aspects of mechatronic techniques or applications will emerge, which may them be gathered into a number of 'themes', for example the use of free ranging automatically guided vehicles (FRAGVs) in manufacturing plants and warehouses.

10.1.1 The job shop: *'We can make anything'*

Operational characteristics

- versatile processes, to meet unknown demands, typical of contractors;
- multi-skilled workforce, to handle an unpredictable variety of tasks;
- groupings by traditional process type, use of capacity charts;
- lead times are traditionally long, but CNC has improved this;
- work-in-progress is often high, but backward scheduling can reduce this;
- the environment often is untidy;
- small batches and one-offs predominate;
- fixed/central inspection stations are often used.

Tools and techniques

- elementary work study, cost estimating is generally experience-based;
- material offtake analysis, to generate base stocks;
- synthetic data based on predetermined motion time systems (PMTS) and standard machining data may be widely used.

Challenges

- vulnerable in recession, clients keep their work in-house;
- much unused capacity is evident;
- material and work identity presents a major problem.

Within this environment the impact of computer numerical controlled (CNC) machine tools has been profound. Historically, work by Koenigsberger, Schlesinger and others demonstrated that the majority of components processed in metal-cutting machine shops were in effect discs, as 80% of components were found to have length/diameter ratio of less than 1. The effect of this on machine tool configurations has been to alter the proportions of machine tools away from the long bed length, much of which was rarely used, of traditional centre lathes towards a more stocky configuration in which the three orthogonal axes of motion were generally of equal magnitude.

The capability of CNC machines to perform a variety of tasks has enabled the need for some distinct categories of machine tools, such as drilling machines, to be very much reduced, since milling and drilling operations are combined in a single machining centre of either the horizontal or vertical spindle type. The co-ordinated drive of three or more axes enables the production of sculptured surfaces, while the high and steplessly variable spindle speeds, provided by modern motor control systems, together with the use of advanced cutter materials such as cubic boron nitride enable surfaces finishes to be produced of a quality which has in many cases eliminated the need for further grinding operations. Computer numerically controlled turning machines are employed for predominantly rotational as opposed to prismatic parts, but will frequently be equipped which auxiliary powered tools which may be deployed on the cross feed axis, for example for the production of key-ways in shafts.

In machining processes, an important aim of design-for-manufacture (DFM) is to design parts on which as many operations as possible can be carried out on a part without re-positioning. Once on the table or in the spindle, it should not come off until it is finished. This can have consequences at the design stage when decisions are made about the part the product geometry.

Quality control and inspection are as important in one-off or small batch manufacturing operations as in large batch or flow line manufacture. The use of traditional measuring instruments such as vernier calipers, surface plates and height gauges and GO NOT-GO snap, plug or ring gauges has been much reduced if not displaced entirely by the use of free-standing co-ordinate measuring machines (CMMs) employing touch-trigger probes or probing systems within the machine enclosure itself. In-process gauging considerably enhances the total capability of a machine tool since the probe may simply be deployed from the tool magazine and the existing machine axis drives and position measurement system used to register the dimensional information.

Because of the high first cost of CNC machines together with their ability to run unmanned, there is an economic imperative to reduce all non-process time, such as tool movements, tool change time and parts handling time to the minimum. Unscheduled stoppage due to unforeseen tool failure and other causes is therefore to be avoided. Such machines therefore normally include a variety of sensors to anticipate failure and enable plant supervisors to estimate the time for which in-tolerance manufacture may be continued before maintenance becomes necessary. It is observable that the job-shops and contractors who have invested in CNC and CMM are those which tend to stay in business.

10.1.2 Batch (or flow) manufacture: *'We have standard products, but we can do specials!'* (but too much becomes special!)

Operational characteristics

- some layout is by process, with bays of machines with similar function;
- some layout is by product with unique features; e.g. a crankshaft line;
- increased use of special jigs and fixtures;
- materials handling aids; conveyors and automatic guided vehicles;
- formally assigned areas for work-in-progress (WIP), if just-in-time (JIT) or Kanban is not used;
- workforce more demarcated by skill level.

Tools and techniques

- formal work study and industrial engineering techniques;
- wide use of synthetic data and pre-determined motion time systems (PMTS);
- queuing algorithms to assist with planning and scheduling;
- route optimisation; string diagrams, manual or software-based;
- demand forecasting and stock management;
- product classification techniques;
- work sampling;
- quality and statistical process control (SPC).

Challenges

- material control at all stages;
- accommodating/controlling the effect of product variations;
- reducing the economic batch size, i.e. lean and agile manufacturing;
- 'getting it right first time' through design for manufacture and assembly (DFMA).

There is still a place for special purpose machines in flow production. For example in the drilling of oil holes in crankshafts, the same gun-drilling operation will be employed for all crankshafts including both 4 and 6 cylinder varieties to produce intersecting deep holes at 45° to the crankshaft axis. However, much of the technology applicable to CNC machine tools will be applied to special-purpose machines including spindle power monitoring (for the detection of tool wear) and the use of in-process gauging.

With this level of manufacture, materials handling becomes more significant. Powered conveyors located generally at the level of the machine worktable will be apparent as may handling systems such as pick and place robots, if not full six axis machines, to place and remove parts from the work area. The less specialised, more versatile parts of such a production environment will have a number of flexible manufacturing system (FMS) cells comprising a CNC machine, tool magazine and automatic tool-changer, pallet shuttle to allow parts to be set up outside the actual machining environment and a robotic handling device. Although versatility is the keynote of CNC-based manufacture, parts classification techniques will probably have been used to establish parts families of broadly similar size, weight and geometry for working on by machines of matching capability.

10.1.3 Mass production: *'Just keep 'em rolling off'*

Manual type

Operational characteristics
- labour intensive (Ford/Taylor, 'Detroit type automation');
- repetitive tasks;
- jobs finely subdivided;
- fixed workstations;
- extensive (paced) conveyors;
- dedicated jigs.

Tools and techniques
- high-level work measurement, PMTS;
- ergonomics;
- queuing theory;
- line balancing.

Challenges
- human motivation and limitations;
- materials management, especially supply to workplace;
- statistical process control (SPC).

For obvious reasons of human endurance and motivation, this type of manufacture is becoming less common in advanced industrial economies. Its raison d'être, the belief that human beings can reach their highest skill levels by being regarded as automata, has been effectively discredited by the work of industrial behavioural scientists such as Hertzberg, Maslow and others. Mass production is increasingly moving over to automated systems for reasons of reliability and product consistency as much as human ergonomic and welfare considerations.

Automated assembly or process

Operational characteristics
- capital intensive;
- inflexible layout;
- single-purpose automatic machines;
- extensive use of robots.

Tools and techniques
- SPC intrinsic to operation;
- adaptive control;
- plant condition monitoring, trend analysis.

Challenges
- maintenance must be good;
- difficult to change product, incentive to use a flexible manufacturing system (FMS).

The application of mechatronic techniques is most pronounced at this level of manufacture with extensive use of robotics, automated assembly and parts feeding systems. Assembly systems will make use of machine vision and force or tactile sensing on robot end effectors. Parts handling will make use of synchronised conveying systems and automatic guided vehicles which can operate in partially structured environments in which a map of the plant layout may be held in memory but the vehicle has to avoid objects or people which may come into its path.

The production environment at this level may fairly be described as sensor-based manufacturing. The effect is to replicate in many aspects the sensory and perceptual capabilities of the human being. While recognising that the adaptability and reasoning capacity of the human being far exceeds that of the machine in many respects, in consistency and tireless repeatability and the ability to operate in hostile environments the machine has the advantage.

10.1.4 Continuous or process: *'If this packer breaks down, we will be over our head in peanuts!'*

Operational characteristics

- process dealing with homogenous materials including liquids, e.g. oil, foods, cloth;
- layout consists largely of pipes, pumps, vessels;
- highly capital intensive;
- labour is primarily on maintenance.

Tools and techniques

- demand forecasting;

- scheduling;
- CPM for maintenance planning;
- condition monitoring;
- materials optimisation, linear programming.

Challenges

- inflexibility;
- high throughput needed to maintain profit;
- sensitive to breakdown;
- potential for large-scale hazards;
- problems with product contamination.

In the process industry, the aspects of mechatronics represented by control systems have a long history. The layout of process plants dictates that much actuation, for example of fluid flow control valves, is carried out remotely using hybrid electro-pneumatic systems and position sensing based on pressure signals. Control of parameters such as flow rate, pressure, temperature and pH are all based on the implementation of sensor-based control loops and implementation of algorithms based on well-established principles such as proportional-integral-derivative (PID) control or more advanced variable structure control techniques. Because of the capital value of process plant and the large outage costs resulting from breakdown, this sector of manufacturing has been in the forefront of the development and application of sensor-based on-line condition monitoring techniques.

In this review of manufacturing operations, it is apparent that the mechatronic principles as embedded in machine tools, co-ordinate measuring machines, control systems and data management systems are present at all levels, but as the product volumes increase and the organisational scale becomes larger, so there is an increasing emphasis upon distributed data systems and communications networks.

10.2 COMPUTER NUMERICALLY CONTROLLED MACHINES, AN OVERVIEW

The early development of numerical control (NC) was from plugboard sequence control systems in the 1960s, with the control units following a logical progression from pre-set lathes and milling machines. Computer numerical control (CNC) uses a dedicated microcontroller at each machine which can contain a repertoire of part programmes with editing features, interpolation subroutines and diagnostics. Any consideration of CNC machines should begin with consideration of how many of the possible *axes* of motion are to be controlled (Fig. 10.1).

For example, lathes are normally 2-axis and drills often 2 1/2-axis, the '1/2' referring to the z motion being under up/down control between stops. Machining centres, either horizontal or vertical, will have at least 3 axes under CNC control, and often up to 5 for vertical spindle machines.

Datum systems for the location of workpieces are of two types; *fixed* with respect to the machine table, usually the SW corner of an *x–y* table or *floating*, in which datum can be set anywhere on the machine table. The second is now more common and much more useful, provided a datum is available on the piece part itself. A probe may be used to establish the datum and offsets

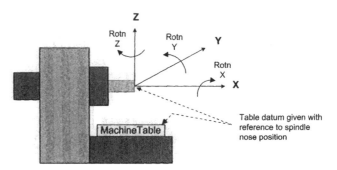

Figure 10.1 CNC machine axes.

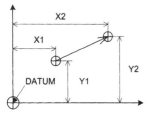

Figure 10.2 CNC machine absolute positioning.

calculated in the programme as described in the section on co-ordinate measuring machines.

Positioning is either absolute or incremental. In absolute positioning, reference is made automatically within the programme to the datum at each move (Fig. 10.2). This is the preferred technique in higher specification machines with DC servo drives and position feed back. Cheaper machines have tended to use stepper motor drives with incremental positioning in which the machine controller makes reference to the last position (Fig. 10.3). There are three common types of positioning system; in ascending order of sophistication:

1. *Point-to-point*. The objective is to move the tool or workpiece to a set of pre-programmed locations. Cutting does not take place during the *x–y* motions, but only at stationary points, for example while drilling. This simple function can be carried out by a programmable logic controller (PLC) for an application such as drilling dowel holes in 'flatpack' furniture. It is not used in machine tools proper.
2. *Straight cut NC*. This may include circular profiles if a rotary table is included. The cutting action takes place during traversing, but only in directions which are parallel to or about main slideway alignments of the machine. The traverse rates and spindle speeds can be controlled – and hence so can the metal removal rate.
3. *Continuous or continuous path NC*. Now the most common form and the most capable. Several systems exist within this category according to the interpolation system employed :
   (a) *Linear*. The simplest form, curved or circular paths are defined by a series of short straight segments; the shorter the segments the more nearly the path approaches a smooth curve. Routines are contained within the machine software, whereby the machine tool programmer sets end points and tangent commands.
   (b) *Circular*. Defines profiles in a succession of arcs with end points/centre/radius. These can be blended to form a smooth curve with no visible discontinuities.
   (c) *Polynomial*. Uses cubic and higher-order curves to generate paths which are effectively continuous within the resolution of the drive and sensor system of the machine. Machines which have this form of control can move most if not all axes simultaneously to generate complex three-dimensional shapes and are commonly used in toolmaking.

Drive systems are of two types:

1. *Stepper motor based*. This is the least expensive system but has some disadvantages as the resolution is limited by the step angle as reflected through the drive train to the table The control systems are usually open loop, so if the machine is overloaded due to an excessive feed rate being demanded and the pull-out torque of the stepper motor is insufficient, it will miss steps and the datum will be lost. Machine 'crashes' have resulted from this problem.
2. *DC servomotor based*. This is the most common system used on machining centres. Resolution is usually finer (e.g. using 7 or 8-bit encoders) and the use of position feedback means that the stalling of an axis drive can be immediately detected and action taken. In higher specification machines the position measurement is on the table itself, rather than on the leadscrew, using forms of LVDT or optical gratings. In all cases, measures are taken to reduce errors due to flexibility; for example from torsional wind-up, by stiff design and backlash eliminators, such as pre-loaded split leadscrew nuts or divided gears.

The structures of modern machine tools are determined by the needs of high metal removal rates which require high cutting speeds, creating a potential hazard from hot high velocity particles and the need for copious coolant flows. Hence the structures are intrinsically stiff, often designed to stand kinematically on three points only without deriving stiffness from the foundation, fully enclosed and configured to allow the free fall of swarf into conveyors.

The CNC spectrum is now very wide with many applications outside metal cutting and with a wide range of parameters such as force, range and rate of movement, accuracy and resolution. Among these are:

- co-ordinate measuring machines (CMMs);
- gantry robots for loading vertical furnaces in heat treatment facilities;
- laser cutting of sailcloth in sail lofts;
- flame cutting of shapes in steel plate in structural fabricators and shipyards;
- roll forming of dished ends for pressure vessels;
- bending of plate;
- stitching shoes or applying beads of adhesive ;
- component insertion to printed circuit boards.

10.3 CNC MACHINES AND PROGRAMMING

Computer numerically controlled machining centres may purchased for use in a stand-alone mode of operation, but will often be part of a system including an automatic toolchanger with a tool magazine or carousel, pre-set or 'qualified' tooling and a pallet shuttle for the set up of the next job, Fig. 10.4. They may also include in-process gauging and automatic part positioning. Toolchanging and pallet movement is included in standard programmes.

Computer-aided part programming (CAPP) is now universal and employs a high-level language to simplify the task of making the part program. The commonest languages are based on the APT (automatically programmed tools) language which contains subroutines to blend curves into straight lines without discontinuities and to perform a large variety of supporting functions.

Manual data input (MDI) is usually available at the machine controller through a keypad and is normally used for editing part programmes resident in memory. It is not a realistic method of entering complete programmes, though that is possible.

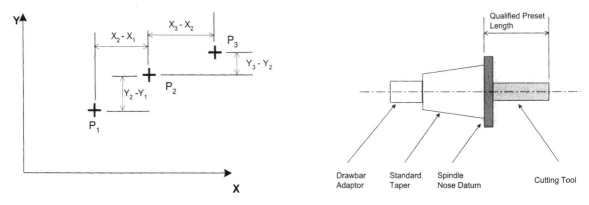

Figure 10.3 CNC machine incremental positioning.   **Figure 10.4** Qualified tooling.

In conversational programming, the controller will ask for information such as material/size of blank/roughness tolerance/tool type/number of finishing cuts. Macros resident in the controller will then perform functions such as the calculations required to optimise the machining conditions including speed, feedrate, depth of cut, number of roughing passes, cutter paths and co-ordinates and will generate the part programme block.

The programme can be viewed on the controller screen together with a representation of the component shape and tool paths before use. CAD–CAM software provides the interface between surface and solid modelling or 2D and 3D drafting packages and the CNC machine tool. The post-processor converts the dimensional information in the CAD program into a part programme usable by the machine tool. Tool data is entered at the CAD workstation. Part programmes are loaded directly to memory at the CNC controller via a serial data link.

In the ideal, the use of comprehensive software systems such as CATEA and Computervision make possible the 'factory without drawings'. All these systems result in formalised lists of instructions which may either be incorporated into common languages containing a core of commands which are almost universal, or be machine specific. Program blocks are constructed within standardised formats.

Commands are made up of a capital letter (address) and a numerical code. There are number of standard command words prefixed by letters thus:

- N, block sequence number. Usually at intervals of 5 or 10, e.g. N020 etc., up to 6 digits;
- G, preparatory function letter. Family of instructions to change mode of operation of the controller, e.g. G71 'work in metric';
- X, Y, Z, axis motions. Prefix co-ordinate information in accordance with the designated axis motions of the machine tool. +ve is assumed unless -ve is entered. Supplemented by A,B,C for rotational axes 'right hand rule applies';
- F, feed rate. Usually mm/minute;
- S, spindle speed. Usually in rev/min. Needs manual interpretation, linear cutting speed;
- T, tool function. Identifies the tool to be loaded from the tool changer. Is uniquely identified with tool location and tool length offset;

- M, miscellaneous. Covers functions such as starting and stopping the spindle, M03, M05, etc.

The programs are constructed in blocks of words, each block terminated by an end-of-block (EOB) statement. In computer assisted process planning, the machining programme may call on subroutines written in 'G-code' type format or use a format specific to the machine control unit (MCU). Statements will include part geometry, toolpaths, arithmetic calculations and cutter offsets. The post-processor will download the programme to the machine controller. The common languages based on APT (US) and EXAPT (Germany) can control up to five axes and have about a 400-word vocabulary using English-like statements. For example:

1. *Position* . General form is:
   display, layout and fonts as shown
   Name = Type/Description
   P1 = POINT/1.0,2.1,0.5
   L1 = LINE/P1,P3
   P1 = POINT/INTOF,L1,L2

Commas are used to separate. The axis order is always X, Y, Z and names used must be previously defined.

2. *Motion*. General form is:
   display, layout and fonts as shown
   Motion Command/Descriptive Data
   GOTO/P1   (move absolute)
   GOT0/-2.0,3.0,0.0   (move absolute)
   GODLTA/0,0,-2,3   (move incremental)
   Contouring has six commands:
   display, layout and fonts as shown
   GOLFT   GORGT
   GOFWD   GOBACK
   GOUP   GODOWN
   and four modifiers:
   display, layout and fonts as shown
   TO   ON
   PAST   TANTO

3. *Postprocessor*. Tool diameter, coolant commands, spindle speeds, etc., are inserted into the APT program at suitable points, e.g.
   display, layout and fonts as shown
   FEDRAT/2.29
   RAPID

4. *Auxiliary statements*. Cutter identity,
   Part No. ...... to complete program.

Various means are employed to reduce programming time and shortening programmes using 'canned' or fixed cycles in which a sequence of operations, inbuilt to the programme, is brought in by a single command. Whether manually programmed or inserted through computer-assisted programming, specific G-codes are reserved for particular types of canned cycle. For example, G80 will cancel the previous fixed cycle and return the tool spindle to the gauge height. G81 is used for a fixed drilling cycle as in the following example of drilling a hole at location X100Y100 from gauge height Z-50 to depth 5 mm.

The full sequence would be:

N050 G00 Z-50 M03	: *spindle to gauge height, rapid, spindle ON*
N055 X100 Y100	: *rapid to hole co-ordinate position*
N060 G01 Z-55 M08	: *feed to depth, coolant ON*
N065 G00 Z-50	: *rapid out to gauge height*

Using a fixed drilling cycle, G81, the programme becomes:

N050 G00 Z-50 M03	: *spindle to gauge height, rapid, spindle ON*
NO55 G81 X100 Y100 Z-55 M08	: *move to co-ordinate position, drill to depth, retract to gauge height, rapid*

This fixed cycle reduces this part of the program by half and is useful if many holes have to be drilled. Fixed cycles automatically perform a number of discrete operations as designated by the appropriate G-code. Codes G81 to G89 commonly reserved for drilling cycles, for example using 'peck' drilling to clear deep holes. They may be regarded as a 'shorthand' for very commonly used cycles.

Loops, subroutines and macros are used where the same, or similar, features are incorporated within the same component to alter the flow of the part program according to pre-defined rules with the object of shortening the program and its development time.

Loops allow the programmer to jump back to an earlier part of the program and execute the intervening blocks a specified number of times. This is especially useful in incremental programming, for example in drilling a line of holes:

N040 G91 Y30	: *set increment, move up Y30*
=N050/6	: *repeat up to N050 six times*
N050 G81 X30 Z-5 F400	: *move 30 in X, drill Z-5 deep, retract*
N055 G80	: *cancel fixed cycle*

(= sign prefixes start of loop, / indicates number of repeats)

Subroutines are used for longer repetitive elements which may not be required in every part; for example a casing which has common external or principal dimensions but different internal features, and are conventionally placed at end of the main program:

CALL command is used to access subroutine

START of subroutine shown by #

END of subroutine shown by $

The EOP ('end of program') marker is used to separate main program from subroutines.

Macros refers to a macro command or macro sub-program; a single command which generates a series of tool path moves. It is a subroutine with the ability to pass values or parameters. For example, in milling a square profile, the sequence of X and Y moves may be the same for all such profiles, as may be parameters such as spindle speed. The parameters which vary from part to part may be those relating to dimensions only. Thus in the main programme the undeclared values are indicated by *. Thus:

N105 G91 X* : *set incremental, traverse X+ value to be supplied*

N110 Y* : *traverse Y+ value to be supplied*

N115 X* : *traverse X- value to be supplied*

N120 Y* : *traverse Y- value to be supplied, back to start*

For a 50 mm square, this would be shown as a calling line in a specific order in a technique sometimes termed 'parametric programming':

= #1 X*50 Y*50 X*-50 Y*-50

Other common G-Code commands in CNC include:

- *reflection / mirror imaging* to deal with handed pairs;
- *rotation* to specify lightening features in a disc, for example;
- *translation* to enable the repetition of features by shifting the datum start point by a specified amount;
- *scaling* to provide constant proportionality for different sizes.

CNC programming is now largely PC-based and enables the user to view sophisticated 3D simulations of the tools paths, checking for clashes and interference and automatically calculating cycles times. This the programme may be substantially proved within the software.

10.4 ESSENTIALS OF CELL MANUFACTURING AND FMS

A flexible manufacturing system (FMS) is based on one or more CNC installations. At its most elementary an FMS installation consists of:

- a CNC machine;
- its tooling and associated tool handling;
- the on-line inspection system, if fitted;
- the immediately associated parts handling equipment.

Computer Integrated Manufacture (CIM) includes the more extensive systems which links the FMS into CAD and into the MRPII and scheduling systems. FMS fits in the area shown on Fig. 10.5 in terms of manufacturing volume and part variety. At the highest volume, the linear conveyor-linked arrangement of Fig. 10.6 with many of the machines being specialised or dedicated to a particular type of part is common. A low/medium volume FMS is more likely to have a group arrangement as in Fig. 10.7, with designated areas on the periphery for tool delivery and CMM if in-process gauging is not used.

In setting up an FMS installation, the following factors will be taken into account:

- *part size*, overall scantlings to be accommodated;
- *part shape*, complexity and number of disparate features;
- *part variety*, in relation to parts families;
- *product life cycle*, in relation to quantity;
- *future plans*, in that these affect part size, shape, materials;
- *ancillary operations*, non-machining operations, such as heat treatment.

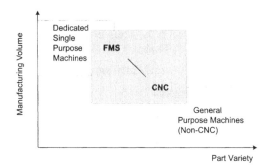

Figure 10.5 Scope of flexible manufacturing systems.

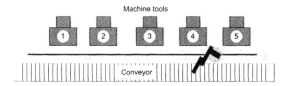

Figure 10.6 FMS with linked conveyor and rail mounted robot.

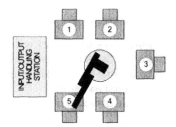

Figure 10.7 FMS with U-configuration and robot.

Parts handling in FMS may be divided into primary and secondary types. The primary system links the individual CNC machines and should:

● be configured for computer control;
● allow independent movement of parts (pallets) between machines;
● provide limited buffer storage between machines;
● allow access to the individual machines for servicing;
● interface with the secondary work handling system.

Examples are powered conveyors, rail-guided robots, in-floor towline carts and automatic guided vehicles (AGVs). The secondary system must load/offload parts to the individual CNC machines and should:

● provide automatic transfer between the primary and secondary system;
● present parts in the correct orientation at the workstation.

Examples are shuttle pallets, dedicated robots or pick-and-place devices. In a small cell, robots can often perform an intermediate role, covering several machines within their arm sweep.

10.5 PARTS CLASSIFICATION SYSTEMS AND DESIGN BY FEATURES

The use of a system for parts classification is a prerequisite for setting up an FMS environment. In the 1950s, there was a recognition that '80% of turned parts are discs', i.e. most have a length/diameter (L/D) ratio of less than unity, but manufacturing plants were full of long space-consuming centre lathes whose full capacity was rarely used. Hence emerged the beginnings of *group technology* (GT), for classifying parts by shape at the process planning stage. Over a period, this has had a significant effect on the type of machines employed in general manufacture. Most CNC machines are short, or vertical, reflecting the proportions of common machined components.

The initial purpose of parts classification was to assist in shop loading so that by assigning numerical codes to parts in relation to their shape and features, they could be assigned to process routes through appropriate machines.

Early coding systems contained fairly broad descriptors, for instance for turned parts, specifying L/D ratios and the number of steps, or for prismatic parts the number of holes and whether the holes were through or blind.

These coding systems were used with manual data input by planning engineers who would scrutinise originally manual and later CAD piece part drawings, and make judgements about the features, code them accordingly and then enter the codes into the planning or MRP system along with their unique identifiers, such as the part numbers.

Such work is evidently *retrospective*. It had no influence on design or design policy and there would not be any explicit intention when creating new parts to make them fit into categories which would conveniently lie in existing GT groupings. Current systems are derived from early work by Opitz and Brisch and further details may be found under any reference on 'group technology'.

The Opitz system, summarised below, uses a digit sequence of the form shown below:

12345	6789	ABCD
(Form Code)	*(Supplementary Code)*	*(Secondary Code)*

- The form code defines the primary design attributes, L/D ratios, and whether it is a rotational or plane shape.
- The supplementary code defines dimensions, material, stock shape, machining allowances and tolerance ranges.
- The secondary code defines the production operation type and sequence. This code is usually special to an organisation, for instance one letter may embrace several operations.

The Opitz and similar systems are limited in that the descriptors relate only to ranges and not to actual dimensions. The only unique identification of a part is its part number, and possibly its serial number, if traceability is required. Part numbers normally have no relation whatever to dimensions or function.

Much effort is now being directed towards a greater refinement of parts classification and its extension in fully automated process planning so that at the time the actual part geometry is created in a CAD system, a means is provided of accessing the data in order to extract the features which relate to processing. From this can be obtained the appropriate process routes and the part/batch cycle times. This can be the front-end of a fully automated MRP system in which, as soon as a CAD drawing is issued, its effect on the production schedule can be assessed concurrently. Wright and Bourne suggest that an automated system would somehow have to 'think' like an expert machinist or production engineer when confronted by a part drawing. Knowledge elicitation experiments were carried out with expert machinists of which the following is an example of the general questioning sequence:

Size:	*How large is the stock to contain this part?*
	Will the stock fit the fixtures we have?
Machine characteristics:	*Can this geometry be made on the available machine tools?*
Tools:	*Will the available tools cut all the part features?*
Fixtures:	*Can the part be held in the available fixtures?*

| Machinability: | *Do the materials raise any new problems?* |
| Tolerances: | *Are the machining processes capable of meeting them?* |

There are well-established principles for setting up stock in terms of using or creating reference surfaces and these have been incorporated in knowledge bases. For example, with a casting it would normally be clear from the functionality of the component which were to be the primary location faces and jig spots would probably be provided. Co-ordinate measuring systems and vision systems can be used for optimising the placement of components on a work-table and for identifying the areas for metal removal. This addresses the potentially expensive question of 'Will this casting clean up ?'

The CAD–CAM environment has now conditioned designers to think more readily in terms of constructive solid geometry in which complex representations are created by means of Boolean operations on primitive objects such as planes, cylinders and prisms. Systems development activity world-wide, continues to seek to create a usable feature-based design system which is necessary for fully automatic CAD–CAM is to be achieved.

Features are defined as:

'Geometric and topological patterns of interest in a part model and which represent high level entities useful in part analysis.'

They may be classified in a hierarchy, or taxonomy. For example, sheet features are:

prismatic	*or*	rotational
flat	*or*	formed
depressions	*or*	edges
localised	*or*	general
	and	

These may be:

1. *explicit* in which case the geometry is fully defined with simple descriptors such as through hole, blind hole, protrusion, depression;
2. *implicit* in which case the shape is defined, with scale factors and parametrics in which numerical values have to be calculated when required.

The use of such systems must take account of the viewpoints of people in different parts of the system or organisation where those in design may 'see' things differently from those in manufacture because of their different objectives. For example, in considering an aircraft wing rib machined from the solid in order to optimise material integrity, the designer sees *ribs* (a positive volume), whereas the process engineer sees *pockets* (a negative volume or void).

Expert process planning systems such as XCUT use object-oriented descriptions based on features and these are linked with process constraints such as external access directions (EADs) from which the feature volume may be machined with cutting tools. Gindy has set these out in a taxonomy by which feature objects are usually stored as 'frames' in knowledge-based systems and are linked in parent/child relationships to other frames representing feature geometry, topology, dimensions and tolerances. The link into CAD is through solid modelling techniques such as constructive solid geometry (CSG) or boundary representation (BRep).

The 'design by features' approach attempts to capture the designer's intent rather than 'second guess' it and would generally constrain the designer to construct the component design from a limited set of pre-defined features. Geometric reasoning techniques need further refinement if the freedom of designers is not to be over-constrained, but the eventual prize of an automated compliance with standards and manufacturing objectives is considered to be worth striving for.

The primitive features shown in Fig. 10.8 represent the primary ones from which most complex components can be built up using constructive solid geometry (CSG) or Boolean operations. Wright and Bourne, in trying out a set of 40 components, found that expert machinists and production engineers had evolved 'plan templates' of generic patterns of operation sequences to deal with particular types of part geometry. Figure 10.9 shows examples of such templates for rectangular blocks which project from the top of a vice and which overhang the ends so that the vertical faces can be reached by the cutting tool.

By the formalisation of such templates, a means can be created whereby the problem of the construction of a sequence has been replaced by one of recognition of a plan. Thus part geometry may be linked directly in code to machining sequences.

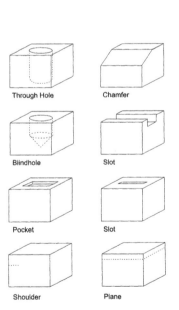

Figure 10.8 Primitive features.

Through Hole

Chamfer

Blindhole

Slot

Pocket

Slot

Shoulder

Plane

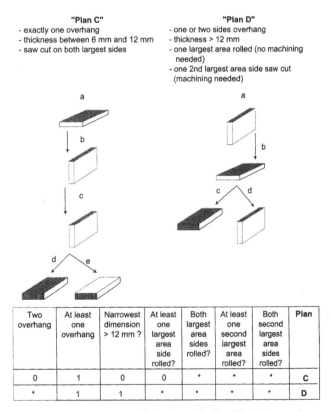

"Plan C"
- exactly one overhang
- thickness between 6 mm and 12 mm
- saw cut on both largest sides

"Plan D"
- one or two sides overhang
- thickness > 12 mm
- one largest area rolled (no machining needed)
- one 2nd largest area side saw cut (machining needed)

Two overhang	At least one overhang	Narrowest dimension > 12 mm ?	At least one largest area side rolled?	Both largest area sides rolled?	At least one second largest area rolled?	Both second largest area sides rolled?	Plan
0	1	0	0	*	*	*	C
*	1	1	*	*	*	*	D

The example codes describe a machining plan as a function of the stock preparation
1 = Yes, 0 = No, * = not significant

Figure 10.9 Plan templates

10.6 THE EXTENSION OF FMS TO CIM

Having established parts families using a parts classification or group technology (GT) approach and having determined the cell layouts, the use of FMS is the logical next step if it can be justified by product variety, product complexity and production quantity. Group technology is generally applied to medium batch production, but may used for one-off cases of highly complex piece parts. A quantitative justification is possible on the grounds of a reduction in process time – the benefit depends on what is done with the resulting 'spare' capacity – a reduction in inventory, a reduction in programming and changeover time and reduction in machining errors and inspection failures. Many examples have been reported of very large productivity gains as, for instance, in a case study by Yamazaki-Oguchi and set out below:

	FMS	CONVENTIONAL
Number of machines	13	54
Production time	528	176
	h/month	h/month
Availability	93%	67%
Number of operators	3	27
Floor area	3000 m²	9000 m²
Piece lead time	3 days	80 days

It is useful to distinguish FMS from a simple GT cell and from CNC. A GT cell can be entirely manual in control and operation but its distinguishing feature is the use of *parts families*. Computer numerical control can be stand-alone and often is. The FMS embraces both the GT family concept and is usually made up of an aggregate of CNC elements under central control. The FMS installation will comprise CNC machines and their immediate tool changers and pallet shuttles, automated material handling and storage systems for both parts and tools and a supervisory computer to manage the whole system. Usually four or more CNC machines are viewed as constituting an FMS environment. Groover suggests shop floor manning levels, excluding parts programmers and computer systems mangers, appropriate to an FMS as being typically:

1 Loader	per 5 machines
1 Toolsetter	per 10 machines
1 General maintenance person	per 10 machines
1 System manager overall	

The functions of the overall computer control system are to:

- start/stop individual machine cycles. The part programme is however executed by the machine's own CNC controller;
- manage downloading of parts programmes to CNC controllers and provide the means of entering/editing programmes;
- interface with production control dealing with how many parts are to be machined, on which machines, in what sequence and schedules the material flow;
- control traffic to run the workpiece transport system in real time. Indexes

conveyors, operates robot load/unload stations and AGVs;

- interface to devices such as pallet shuttles, which may be under the CNC machine's own controller;
- monitor the work handling system. Often the part identity in an FMS system is made possible only by its identification with a specific pallet or fixture location;
- control tool use, keeping track of location, life status, regrind, entering and leaving the system for new tools and tools which are written off;
- monitor and report upon overall system performance including machine condition and SPC data.

Data files will be held on part programs, routes, production parameters, pallet references, tool locations, tool-life and reports made on machine utilisation, production rate, status or activity and tool performance. The satisfactory operation of an FMS requires the application of line-balancing techniques with a clear identification of bottlenecks to assure a smooth production flow in relation to cycle times and quantities.

10.7 CIM ARCHITECTURES

CIM involves the whole process from the definition of a product to its shipping and maybe even its support afterwards. Figure 10.10 shows the basic activities within CIM, and how it is much more comprehensive than CAD–CAM and FMS. However CAD is the initiator for everything else via the Bill of Materials. Most products are assemblies and the planning of their manufacture requires a 'bill of materials (BOM) explosion'.

The BOM is a static document which resides in the *product database*. However, when orders are received, the BOM for each product has to be amalgamated with all the other BOMs for different products into a production

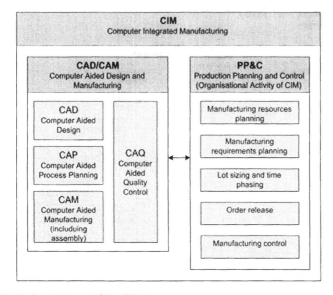

Figure 10.10 Activities within CIM.
(*After Rembold, Naji & Storr*)

schedule for parts and assemblies. This then becomes a dynamic requirement, operating in several layers down to the loading and control of individual machines. In various of the works cited in the bibliography, descriptions are given of a number of CIM models which have been found to work in practice, some specific to companies (Siemens), some from academic establishments (Karlsruhe) and others as combinations of best practice put together by national standards bodies (NIST). All can be examined as a series or hierarchy of functional blocks, in each of which a transformation occurs linking inputs, activities and outputs.

For example, Fig. 10.11 shows the flow diagram for a typical hierarchical organisation. It can be seen that as well as physical transformations involving raw parts, processes and finished parts there are parallel information transformation processes attached to these consisting of manufacturing instructions, product data and status reports. In the old manual systems, the second group would have been represented by paper in the form of route cards, batch tickets, drawings and process instructions. Now it is dependent upon electronic communications in which the establishment and maintenance of product databases is the key issue.

A practical example is the manufacture of white goods in the factory of LG Electronics in Changwon, Republic of Korea. Domestic refrigerators are manufactured in the modern plant in four basic sizes: 10 litre (student), 30 litre (small domestic) and 45 litre and 70 litre. The total output is 1.5 million units

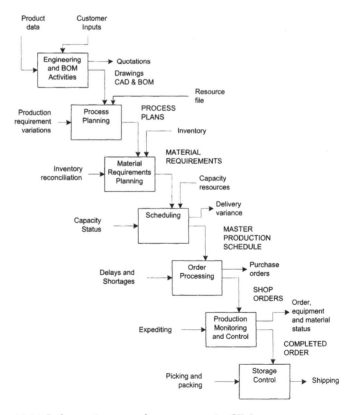

Figure 10.11 Information transfer processes in CIM.

per annum of which 50% are in the 45 litre size, the remainder being distributed roughly equally. The proportions change by up to 8% either way in each band.

In considering how to set up a CIM system the manufacturing engineer would need to address a number of questions:

- What sort of parts configuration is to be found in this domestic refrigerator?
- What visual information is available to help understand the BOM?
- How could the parts be classified as to process?
- What number of sub-assemblies and parts might be involved?
- How much would be made in-house?
- What might the manufacturing operation look like physically?
- At these volumes, where might process automation be used?
- What would be the shop floor sensor, data collection and other hardware requirements?
- What are the most significant information management issues in this environment?

10.8 IDENTIFYING INFORMATION FLOW IN THE CIM SYSTEM

These considerations of the manufacture of domestic refrigerators illustrate the need for information management in an environment where manufacture is based on sub-modules (e.g. compressors) which come together on a flow line on which is placed or erected the 'base component', in this case the carcass of the unit consisting of pre-painted, pre-insulated panels. Rembold sets out a number of CIM models which describe the data flow in such a manufacturing organisation. The aim is to be generic: the model is intended to be applicable to any manufacturing operation that involves piece parts but it would be less applicable to a continuous process.

The model described in most detail by Rembold is that based on industrially-linked work carried out at the Universities of Amhurst (USA) and Karlsruhe (Germany). It envisages a hierarchy of information systems which also links in with the layered communications manufacturing automation protocol (MAP) standard originated by General Motors, which is designed to facilitate the intercommunication of hardware modules and systems, such as machine tools and robots, whatever their source. The start point for all factory-based and inter site information exchange systems is the product database comprising:

1. a *geometric layer*, CAD, contains all the information pertaining to the shape of the product;
2. a *process planning layer*, contains the operation sequences for standard procedures and also descriptions of the manufacturing equipment;
3. a *'variation' layer*, provides the user interface by which the particular product parameters, not otherwise defined in the geometric model, such as surface finishes, can be input;
4. a *communications (interconnection) layer*, enables information to be exchanged through standard interfaces and format translators, such as IGES.

The CIM models developed by Rembold all have a recognisable format and are based on the IDEF0 representation shown in Fig. 10.12. This technique has been used in many companies to analyse their communications requirements and an outline is given here.

The acronym IDEF stands for 'integrated computer-aided manufacturing definition methodology'. Essentially, it consists of a linked series of functional blocks each set out in a particular format as shown in Fig. 10.13.

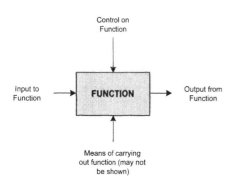

Figure 10.12 IDEFO model block.

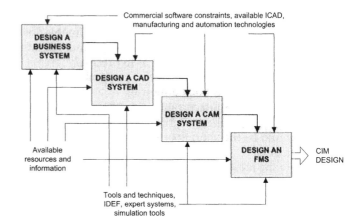

Figure 10.13 Flow diagram for the design of an FMS. (*After Ranky*)

At the top of the tree, there is a high-level description. Progressing down the layers, the individual parts become more detailed. The relationships between the layers of 'parent–child' form. For example, the parent diagram of a machine shop which converts an input of an engineering drawing and some raw material into the output of a finished product. The parent diagram 'owns' or can be decomposed into the child diagrams involved in manufacturing the component. For example, interpreting the drawings and determining the processes required. So the lower diagrams detail the higher.

The procedure can be applied to any activity in which something is changed by a process or activity, shown in a box. Around each box the following question is asked:

What are the inputs, outputs, controls and mechanisms?

The connection of the arrows to the activity box is always in the same sense:

- inputs/outputs, left/right;
- controls/mechanisms, top/bottom.

The IDEF0 model can therefore be a rigorous and highly detailed way of describing a manufacturing system and identifying the information flows required to execute the activities, that is the transformations which comprise the whole process. A simple procedure in essence uses a block diagram approach familiar to control engineers and can be used as a specification tool for FMS/CIM systems development.

10.9 DATABASES IN MANUFACTURING

Most of the functions that have an involvement in the CIM database have now been identified. A summary list for which is shown in Fig. 10.14.

An objective in efficient manufacturing may be to offer the maximum amount of product variety from the smallest possible number of individual piece parts. This is particularly the case where a product is available in a range of sizes, or different materials of construction, for example, a range of hydraulic valves, electric motors or process pumps. A fundamental question in constructing a product database is how it is structured. There are three basic approaches.

10.9.1 A hierarchical data model

This is set out as a parent–child diagram in which there are no links on the same level. The links between levels are of a 1:n relationship, e.g. a stool top:legs is 1:3. There is no linkage between objects at the same level. In Fig. 10.15, the boxes containing the objects are 'nodes', the relationships are shown by arrows indicating 'part of'. The model is simple and easy to understand, but because every product is represented in this way, the database is highly redundant. Many similar product representations will exist which only differ in some small part.

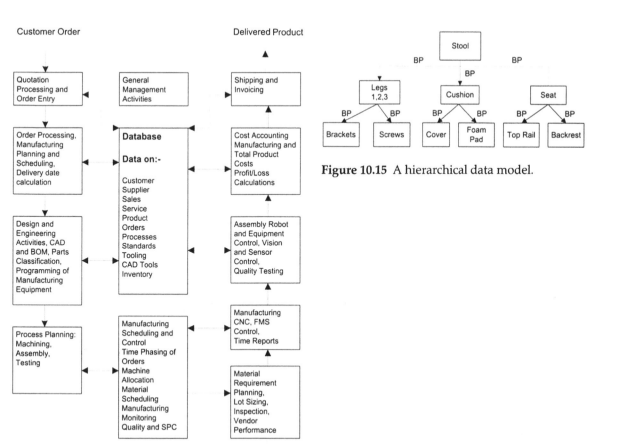

Figure 10.15 A hierarchical data model.

Figure 10.14 Functions of the CIM database.

10.9.2 A network model

The network model looks similar to the hierarchical data model but as well as the vertical (hierarchical) relationship contains information about the relationship between objects at the same level. For example, the statement:

'part a is the base of part b'

contains links between objects on the same level expressing the order in which parts are assembled. For instance, MB = 'mounted before', MO = 'mounted on'. This can describe the topology of a product in terms of how it fits together, thus giving the relationship between the parts. The database therefore has $m:n$ relationships and evolves into a ring-type structure.

There is less redundancy in this database, but much greater complexity, making it more difficult to change, since the changes are propagated throughout the network.

10.9.3 A relational model

This does not deal with the individual record of a product, but contains the data sets which make it up. The product does not exist as a complete description until the data sets have been called up. It is visualised in tabular form which might contain a list of five motors with a range of three powers and three speeds used in three products. Queries which one might use with such a database could include:

- 'Find all the motors used in product A.' (*We have just had an order for 200 of A, so how many additional motors of H power and N speed do we need from the supplier?*)
- 'Find all the 2 kW motors in our product range.' (*Aggregate this with our order schedule - isn't it time we negotiated a further bulk discount?*)
- 'Find all the 1500 r.p.m. motors used in products A and C.' (*Can we improve our use of standard motors, use common parts across more products and reduce the variety of our purchases?*)

Relational databases are becoming widely used because of their ability to provide answers such as the above. However, their implementation is computationally expensive and needs a large memory.

10.9.4 Database management

A major issue in database management is where the authority resides for data entry and editing. The principal argument lies between a totally centralised structure and one distributed in ring or star form as in Fig. 10.16. Many functions in an organisation need access to a database, some only to read it, but others play the major role in generating the data in the first place. Design engineers have always tended to proliferate varieties, whereas the purchasing function wants to minimise varieties to make economies of scale in buying. How can these conflicts be resolved? Who in fact 'owns' the data?

10.10 IN-PROCESS GAUGING AND CO-ORDINATE MEASUREMENT

The process of dimensional checking of components has been revolutionised by

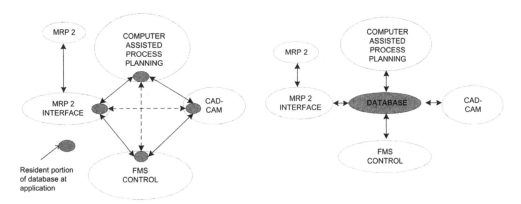

Figure 10.16 Database structures.

the wide adoption of touch trigger probing following its invention by McMurtry and its development by Renishaw plc. The principle of the touch trigger probe is as shown in Fig. 10. 17. A jewel tipped probe projects from a plate located kinematically on three contact points. In the event of the probe touching a surface, one of the contacts will be broken as the plate pivots about the line joining the other two contacts. This then sends a signal to the processor within the 3-axis co-ordinate measuring machine and the axis positions are then recorded.

Such free-standing machines of the moving gantry, as shown in Fig. 10.18, or vertical pillar type may be used for the routine checking of workpieces brought to a central inspection area. The probing head itself is now frequently employed to carry out a variety of operations within the working envelope of a CNC machine for checking a part while it is still on the table or in the spindle. In this use, the probe may be mounted in a standard tool-holder and deployed from the CNC tool store. When used in this way, the probing system presents a number of powerful aids to machine utilisation and productivity, which are described in the following sections.

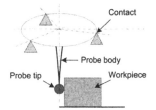

Figure 10.17 Principle of the Renishaw probe.

10.10.1 In-cycle gauging

The probe is mounted onto the machine spindle, or on an auxiliary slide, and is used to check dimensions, for example of critical diameters, before the component is released. The resolution of the probe system is of the order of 0.001 mm or finer and is thus well within the resolution of most CNC machines. At each contact, the position of the machine axis drives is recorded and stored. This then enables the effect of tool wear to be automatically compensated by the calculation of tool wear offsets.

10.10.2 Automatic job set-up and fixtureless machining

Traditionally, in order to machine a prismatic part, it was necessary to locate it in a fixture with specified locational tolerances with respect to a defined datum and alignment. Using probing it is possible to hold the part to the worktable using simple mechanical or magnetic clamps and use the machine axis drives to drive the probe to a minimal number of contact points, usually three, on the workpiece. In this way, as illustrated in Fig. 10.19, a frame of reference can be set in the space owned by the workpiece.

If using a rotary table, Fig. 10.20, the position of a datum feature such as a mounting face on the part can be automatically aligned to the orthogonal machine axes. Alternatively, the part programme co-ordinate information may be automatically aligned to that feature.

10.10.3 Optimised machining by automatic location of centre

A difficult problem in setting up for the machining of castings and forgings has been the determination of the 'best centre' in order to ensure a casting cleans up. For instance, irregularities and errors may occur in a casting due to the displacement of cores and it may be important to ensure that the external features of the casting are machined with regard to a datum set on the cored hole. In this way a uniform wall thickness for a valve body, for example, may be obtained. The probe system may therefore be used to provide a feature shift in which centres may be determined by contacting at three or more points and averaging and then the requisite offset used to modify the machining programme.

10.10.4 Tool setting

Generally, CNC machine tools operate with 'qualified tooling' as in Fig. 10.21, in which the tool lengths are pre-set, usually in a pre-setting fixture, to ensure that the stand-off or length from the mounting face on the tool holder, to the cutting face of the tool is known and can be programmed. With probing, this becomes less important, as a cube-ended stylus can be mounted on the machine table, as in Fig. 10.22, and used to measure the tool length.

Alternatively, the tool-setting probe may be used to check tools from the magazine after their insertion in the spindle, but before cutting commences. In this way the actual length may be compared with that stored in the tooling file. In a similar way, gross errors, such as those due to tool breakage, may be easily detected.

10.10.5 Other uses of probing in CNC

Probes may be able to perform some of the functions of a vision system to a limited extent where part geometries are predictable. Parts which are mis-located or placed end-for-end in fixtures may be detected. The amount of stock on a part presented for machining may be checked to determine if it is between the upper and lower limits. The CNC machine itself may be able to perform self-checking routines, for example to determine the extent of distortions affecting the relative position of the spindle and worktable during warm-up. Probing is also used very widely for digitising prototypes.

In spite of the advances in rapid prototyping techniques to generate hard 3D shapes directly from 3D CAD solid models, many products still make use of 3D studio models in wood or clay which are then digitised using pillar or gantry-type CMMs. These may be very large with plan areas up to 64 metres.

10.11 RAPID PROTOTYPING AND MODELLING

Rapid prototyping (RP) is a technology which enables product developers to produce solid models, or even master patterns, as soon as the design is ready

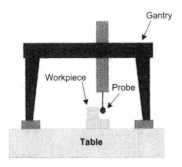

Figure 10.18 Moving gantry co-ordinate measuring machine.

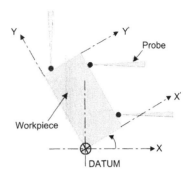

Figure 10.19 Setting of datum relative to part.

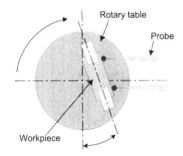

Figure 10.20 Automatic alignment of job.

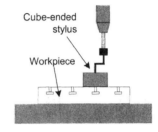

Figure 10.21 Use of cube-ended stylus for tool setting.

on the computer screen. It produces, relatively quickly, life-size models that can be handled and used as the basis for design discussions in interdisciplinary teams, some of whom may not be technologists. The prototype piece parts may be used for early assessment of how subassemblies will fit together. A large number of RP processes have been developed of which a few are emerging as the principal contenders for wider commercial application. These will be reviewed briefly.

10.11.1 Stereolithography

Stereolithography (SL) is the best known and most widely used rapid prototyping process and is shown in Fig. 10.23. In this system a narrow beam of ultraviolet light is focused on to the surface of a ultraviolet light sensitive polymer contained in a bath. A process of photopolymersiation takes place in which the liquid solidifies to a shallow depth, typically 0.5 mm. By scanning the beam in x–y across the surface over an area representing the product section at a particular depth z, a solidified slice is produced. By incrementally increasing the depth of the polymer liquid or lowering the platform on which the emerging prototype is created, thus allowing a film of liquid to spread over the previously hardened slice, a succession of slices can be built up, each bonded to the one below it, until a full 3D object is 'grown'.

10.11.2 Selective laser sintering

The selective laser sintering (SLS) process of Fig. 10.24 employs a container of

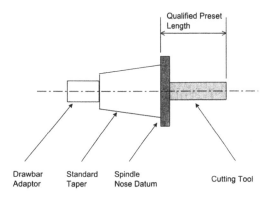

Figure 10.22 Qualified tooling.

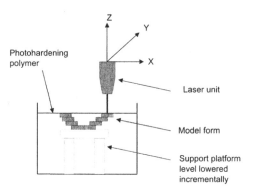

Figure 10.23 Stereolithography.

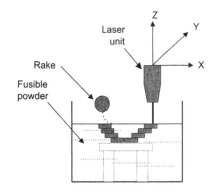

Figure 10.24 Selective laser sintering.

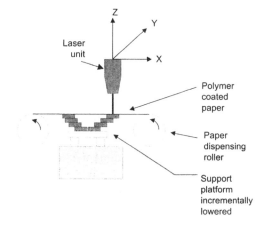

Figure 10.25 Laminated object manufacturing.

heat fusible powder normally a thermoplastic such as polycarbonate or a wax. A rake or roller sweeps the surface level and a carbon dioxide laser scans the section to be solidified. Under the influence of the laser, the particles fuse to produce a solid slice. The work platform is then lowered and additional powder is added and rolled level in preparation for the next scan. In this way, new slices are generated and fused to the layers beneath.

10.11.3 Laminated object manufacturing

One of the simpler and less expensive rapid prototyping processes, laminated object manufacturing (LOM), as shown in Fig. 10.25, employs a focused laser beam above a table across which is dispensed from a roller a sheet of heat-activated polymer impregnated and adhesive coated paper. In contrast to SLS and SL, the laser beam does not scan a whole section area, but incises only the section outline. The slice is bonded to a descending work table which is lowered at each slice increment for a new portion area of paper to be deployed across the preceding cut slice. The 3D object is then built up by successively cutting and bonding the sheets of material.

10.11.4 Fused deposition modelling

Fused deposition modelling (FDM) extrudes heated thermoplastic material from a nozzle positioned above an x–y table, depositing it as thin layers to build parts from the bottom up. This is considered to be a relatively low-cost process because it requires no laser, needs no post curing, and can use a variety of materials.

10.11.5 Ballistic particle manufacturing (ink jet manufacturing)

This process produces parts by projecting droplets of molten material on top of each other in a discrete manner. The droplet gun creates the required full section slice in the x–y plane on an elevator which moves downwards after each layer is formed. Typical jet systems have multiple nozzles and operate at up to 10 kHz with high layer deposition rates. The generated droplets are about 50 μm in diameter and are normally of investment casting wax. The process is being developed to deposit low melting point alloys of aluminium and zinc.

10.12 CAD–CAM INTEGRATION AND RAPID PROTOTYPING

The 'soft modelling' of products on CAD systems has now entered the stage where PC platforms with 3D surface and solid modelling software such as AutoCad 14 are capable of handling RP processes. A typical procedure takes a CAD file and converts it into a standard STL file format which most RP systems can accept. Either a full solid model, for slice-based systems such as SL, or a fully enclosed surface model for LOM, is required. The quality and completeness of the CAD model will be reflected in the quality of the physical prototype.

A typical procedure is for the software to first 'slice' the CAD model into discrete cross sections or layers contained in STL format. The file is then downloaded to the RP system which controls the deposition and forming of material and the incremental movement of the work table. Rapid prototyping is thus a further example of CNC-based systems dependent upon sound mechatronic design principles in the configuration of axis drives and their control systems.

10.13 RAPID TOOL MAKING

A large question is how far RP processes can actually produce models that represent the real product. With stereolithography, very fine surface finishes are possible in which the incremental steps are barely discernible but the models themselves may lack the robustness of those produced by the later ballistic particle techniques. Rapid prototyping is normally regarded as the first stage of a rapid tool making process for the production, for example of injection mould dies. Several steps are normally required to create the tool or die after the creation of the first RP model.

The RP model may be used to generate tooling in a number of ways. Limited life prototype moulds for low pressure injection moulding may be generated by metal spraying the model with an aluminium or zinc alloy until a sufficient thickness has been built up for the metal shell to be carefully detached from the model. The shell can then be supported by backing with an epoxy resin and contained in an aluminium die block forming a die cavity suitable for 1000 to

2000 injections. Alternatively, the model can be used as a sacrificial pattern in a lost-wax casting process. Ballistic particle RP, because it forms a model of higher strength, can be used to produce tools directly, using subsequent processes of sintering and infiltration to fill the voids between the particles.

Rapid prototyping is reducing the time required to achieve prototypes parts in complex shapes in a very dramatic way. In the automotive industry, models of whole cylinder heads and exhaust manifolds have been produced in three weeks from the start of CAD, about 20% of the time taken for traditional means. The overall cost savings have been even more dramatic, with claims of the overall cost to obtain a prototype being cut to 10%. The entry price for owning this technology is however high, with in 1999 the simpler industrial LOM systems costing in the region of £50 000 and the larger SL systems up to £250000. The costs of some of the materials, including the resins used in SL, are substantial.

As the capability of the systems to generate models is generally faster than the ability or need of small and medium companies to generate new product concepts, many such systems are located in RP bureaux offering services to clients on a contract basis. One aspect of RP is a possible modification of emphasis in the design culture from 'do it right first time' to 'do it again'. While the first approach attempts to eliminate mistakes, the second exploits the 'mistake' to accelerate learning, thus presenting a dilemma in which the time savings available from RP may be compromised by the tendency of design engineers to want the perfect rather than the very good.

10.14 PRINCIPLES OF DESIGN FOR EASE OF (AUTOMATED) ASSEMBLY

With the automation and handling of piece parts production now well established, the emphasis in manufacturing is turning even more to the assembly process, which accounts for 40–60% of the cost of manufacture for many products in manual processes, but may be less if extensive automation is used. Assembly techniques and design for manufacture and assembly (DFMA) must be considered together if the benefits of this aspect of concurrent engineering (CE) are to be realised.

Three main stages can be identified in manufacture:

- primary processes for raw materials production, such as casting and forging;
- secondary processes for piece part finishing, such as turning or broaching;
- composing processes for the assembly of piece parts.

Andreasson uses the term 'composing' to describe the process of assembly. Although the cost of automated assembly may be much less than for manual processes, the intention to assemble automatically must be taken into account at the very start of the design stage. Some ways in which an assembly may be composed include:

- joining: enclosing, sliding in, inserting;
- filling (of void): with gas, liquid, powder;
- interference: wedge, screw, squeeze, shrink, force, rivet, nail;
- phase change: cast, forge;
- change of form or shape: fold, swage, clench, roll;

- material interface effect: weld, solder, glue, plate, diffusion bond;
- other: sew, tie, wrap, splice, weave, braid.

Before a product can be assembled, the individual parts must be handled in a sequence in a series of operations which will typically include steps to *store, move, sort, index, orientate* or *merge* the components and afterwards to check that the assembly has been successful. Composing thus lies between the two sub-processes of handling and checking:

Handling	**Composing**	**Checking**
Putting two or more parts into a particular mutual position.	The actual movements involved in assembly.	Establishing whether the assembly has been completed to tolerance. Are all the parts there?

The most challenging part of an automated assembly process may be the initial storage and transportation of the parts to ensure their correct presentation to the composing process. These may include actions to:

- move: vibratory rails, bowl feeders, chutes, pipes;
- separate/merge/orientate/rotate: flights, holes, ramps, pins;
- allocate/select: starwheel, escapement, pick-and place simple robot, SCARA robot;
- insert/extract: gripper, sucker – end effectors of various types.

Equipment design for automated assembly is something of an 'art' in which there remains great scope for pure invention. For example, the separation and feeding of helical springs is particularly difficult. It is often possible to buy generic base units for items such as vibratory bowl feeders and then fit them with tailored flights and gates to handle specific components. The cost of such equipment must be considered carefully. For example, a bowl feeder can usually handle only one type of part at once and will cost £2000 to £5000 per machine (1999 pounds). The general principles for DFA, which can apply equally to manual or automated processes, may be summarised as follows:

- minimise the dexterity requirement: needed to grasp or position parts;
- avoid 'nuts and bolts': use captive nuts, integral nuts or snap-fits;
- modularise the assembly: build up the product by toleranced, testable sub-assemblies;
- select one part to be the base part: to provide a self-jigging configuration;
- modularise the design: obtain product variety by using the smallest number of standard components;
- reduce the number and aggregate components: exploit mouldings, powder met. etc.;
- limit the required directions of access: perform all operations from the top if possible;
- use the minimum number of fasteners: compatible with securing the parts against imposed loads;
- use generous leads for insertion: avoid sharp corners and use chamfers;
- differentiate materials and functions: for example, use brass thread inserts in a polymer;
- use principle of least constraint: for example, provide axial location of shafts from one end only;

- aim for 'hopperability': where small parts are used ensure they will feed without tangling

It must however be stressed that the tolerances necessary for successful automatic assembly will normally be tighter than for manual assembly because the adaptability of automated assembly equipment including robotic end-effectors is less than can be achieved by human perception and intelligence. Further aspects which require caution include the possibility of jamming and wedging conditions resulting in a failure to assemble at all or, possibly more seriously, the passing on of damaged assemblies which may be very difficult to detect once they get into processes further downstream. Supplier quality assurance is therefore extremely important.

Formal approaches have been developed to assist the process of design for assembly DFA or design for manufacture and assembly DFMA. Boothroyd and Dewhirst set out the first well documented systematic approach on DFA and many companies have followed this, especially in the UK by Lucas-Varity plc (now TRW) which is a leading exponent. Formal DFMA takes into account Andreasson's principles mentioned above, for example:

- parts count reduction by looking at the essential tasks/functions each part has to perform. Using mechatronic approaches whole mechanical modules can be replaced, for example by replacing cams and link- ages by software profiles and simple linear actuators;
- using a base component as the assembly jig;
- using standard methods where possible, for example by standardising on simple grippers rather than using suction pads and therefore, if needed, including a grippable feature such as a groove or upstanding rim on the part;
- using common modules across a range of products, for example common bearing assemblies;
- using parallel assembly in a 'tributary and river' arrangement.

In the composing stage, the concern is with the joining methods employed. Here, the application of DFA is sometimes the 'enemy' of design for mainte- nance if not for design for recycle-ability. For example, consider the methods available for joining parts:

- nuts and bolts;
- screws into captive nuts;
- screws into cut threads or threaded inserts;
- rivets;
- welding (continuous);
- gluing/soldering;
- spot weld/RF or ultrasonic weld;
- clip;
- fold or clench.

The most expensive in production is probably *nuts and bolts*, but any method below *screws into cut threads* will almost certainly be non-user maintainable and perhaps not maintainable at all such that the product will be destroyed by any attempt to reverse the assembly process. For example, integral snap-in clips are often used in assemblies which are driven to a fixed stop, say of a spigot into a recess. The tapered tang on the clip deforms elastically while being inserted by an air cylinder (say) of sufficient force, but springs out on passing through the aperture in the sheet. There is usually no means of subsequently accessing the tang of the clip to release the assembly. However, the assembly time may be milliseconds and so is attractive from the point of view of manufacturing costs and competitiveness in product market price. Liquid metal injection and gluing have similar advantages and disadvantages.

For successful automated assembly, the use of close tolerances but generous clearances and lead angles is needed to ensure a 'right first hit' performance with uncertainty eliminated as far as possible. The checking process can be accomplished by a number of means – weight of finished product; checking a key dimension; checking a function, for example of a PCB and its assembled components following a programmed sequence, using vision systems and comparing with a stored image.

The techniques available for DFA are well proven and their creation has proceeded in parallel with developments in robotic technology and assembly systems. To assist in this, robot architecture has been configured specifically for assembly tasks where the parts have been designed to assemble from a single access direction into a base component. The selective compliance assembly robot arm (SCARA) machine design has been created to fulfil this requirement.

Since robot drives and structures often cannot achieve positional accuracy's within commonly required assembly tolerances, generous leads are needed for the insertion process, but this alone may not be enough to prevent jamming or wedging of the components if a misalignment or offset is present.

Williams analyses a number of cases of this condition which could be referred to idiomatically as the 'sticky drawer' problem and which is frequently met during peg-into-hole assembly tasks. Whether the outcome is a successful assembly or not is determined by the initial misalignment, the frictional conditions and the degree of relative compliance of the mating parts in the relevant degrees of freedom. Usually the base part has negligible compliance with respect to the assembly fixture, so it is often the practice to deliberately introduce compliance into the robot end-effector.

10.15 SENSORS IN MANUFACTURING

All applications of mechatronics in manufacturing depend upon the use of sensor technologies of which the following are the principal types:

10.15.1 Contacting sensors

- touch probes, as used in the Renishaw system, for CMMs and integration into CNC machine tools for the dimensional checking of machined components;
- linear variable displacement transducers (LVDTs) for measuring the displacement of machine tables;
- micro-switches for the simple contact sensing and detecting the presence of objects, as in automated assembly procedures;
- strain gauged diaphragms, beams and torque tubes used variously in the measurement of pressure, force, weight and torque. Silicon based systems now make it possible for these devices to be fabricated with plan dimensions of a few tens of microns;
- piezoelectric sensors as an alternative to strain gauge sensors and which are particularly suitable for the measurement of acceleration;

- stylus-based inductive or capacitive sensors used for the measurement of surface finish;
- tactile sensors based on a matrix of pressure pins acting on variable resistive or capacitive devices. These are particularly useful where vision systems cannot be positioned to view the rear of objects.

10.15.2 Non-contacting sensors

- vision systems for parts identification and shape checking by template matching, surface finish and reflectance evaluation;
- ultrasonic for flow rate, displacement, distance to an object, crack detection and level detection;
- infrared for temperature and temperature distribution and also for object sensing;
- laser and optical for position, displacement and object sensing and surface finish evaluation;
- bar code reader for component identification and for confirming the location of automatically guided vehicles;
- magnetic field and Hall effect sensors for the detection of the presence of objects, position of actuators and measurement of rotational speed. Multiple magnetic sensors may be used in non-invasive inductive tomography to determine the distribution of immiscible liquids or to check their homogeneity within pipes or vessels;
- optical shaft encoders and optical gratings to measure the position and speed of rotating shafts and linear slides.

Vision systems are now widely used as part of integrated inspection systems as more fully described by Wright and Bourne. A simple system observing a process is represented in Fig. 10.26. The system observes the parts to determine if they are within specification according to the criteria selected, such as dimensions, orientation or colour. The image acquisition system includes structured lighting, a camera which is normally based on a charged couple device or CCD array and a frame grabber to enable the image to be processed if the parts are in motion on a conveyor, as is frequently the case. The use of image processing is computationally intensive and may still present difficulties when attempting

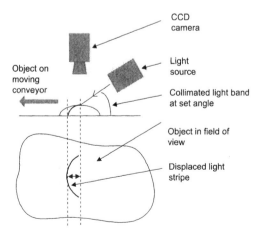

Figure 10.26 Use of structured light striping.

to detect small faults in a very large visual field, for example in detecting 0.5 mm diameter bubbles in a 2 m square transparent acrylic sheet.

In sensor integration in a machine system, close attention must be given to good principles of structural design, location and alignment to achieve reliable determination of the measurand. Errors often arise because of incorrectly positioned sensors and poor mountings. An example of a linear slide with position measurement is shown in Fig. 10.27. For example, linear and rotary position sensors should be located to:

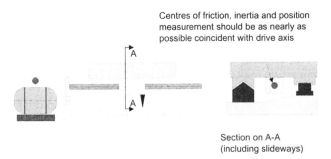

Figure 10.27 Linear slide.

- minimise parallax offset errors;
- be insensitive to thermal variations.;
- isolate from vibration and external disturbing forces;
- be located to avoid natural frequencies, including torsional excitation by pulsating inputs;
- provide a means of access for initial setting and adjustment;
- avoid distortion of mounting alignments due to clamping forces;
- avoid distortion or deflection due to loads from the leads;
- avoid spurious signals and 'cross-talk' due to electromagnetic interference (EMI) and electrical noise.

If the sensor is in the drive loop, then the backlash and elastic deflections must be minimised. Angular position sensors are often attached to power transmission elements such as ballscrews and thus incorporate errors through measuring the elastic torsional deflection of the screw. The further away from the required point of measurement the sensor is located, the more chance there is of introducing backlash and torsional wind-up errors. However, this must be set against the very considerable cost advantages in using compact inexpensive rotary transducers and mounting them on the fastest moving parts of the system, for example on the drive motor shaft. The gearing effect then enables a low resolution transducer to give a high-resolution linear position, but only if all the elastic and backlash errors are eliminated in between.

Slocum sets out a very comprehensive approach to the principles of sound mechanical design without which the benefits obtained from the intelligence and self-compensating attributes of mechatronic integration will not be realised. Leading points and questions to which attention should be directed include:

- *Speed and acceleration limits*. High speed motion through repeated cycles may give rise to unacceptable heat inputs in plain sliding bearings. Rolling bearings have less friction but may be less stiff in the out-of-plane direction.
- *Range of motion*. Flexural elements may be good for short movements only,

but the length of some elements such as ball screws dictates that they be used on long travel applications only.

- *Applied loads.* Roller bearings have higher load capacity than ball bearings, but create more friction. Oil hydrostatic bearings have much higher load capacity than aerostatic bearings. The consequences of off-design operation and failure need to be assessed.
- *Accuracy.* Can the motion be servo-controlled to the required position? Are the lateral deviations from the ideal motion path acceptable?
- *Repeatability.* Once any deviations from required accuracy have been mapped, can they be repeated?
- *Resolution.* Can the bearing allow for a small increment of motion? How is it affected by the friction level and smoothness of motion determined by the demanded acceleration profile or the external disturbances? Distortions and surface finish errors may promote 'stick-slip' and create a 'force hump' and a jerky start from rest.
- *Preload.* Many structures experience non-linear deflections in the first stages of loading, becoming linear as the load increases. By preventing stress reversals, fatigue performance may also be enhanced. It is beneficial in both cases to apply a carefully determined pre-load to the bearing, especially the case with rolling contact bearings. Too much preload however increases the friction and may worsen resolution and repeatability.
- *Friction.* High static friction with lower dynamic friction results in stick-slip or 'stiction' which leads to limit cycles in servos. Rolling bearings may prevent this problem. Some dynamic friction is desirable because it assists damping; however it also generates heat.
- *Other factors.* These include out-of-plane stiffness, thermal performance, environmental sensitivity, sealability, size and configuration, weight and support arrangements and ancillary requirements for maintenance, material compatibility, required life, maintainability and cost.

As well as the very evident employment of mechatronic principles in metal-cutting based manufacturing and statistical process control (SPC) based on CMM techniques and in MEMS, Chapter 12, there are some areas where applications continue to develop.

10.16 ON-LINE MONITORING OF MANUFACTURING EQUIPMENT

The application of on-line monitoring techniques has received increasing attention over the last three decades from the large-scale process and petrochemical industry because of the consequential costs and potential hazards arising from unpredicted breakdown. Vibration signature analysis for the detection of gear tooth wear and changes in shaft centre distance due to thermal effects or bearing wear is now the most commonly used technique.

Other methods using permanently installed transducers include capacitive and inductive transducers for determining changes in the axial position of shafts, embedded thermocouples to monitor bearing conditions and optical and thermal imaging techniques to observe, for example, the deterioration in the refractory linings of high temperature process equipment such as kilns. As general engineering manufacture has become increasingly automated with the use of capital intensive equipment and continuous operation, so the costs of loss of performance or breakdown have become less acceptable.

More comprehensive use of installed instrumentation is also used to make processes intrinsically more capable. An interesting example mentioned by Hong is the drilling of holes in aircraft structures for the fixing of external skins. In this case, a robotic arm carrying an air-powered drilling unit performs the drilling process. A precision ball screw driven by a stepper motor provides the feed motion on to the drill spindle. Because of the large number of holes to be drilled and their inaccessibility to a human observer, drill failures were not acceptable. Hence the system has to replicate to some degree the 'feel' and intelligence of a human operator. Thus, because of the flexibility of the robot arm at its maximum sweep, a compliance mechanism is incorporated in the drill unit mounting to allow proper insertion of the drill nose into the template holes.

Monitoring the drill condition is a difficult task because while drilling, the drills cannot be inspected visually as they are engaged in the material and there are a large number of ways in which a drill may fail including corner wear, asymmetry, lip height wear, crater wear, chisel wear, chipping and others. Failure will be defined by its effect on the product such that holes are drilled oversize, out-of-round or with a ragged internal surface. The different types of wear may give rise to different effects, for example corner and chisel wear, in effect 'blunting', will increase the drilling torque whereas loss of symmetry will cause vibration.

The drilling unit therefore is equipped with five sensors which measure thrust force, displacement, vibration amplitude, speed and air pressure. The thrust sensor is a strain-gauged load cell mounted between the feed mechanism and the carriage, whose deflection is a measure of the thrust force. The speed sensor and the pressure sensor are both used to measure the torque in the drilling process and the lateral vibration of the drill head is measured by an accelerometer whose signal is converted to displacement. When a drill starts to wear the changes in the measured parameters are used to determine when the process should be stopped for drill replacement.

Adaptive techniques are now widely used with CNC machine tools in order to allow machining to continue in the presence of limited amounts of tool wear, or to maintain tolerances in thermally unstable conditions. In the first case, spindle power and vibration monitoring detects the onset of wear which would eventually result in the product going out of tolerance. When used in conjunction with in-cycle gauging with a touch-trigger probe, a progressively increasing offset can be applied to compensate. Multiple sensor inputs can make use of neural net and fuzzy logic techniques, as described in Chapter 6, to take decisions as to whether and when the process should be interrupted for a change of tool. In the second case, temperature sensors are placed in the machine environment or embedded in critical parts of its structure. A prerequisite is that there should be a thermal distortion map of the machine showing the time-history dependent behaviour of the spindle and table positions over the working envelope for a range of power inputs.

10.17 AUTOMATIC GUIDED VEHICLES AND PARTS HANDLING

An important aspect of the sensor-based manufacturing process is how to move the raw materials, for example to an FMS cell, and how to move the components on to the next stage. Most systems which lend themselves to full automation are based around bar codes and programmable code bars, which are read by either laser scanners or magnetic readers. This technology facilitates automatic material flow control, supervised by networked computers, and allows full optimisation of the FMS.

In some cases it is possible to attach or imprint an identifier directly on to a part. For work-piece identification four basic methods are used:

- *Mass flow tracking*. With this system, a simple light beam or a camera identifies the product. Often, only one parameter such as height is checked to distinguish one part form another. This method can give information about the product type but not about a specific part.
- *Alphanumeric characters*. The information is either printed on the product or attached to it with labels. The print has to be held in a well defined position to enable it to be read by an optical character recognition (OCR) system, and is therefore not commonly used.
- *Bar code*. This is the most popular and versatile system for part identification and can be attached as a label, or printed on the surface. Laser scanners can work at conveyor speeds of up to 3 m/s and tolerate considerable angular inaccuracies.
- *Special coding*. Codes can also be applied to a product by cast-in dimples, punched hole patterns or other markings which can be recognised by vision or tactile sensors. Such markings have the advantage that they are more permanent and are less likely to be damaged or washed off by subsequent processes.

In many cases, the identity of a part or assembly cannot easily be maintained on the unit itself and its identity is therefore linked to the pallet or other storage or handling device which contains or transports it. In this field the use of automatic guided vehicles (AGVs) is becoming common with increasing interest being shown in the possibilities for the more intelligent free-ranging automatic guided vehicles (FRAGVs) and the prospects for more truly autonomous vehicles. The range of capability can extend from the simplest line following vehicle to vehicles which can choose an optimal route to a series of target locations, avoiding dynamic obstacles and mapping changes in the environment on the way.

All AGVs employ some form of passive or active guidance system. Early path-following AGVs and many current examples follow fixed routes determined by an energised wire loop embedded in the floor. Docking stations, for example at pallet offtake points to CNC machines, are recognised by magnetic sensors or limit switches. Two search coils mounted on the vehicle sense the guide wire and the vehicle steering is controlled so as to minimise the difference in voltages between the two search coils. The position along the guide path may be determined with a limited degree of precision by odometry using shaft encoders on the drive system. This method however requires frequent updating using a system of reference points. The main limitation with conventional AGVs is the inflexibility of the path although limited loops and crossovers are possible using 'path switch select' or 'frequency select' methods when decision points are reached.

Free-ranging vehicles (FRAGVs) have the ability to navigate without a fixed pre-determined path and thus offer a more flexible transportation network. The sensor fit for a FRAGV needs to be much more comprehensive and may include several of the following:

- a heading sensor such as a flux-gate compass or inertial navigation system combined with odometry for dead reckoning;

- ultrasonic ranging for the determination of distances to obstacles, or to enable the following of walls at a predetermined distance;
- infrared guidance using stationary LEDs mounted at strategic points and a vehicle-mounted camera and triangulation;
- vehicle-mounted stereo vision systems able to recognise and match objects in sight to objects in an on-board database;
- a laser mounted on the FRAGV which can scan barcodes fixed to walls around the workplace and identify particular positions or reference points;
- global positioning satellite (GPS) systems which can be used for location within a circular error of 20 mm.

Systems are now being developed in which the FRAGVs equipped with an a priori stored map of an environment in which gangways, doorways, alleys, stanchions and fixed manufacturing equipment can interrogate their surroundings and determine whether new objects in the planned path are static or moving and take decisions on updating the map after a number of journeys. FRAGV technology holds the promise of many beneficial developments in manufacturing and is perhaps the last step needed for the achievement of the truly automated factory.

11 Future developments in mechatronics

Having begun by considering the evolution of mechatronics, it is probably appropriate at this point to consider its future development. While any attempt to predict futures is likely to be subject to significant error, it is nevertheless possible to suggest mechatronics related developments in a number of areas of technology. Unlike Chapter 12, in which a series of case studies based on current technology are presented, in this chapter the emphasis will be on the development of future systems which have mechatronics at their core including automotive systems, home systems, telecare and telehealth, automation, manufacturing systems and design.

11.1 AUTOMOTIVE SYSTEMS

Automotive technologies have always been at the forefront of mechatronics with features such as engine management systems, traction control, automatic braking systems, active suspension, intelligent four-wheel drive and environmental controls now available on many vehicles. Referring to Fig. 11.1, features that are likely to be included in the near future include collision avoidance systems, drive-by-wire, lane following and route control.

11.1.1 Collision avoidance

Collision avoidance systems involve the use of a radar system mounted in the front of the vehicle to monitor the presence of a lead vehicle. By combining the information from the radar system with the direct operation of the vehicle throttle, it is possible to control the speed of a following vehicle directly in response to changes in speed of the lead vehicle while maintaining the appropriate distance between vehicles.

11.1.2 Drive-by-wire

By replacing the direct mechanical linkage between the steering wheel and the steered wheels with an electronic connection, drive-by-wire operation is achieved. The introduction of drive-by-wire then facilitates the introduction of automatic steering and control options to the vehicle. To be accepted, the performance levels of a drive-by-wire system have to be established and their safe operation proven and verified under a wide range of conditions.

11.1.3 Lane following

Lane following systems based on the use of television cameras mounted on the vehicle to track lane markers are now available and, when used in conjunction with collision avoidance systems and drive-by-wire, enable the vehicle to autonomously maintain its position within the lane. As with drive-by-wire, lane following systems remain to be proven and verified.

11.1.4 Route control

The combination of collision avoidance with drive-by-wire and lane following supports the autonomous operation of vehicles with route control. Perhaps based on the use of global positioning satellite (GPS) technology, a route control system would enable a vehicle to be inserted into a designated lane on a motorway or equivalent road following the establishment by its user of its destination, after which it will be automatically driven until the point at which it is to leave the motorway when control will revert to the user. Such systems have been demonstrated on test tracks and would allow a much higher utilisation of road space as all vehicles using the designated lane would be controlled to achieve optimal speed and separation under all conditions.

The achievement of an effective route control system based on the use of collision avoidance, drive-by-wire and lane following strategies requires a number of mechatronic solutions both in respect of individual systems and in the integration of those systems. Of particular importance is the need to ensure the safety of the final system and to prove its reliability under a wide range of operating conditions. Nevertheless, autonomous vehicles have already proved themselves capable of operating on conventional roads with no modification to those roads and is anticipated that commercial systems will become available in the not too distant future.

11.2 HOME SYSTEMS

The concept of the 'Smart Home' in which individual systems and devices such as environmental and heating controls, washing machines and cookers are integrated with security systems and even the ability to remotely control and adjust features such as curtains and to open and close windows was developed in the late 1980s. The commercial development of a range of system components and appliances based around an integrated bus architecture will require the application of mechatronic concepts to their design and operation.

Such a Smart Home could, as suggested by Fig. 11.2, be accessed remotely by telephone or via the internet, for instance to change the settings on the heating controller, or locally, possibly using the television as an interface option. It is also likely that speech, in the form of natural language commands, will increasingly be used to issue commands to a wide range of domestic appliances such as cookers or washing machines.

Other mechatronic home systems that are likely to become available include voice controlled appliances, intelligent agents such as autonomous robotic vacuum cleaners which will clean during the night and robot lawn mowers, though in each of these cases there remain problems to be solved in terms of providing adequate on-board energy storage, or washing machines which monitor the price of electricity before deciding whether to turn themselves on.

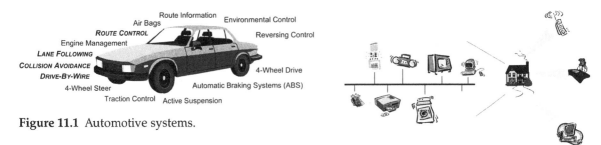

Figure 11.1 Automotive systems.

Figure 11.2 Smart home technologies.

11.3 TELECARE AND TELEHEALTH

Faced with an ageing population in which the proportion of individuals aged 65 or more in the population is going to continue to increase, the option of using technology to support independent living in an individual's own home is increasing in importance. Such telecare and telehealth systems will integrate features of Smart Home technologies with specific support systems such as fall sensors and lifestyle monitoring systems with aid technologies such as drug dispensers. Early release from hospital might also be supported by the use of implanted microsystems or micro-electro-mechanical systems (MEMS) which monitor blood pressure and blood chemistry and provide on-line diagnostic capability.

Constructed around a series of smart nodes as suggested by Fig. 11.3, many of the systems and sub-systems associated with telecare and telehealth, such as the fall sensor referred to above, are mechatronic in nature. Their development will however require not only the collaboration of technologists but the bringing together of medical, clinical and social expertise to ensure the appropriate application of these technologies.

For individuals with problems of access and mobility, mechatronics is also likely to contribute through the development of systems to support remote access. For instance, a small mobile robot carrying a television camera has been used to provide a form of telepresence for a housebound individual. Operating under the direct control of the user, this robot can be controlled to allow them to visit parts of a house, to explore gardens and to participate in a range of activities with other people in a way which would not otherwise have been possible.

Other related systems include adaptive housing where, for instance, the height of a work surface can be adjusted to suit wheelchair users or smart storage units are used to open up access to cupboard space. Smart wheelchairs

Figure 11.3 Telecare smart nodes.

which autonomously adapt to individual user requirements will also support increased mobility by making it easier to avoid obstacles outside the user's line-of-sight or to manoeuvre more effectively in enclosed spaces. Work on walking systems will support the development of powered and intelligent prostheses which will adapt to the user's walking patterns, improving their mobility still further.

11.4 AUTOMATION AND ROBOTICS

Mechatronics will contribute to the development of intelligent automation through advanced robots capable of operating in unstructured environments such as a construction site. Such robots will have the ability to adapt their operation to suit the conditions in which they find themselves and will integrate their operation with user commands, possibly from a remote location. The combination of machine intelligence with human intelligence in a co-operative, mechatronic environment will have a number of advantages for operation in hazardous environments such as construction, sub-sea and space where the use of the intelligent robot reduces the level of risk or hazard to which the operator is exposed. Other areas of application include medical robots, including surgical assistants, automated recovery systems, intelligent agents capable of acting on behalf of the physically disadvantaged and robot explorers.

Developments are likely to include autonomously mobile systems able to plan their own route through an unstructured environments, systems capable of carrying out their own task planning on the basis of defined goals and systems which will learn from their operators about particular task domains.

11.5 MANUFACTURING

Manufacturing technology is likely to become increasingly closely linked to the design process with the integration of knowledge based design strategies incorporating an understanding of the manufacturing processes and of integrating these processes during product design. Thus the process of product design will increasingly be associated with the automatic generation and definition of the related manufacturing processes.

It is also likely that the nature of machine tools will also change through the application of mechatronics. Current machine tools are of necessity large and bulky in order to sustain the reaction forces associated with the machining process with minimum structural deformation. By replacing the existing structure with an active structure incorporating a series of actuators, the overall system could be made smaller and lighter as well as more able to compensate for factors such as thermal expansion or vibration.

Such smart structures have already been used in a number of applications. For instance, the mirror of a large optical telescope was constructed from a series of small mirrors each with its own actuator, allowing compensation for changes in temperature and permitting a much larger stable reflecting surface than would have been possible by conventional means. Other suggested applications include the use of microactuators to change the shape of an aircraft wing to provide control instead of conventional flight control surfaces and active vibration control for complex structures.

11.6 DESIGN

The design of mechatronic systems implies the bringing together of a range of technologies, many of which have over time evolved their own, individual design methodologies and procedures. The increasing capability of computers and the development of enhanced AI systems should support the development of expert system based design support tools which will operate at the functional level and which will have the ability to capture design expertise. Such systems will have the capability of providing immediate support and advice on the impact of decisions made in one technological domain on other areas of technology. Thus, for instance, a decision on the required controller strategy together with information from structural and other analysis can be provided to the designer responsible for the implementation and realisation of the electronic system in the form of an input to the relevant design support tool.

11.7 SUMMARY

Mechatronics has had and will continue to have a major impact on the design and development of many engineering systems and processes. In the above paragraphs a brief attempt has been made to identify how this development is likely to continue in relation to a number of areas of technology. The examples given are by no means exhaustive but are intended to provide an illustration of the likely ways in which mechatronics is going to impact upon engineering in the next few years.

Case studies in mechatronics 12

12.1 MECHATRONIC GUIDED BUS

In recent years, there has been an increasing move to re-introduce mass transit systems in the form of trams and light railways into city centres. Such systems provide an effective means of moving large numbers of people within an urban environment, particularly where their tracks can be separated from other forms of vehicular transport such as cars. However, trams and light railways lack flexibility and offer limited coverage as routes are restricted to where track has or can be laid and the infrastructure costs of opening up new routes is often high.

In comparison, buses are highly flexible and are capable of being moved between routes and of operating outside the urban area. However, even where bus lanes are provided, but not physically separated from other vehicles, they are prone to delays in periods of high traffic density such as the morning and evening peak periods.

It has therefore been suggested that what is needed is a system which combines the route flexibility of a conventional bus with tram like operation in designated urban areas. Such a system would have a number of advantages over conventional light railways or trams including:

- the ability to use normal roads thus enabling the same vehicle to be used outside the designated areas and between towns;
- increased manoeuvrability with respect to conventional light railways or trams;
- the basic infrastructure is already in place in the form of the road network and all that needs to be added are the guides.

When operating in the urban environment, the guided bus has a number of advantages over conventional buses including:

- By having a designated path from which other vehicles are excluded, delays are reduced.
- The clearances required by a guided bus between the bus and the kerb are less than the clearance required by a conventional bus lane and guided buses therefore take up less road space.
- Driver load is decreased as the driver does not have to steer the bus when it is in the lane and can concentrate on controlling speed and ensuring operating safety.
- By combining the guidance with improved communications, scheduling can be improved and user information enhanced.

To date, a number of mechanically guided bus systems have been introduced into cities in Europe in which the buses are guided between two reinforced kerbs by using roller arms attached directly to the steering mechanism which act to centralise the bus between the kerbs as suggested by Fig. 12.1. Such systems, while providing guidance, suffer from a number of disadvantages including:

- the imposition of high loads on the kerb which necessitating a heavy and expensive structure;
- increased wear in the steering pinions due to the imposed loads;
- the need in some countries, for instance the UK, for a special dispensation as the rollers make the bus wider than the maximum legal width;
- it is almost impossible for the bus to reverse in the guideway without becoming stuck;
- a large turn radius must be used when in the guideway to ensure that the tracks of the front and rear wheels follow the same track when turning.

A mechatronic solution to achieving a guided bus offers a number of advantages over the mechanically guided version including:

- lighter kerb structures;
- significantly reduced load on the steering mechanism;
- improved cornering;
- reversing capability;
- ability to interface with other systems.

The case study therefore describes the development of the operating and control strategies required by a mechatronic guided bus including testing on a scale model of a bus and the associated track.

12.1.1 System configuration

In addition to providing a reference for the guidance system, the use of a kerb

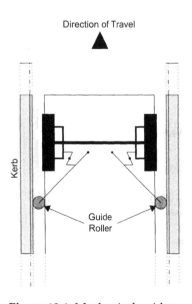

Figure 12.1 Mechanical guidance.

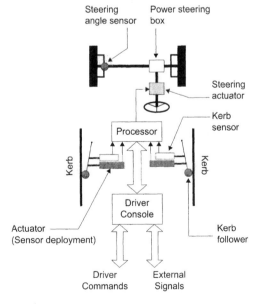

Figure 12.2 Mechatronic guided bus systems.

acts to isolate the guided bus from other road vehicles. However, for the mechatronic guided bus shown in schematic form Fig. 12.2 it can be seen that this kerb can now be a much lighter structure than before as it is no longer required to provide a reaction force. Similarly, the sensory system, using kerb following rollers or whiskers in the first instance, can be made much lighter as they are not required to transmit any force.

Instead, the sensors detect the position of the bus relative to the kerb and transmit this information via the system processor which contains the required information about the steering geometry of the bus to the steering actuator connected to the steering column. Using this configuration, it is no longer necessary for both sensors to be in contact with the kerb in order to guide the bus, allowing one kerb to be removed to enable reduced a turning radius to be used. In addition, as the sensors are much lighter, they can be arranged to extend or retract automatically on entering or leaving the guided sections, thus eliminating the need for any special dispensation with regard to vehicle width.

Steering behaviour

When turning, the track of the rear wheels is inside that of the front wheels with the difference increasing as the radius of the curve is reduced. Referring to Fig. 12.3:

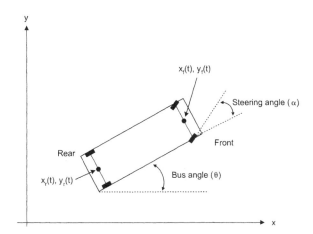

Figure 12.3 Steering geometry and behaviour.

$$\tan \theta = \frac{y_f(t) - y_r(t)}{x_f(t) - x_r(t)} \tag{12.1}$$

Assuming that in a time interval δt the front wheels will move in the direction in which they are pointing. For a constant velocity v:

$$\delta x_f = v \delta t \cos(\theta + \alpha) \tag{12.2}$$

and

$$\delta x_f = v \delta t \sin(\theta + \alpha) \tag{12.3}$$

If the interval δt is small then the rear wheels can also be assumed to move in the direction in which they are pointing, in which case:

$$\theta(t + \delta t) \approx \tan^{-1} \frac{y_f(t) + \delta y_f - y_r(t)}{x_f(t) + \delta y x_f - x_r(t)} \tag{12.4}$$

The new position of the rear wheels $\{x_r(t + \delta t), y_r(t + \delta t)\}$ can then be found assuming a bus length of L_B as:

$$x_r(t + \delta t) = x_f(t + \delta t) - L_B.\cos\{\theta(t + \delta t)\} \tag{12.5}$$

and

$$x_r(t + \delta t) = x_f(t + \delta t) - L_B.\sin\{\theta(t + \delta t)\} \tag{12.6}$$

Using these relationships the curves of Fig. 12.4(a)–(d) can be obtained for various turning radii of 2, 4, 8 and 16 times the vehicle length respectively which show the variation in track between the front and rear wheels that results, from

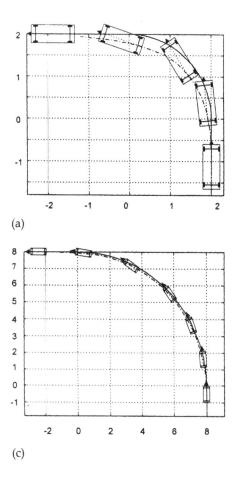

(a)

(c)

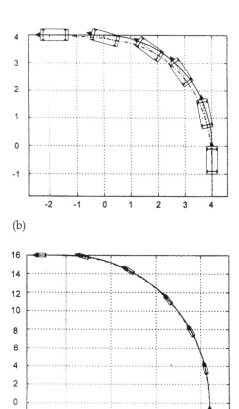

(b)

Front $\leftarrow$ ⊥ ⊥ Track of front wheels ———
Track of rear wheels – · – ·

(d)

Figure 12.4 Turning behaviour.

which the problem of achieving a tight turning radius with a mechanically guided bus is clearly seen.

Control strategies

In order for the mechatronic guided bus to operate the three states shown in Fig. 12.5 and listed below need to be accommodated:

- straight line motion;
- turn right;
- turn left.

The on-board systems for a mechatronic guided bus where evaluated using a 12.5:1 scale model. Using the model, straight line control was achieved by implementing a proportional plus integral (PI) controller with information derived from both sensors with the goal of keeping the bus in the centre of the guideway. Analysis of the controller equations yielded the open loop transfer function of equation (12.7). This corresponds to a pair of poles at the origin and a zero at $s=-v/L_B$, suggesting that the system is marginally stable which corresponded to the observed behaviour of the model with oscillations occurring at high speeds. The control law was then modified to make the system gain proportional to the velocity of the bus and this had the effect of reducing the observed oscillations:

$$G(s) = \frac{v(sL_B + v)}{s^2 L_B} \tag{12.7}$$

To enable the mechatronic guided bus to achieve a small turning radius, the inner guide kerb must be removed in which case the control signal will be derived from the outer guide kerb only. The onboard system must therefore be capable of detecting the current mode of operation and of switching between modes as required. Using the model the following options for detecting the transition from straight line motion to turning motion and vice versa were evaluated:

- monitoring the turn angle of the front wheels;
- removal of the inner guide kerb on entering a turn.

Of these, it was found that while monitoring the angle of the front wheels could detect a corner, there was a tendency for the model to oscillate with the controller hunting between modes on straight sections of the route. Removal of the inner kerb produced the effect shown in Fig. 12.6 as the bus initially attempted to maintain a position equidistant between kerbs. By including a reference to the rate at which the position relative to the guide kerbs changes then a much smoother transition between straight line and turning motion was achieved.

Entry into the guideway

The entry into the guideway will in the first instance be controlled by the driver who will release the wheel following the deployment of the kerb followers. These will then pick up the guide kerbs and take over steering control. Referring to Fig. 12.7, it can be seen that this presents a problem in terms of the control strategy used since if the conditions as set out previously are employed,

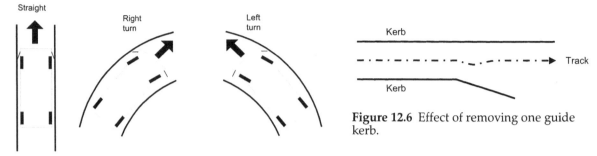

Figure 12.6 Effect of removing one guide kerb.

Figure 12.5 Turning behaviour and sensor contact.

the initial contact by a single kerb follower will be interpreted as a bend, causing the bus to turn towards the guide kerb that is first contacted. A similar condition will exist where there is a gap in the guideway, for instance at a cross-roads.

In order to prevent this type of steering error occurring, the steering algorithm was adjusted so that, following the initial deployment of the sensors or if contact is lost on both sensors, then the first contact is assumed to be at the entrance to a new section of guideway, eliminating the problem and achieving a smooth entry into the guideway.

An alternative strategy would be to communicate with the bus as it was approaching the guideway in order that it would be able to anticipate the fact that it would initially pick up the guideway with a single sensor. This communication could be achieved by a variety of means including an inductive loop in the roadway and radio beacons.

Reversing

The problem of reversing the bus in the guideways is illustrated by Fig. 12.8 which shows the variation in steering commands required. By locating a second pair of kerb followers at the rear of the bus which would be deployed when reverse was selected, with the front followers being simultaneously retracted, then by reference to Fig. 12.9 the relationship with the steering angle becomes that of equations (12.8) and (12.9), enabling the bus to reverse safely in the guideway.

$$\frac{x_f(s)}{\delta(s)} = \frac{v(v - sL_B)}{s^2 L_B} \tag{12.8}$$

$$\frac{x_r(s)}{\alpha(s)} = \frac{v^2}{s^2 L_B} \tag{12.9}$$

Sensor deployment

As the kerb followers for the mechatronic guided bus are not required to transmit a force but merely to detect the presence of the guide kerb, they are much lighter than those used by the mechanically guided bus and need not be permanently deployed. Using techniques such as the inductive loop and radio

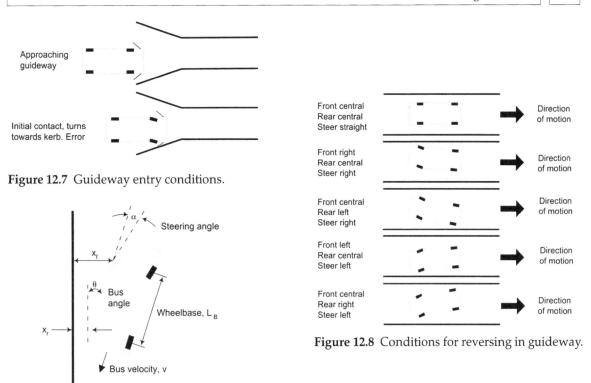

Figure 12.7 Guideway entry conditions.

Figure 12.8 Conditions for reversing in guideway.

Figure 12.9 Steering angle while reversing.

beacons referred to above, the kerb followers would be automatically deployed on approaching the entrance to the guideway and retracted on leaving the guideway, removing the need for any special dispensation on width.

System operation

Once the mechatronic guided bus becomes available then it opens up a variety of different operating strategies including:

- Direct communication between the bus and the bus stop where information would be displayed about the time of arrival of buses with the possibility of passengers being able to signal to the bus that they are waiting at a particular stop.
- The control of a group of buses could be controlled by a lead bus. Using radar based techniques together with inter-vehicle communications the speed of a group of buses could be controlled from the lead bus. This would enable them to operate as a multi-unit vehicle with close spacing in an urban environment but with the ability to split off and follow different routes as and when required, for instance to service a particular area or to travel between adjacent towns.

12.1.2 Conclusions

Guided buses offer a number of advantages over conventional buses in an urban environment retaining route flexibility. Mechanically guided bus systems, such as those currently in or entering service, suffer from a number of disadvantages in terms of the loading on the vehicle steering mechanism and the need for special dispensations with regard to width as well as operational limitations such as the inability to follow small radius curves or to reverse in the guideway. By using a mechatronic approach to achieving a guided bus, these limitations can be overcome while the opportunity for adding additional features is increased.

12.2 MICROSYSTEMS, MEMS AND BUILT-IN-SELF-TEST

Microsystems and micro-electro-mechanical systems (MEMS) are miniature mechatronic systems typically incorporating sensors and/or actuators together with the associated electronics for signal conditioning, signal processing, control, power and communications. This general development is illustrated by Fig. 12.10 which shows the evolutionary path of smart sensor technology from individual devices to complete systems offering improvements in accuracy, functionality and flexibility over their more conventional, full-scale and 'dumb' counterparts.

Microsystems may be realised either in the form of a single chip or as a multi-chip-module (MCM) using thick film or thin-film hybrid technologies. A particular example of the MCM approach is the '3D-chip' concept shown in schematic form in Fig. 12.11 in which the individual silicon dies are bonded layer by layer to a flexible film to form a cubic array with vertical inter-chip connections. Other technologies deployed include the use of micromachining to create intricate structures on chip such as the gear train and mirror mechanism produced by Sandia National Laboratories and shown in Figs 12.12 and 12.13.

Whatever technologies are used in their manufacture and assembly,

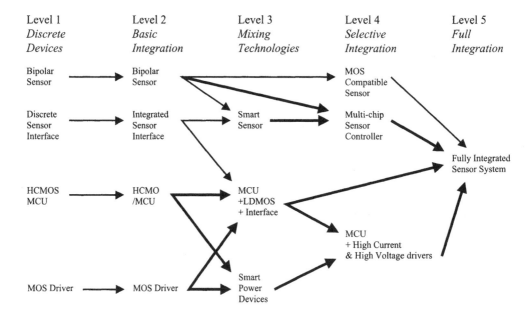

Figure 12.10 The evolution of the fully integrated smart sensor.

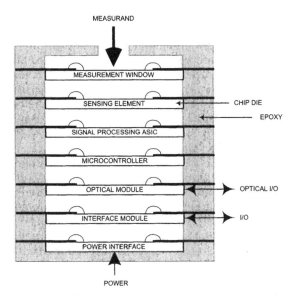

MEASURAND

MEASUREMENT WINDOW

SENSING ELEMENT — CHIP DIE

EPOXY

SIGNAL PROCESSING ASIC

MICROCONTROLLER

OPTICAL MODULE — OPTICAL I/O

INTERFACE MODULE — I/O

POWER INTERFACE

POWER

Figure 12.11 A possible '3D' chip layout.

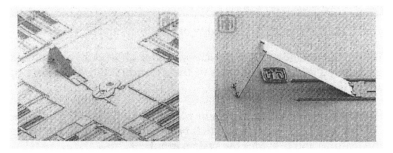

Figure 12.12 Micromachined gear train.
(*Sandia National Laboratories*)

Figure 12.13 Micromachined mirror mechanism.
(*Sandia National Laboratories*)

microsystems or MEMS are characterised by the integration at a nanometre scale of a large number of components and the presence of local and dedicated electronics. The increasing availability of microsystems or MEMS gives significant benefits in a number of areas, for instance the higher resonant frequencies of the on-chip mechanical structures results in improvements in system bandwidth while the reduction in thermal mass means a greater sensitivity and a faster response to changes in temperature. However, care must be taken to ensure that the devices are not too small for the intended task, thus reducing the seismic mass of an accelerometer means that it is less capable of responding to low-g accelerations. The requirement that the mechanical elements of the system are properly matched to the application is therefore even more critical when considering microsystems and MEMS than may well be the case with more conventionally sized systems where a degree of mismatch using standard or 'off the shelf' components can be tolerated.

Applications proposed for microsystems or MEMS range from the highly exotic such as the idea of groups of miniature 'cleaning robots' injected into the bloodstream to 'de-fur' arteries or a micropump embedded in a patient's arm to control the rate of release of a drug directly into the bloodstream in response to the detection by the microsystem or MEMS of changes in their physiology. Other applications suggested include covering the wing of an aircraft with miniature 'flaps' which interact with vortices to reduce drag and improve performance and the use of a series of miniature actuators to control the shape of a mirror to focus a surgical laser more accurately to more prosaic applications such as smart pressure sensors. Indeed, it often seems that the possible scope and application of microsystems or MEMS is only limited by the imagination of those working with the technology.

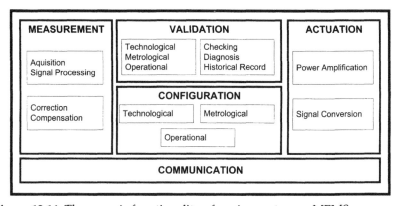

Figure 12.14 The generic functionality of a microsystem or MEMS.

The generic functionality of a microsystem or MEMS can be subdivided as suggested by Fig. 12.14. The roles of the individual blocks making up this figure are then:

1. *Measurement*. The measurement block contains the sensing element together with the associated signal processing, correction and compensation features and provides the system input.
2. *Actuation*. The actuation block provides the output functionality of the system and groups, for instance, the drive electronics for power amplification with the required signal conversion electronics.
3. *Communications*. The communications block is responsible for the connection of the system with the outside world. It would be configured according to the standard being used and is the means by which the system receives and transmits data.

4. *Configuration*. The configuration block is concerned with the operation of the system itself and may be further subdivided as follows:
   (a) a technological domain which is concerned with the internal structure of the system
   (b) an organisational domain which contains information such as the addresses and identifiers of other connected systems
   (c) a metrological domain containing information related to the measurement function being performed
5. *Validation*. The validation block is responsible for checking the behaviour of the system in relation to the technological, organisational and metrological domains.

Many microsystems or MEMS are in use in safety critical applications and there is therefore a need to check and monitor their performance through the application of techniques such as 'design for testability' (DFT) and built-in self-test (BIST). In this context, an important measure of the performance of a system, sub-system or device is the concept of *availability* which may be expressed as:

$$\text{Availability} = \frac{\text{Mean Uptime}}{\text{Mean Uptime} + \text{Mean Downtime}}$$

$$= \frac{\text{Mean Time to Failure (MTTF)}}{\text{Mean Time to Failure (MTTF)} + \text{Mean Time to Repair (MTTR)}}$$

$$= 1 \bigg/ \left(1 + \frac{\text{MTTR}}{\text{MTTF}}\right) \tag{12.10}$$

This implies that a high availability system can be achieved by increasing mean time to failure (MTTF) or reducing mean time to repair (MTTR).

Strategies to achieve high system availability include the perfectionist approach which aims to eliminate the possibility of failure. However, as zero defect designs, devices, processes and systems do not exist, absolute perfection is not achievable in practice. The main alternative fault tolerant strategy uses redundancy to support system reconfiguration and embeds fault detection and isolation procedures within the system. Other strategies such as 'fail safe', 'limp home' and 'graceful degradation' all allow for operation at reduced performance to allow the system to reach a safe condition. Figure 12.15 sets out some of the possible strategies that might be adopted for achieving high availability and the relationships between them.

The complexity of microsystems has led to the development of the concepts and mechanisms of DFT in which the test procedures to check the operation of an integrated circuit are built is as part of the design process and Fig. 12.16 illustrates some of the techniques used for this purpose. The extension of these and other procedures into the in-service operation of the microsystem in the form of BIST strategies and procedures further extends the capability of the microsystem or MEMS. Figures 12.17 and 12.18 and Table 12.1 illustrate the application, impact and importance of developments in this area.

12.2.1 Vehicle airbag controller

A vehicle airbag controller microsystem is shown in schematic form in Fig.

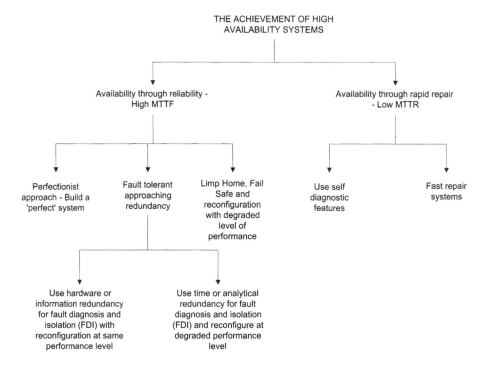

Figure 12.15 Strategies for the achievement of high availability systems.

Table 12.1 Self-test and diagnosis in relation to microsystems and MEMS

The importance of self-test and diagnosis in microsystems and MEMS

- High performance requires on-chip evaluation.
- Manufacturing test times reduced by internal test support
- Increased accessibility, controllability and observability
- Modular design with self-testing library elements
- Self-test and diagnosis supports rapid prototyping and reduced time to market

---

Applications and demands for self-test and diagnosis in microsystems and MEMS

- Increased high-level system reliability through built in self test (BIST)
- Increased testability through design for testability (DfT) and BIST
- Packaging means to self-test and diagnosis is the only viable approach to testing
- Support for the requirements of safety critical systems
- Improved fault tolerance
- Counteracting intermittent measurement fault with testing to improve functionality
- Reduced down-time through preventive maintenance
- Standards and regulations

Impact of self-test and diagnosis on system quality and performance of microsystems and MEMS

- Self-test capability supports increased system complexity
- Self-test and diagnosis is essential for distributed systems
- Diagnostic capability is in some cases available 'free' as a result of a DfT strategy
- Quality improvements products and in applications

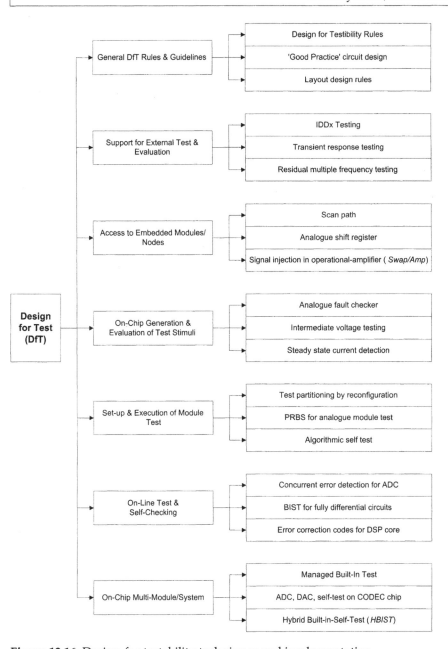

Figure 12.16 Design-for-testability techniques and implementation.

12.19 and comprises a 'smart' accelerometer together with a microcontroller and power supply. Self-test and design features must ensure that the airbag does not cause an unacceptable hazard in its own right or risk to human life or the environment in the case of a fault. Potential faults must therefore be detected, interpreted and corrective action initiated as soon as they occur. Corrective action might take the form of system reconfiguration using built-in redundancy or more likely in the case of the automotive industry the application of a 'fail safe', 'graceful degradation' or 'limp home' strategy. Thus one

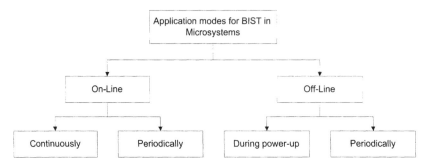

Figure 12.17 The implementation of BIST in microsystems.

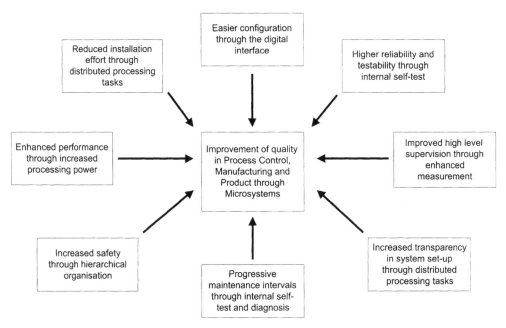

Figure 12.18 The benefits of incorporating BIST on the quality and performance of microsystems.

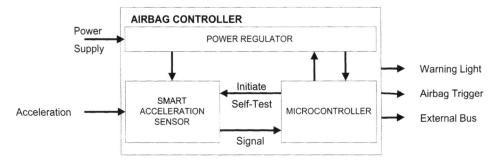

Figure 12.19 Airbag controller block diagram.

possible response to a detection of a fault in the airbag microsystem would be to indicate the fault to the driver, for instance by means of a warning light on the dashboard, so that they know to go to the garage for repair.

A typical structure for a comb type micromachined accelerometer such as could be used to measure acceleration is shown in Fig. 12.20 in which the spacing between the plates is of the order of 1 μm. To reduce the sensitivity of the accelerometer to time and temperature effects, a force-balance system may be used which typically reduces the movement of the seismic mass by a factor of 10 for an acceleration of the order of 50 g to around 0.01 μm. In testing the accelerometer, it is important that the movement of the seismic mass is checked and various techniques have been developed for this purpose including:

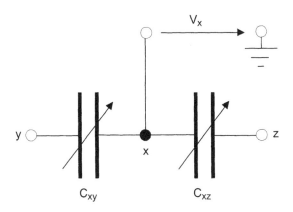

Figure 12.20 Accelerometer sensing element with capacitive readout showing the equivalent circuit.

- electrostatic excitation in which a voltage is applied to a plate parallel to the seismic mass with which it forms a capacitor. Electrostatic forces then result in a movement of the seismic mass;
- thermo-electric excitation in which a deformable element is heated by passing a current through it. This results in a change of shape of the element which in turn causes a displacement of the seismic mass;
- the injection of a test-signal into the feedback path of the servo loop of a force balanced accelerometer to cause the seismic mass to move.

A typical self-test strategy would inject a signal into the accelerometer sensor system to produce an offset in all the signal lines as well as causing a physical deflection seismic mass of the accelerometer to take place. Only a correctly functioning system will respond to the test-signal with a response matching the test pattern stored in the microcontroller.

It is important to note that it is not the sensor which performs the diagnosis and triggers the response but the microcontroller. However, the incorporation of a self-test function in the accelerometer system is important in achieving the required sensitivity in the detection and response to possible faults and degradation of function.

12.3 HIGH-SPEED TILTING TRAINS

When cornering, train passengers are subjected to a lateral force which is proportional to v^2/r where v is the train velocity and r is the radius of the curve which, if it becomes excessive, can cause discomfort. Conventionally, the track will be inclined into the bend as in Fig. 12.21 to counter this lateral force and achieve a resultant force which is as near normal to the plane of the track as possible at the design speed for the curve. However, should the speed of the train be greater or less than the design speed for the curve, compensation for the lateral force will be incorrect and a possible source of passenger discomfort.

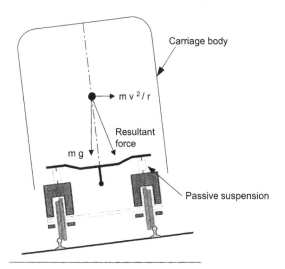

Figure 12.21 Cornering forces on a conventional carriage.

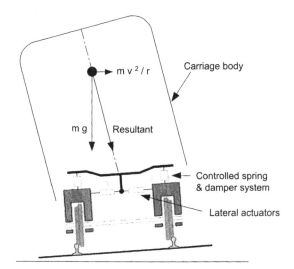

Figure 12.22 Cornering forces on a tilting carriage.

High-speed trains such as the French *TGV* (*Train à Grande Vitesse*) or the Japanese *Shinkansen* or *Bullet Train* use purpose built tracks with large radius curves and appropriately inclined track to minimise the effects of the lateral force while cornering to enable high average speeds and reduced journey times. Such purpose-built track is however expensive to construct and requires that there is the land available for the purpose. In many instances it may not be practical to build a new track, in which case the preferred solution would be to operate trains at higher speeds using the existing track.

In the first instance, some increase in speed may be possible by increasing the inclination of the track at a curve to enable it to be taken at a higher speed. This solution is however limited by the need to ensure that passenger comfort is maintained at train speeds below the design speed for the curve. Indeed, it is possible that if a train takes such a bend too slowly then the passengers will experience discomfort due to the effects of gravity apparently pulling them in towards the centre of the curve. This effect can be seen by reducing the lateral force in Fig. 12.21 to, or near to, zero.

12.3.1 Tilting trains

An alternative to increasing the inclination of the track would be to tilt the carriage into the bend as in Fig. 12.22. This would produce the same effect in terms of passenger perception and ride as increasing the inclination of the track, but would have the advantage that the degree of tilt could be controlled in relation to the train speed enabling the required ride to be achieved over a wide speed range. The result of tilting the carriage would be an increase in cornering speed for the train of up to 40%.

Tilting mechanisms

A tilting train uses an active suspension under the control of an onboard processor. In operation, the bogie rides on the track and provides the reference for the tilting mechanism for the carriage. Referring to Fig. 12.22, the lateral actuators will then act in conjunction to provide the tilting force while the controlled variable rate springs and dampers act to provide additional control over the motion. Figure 12.23 shows the actuator geometry used by the X2000 tilting train produced by Adtranz and operated by Swedish Railways while Fig. 12.24 shows the X2000 in service with the tilting action of the carriages being clearly visible.

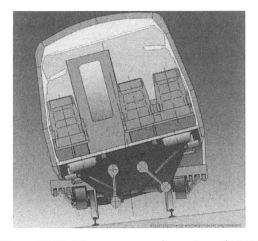

Figure 12.23 Tilt actuator configuration of X2000.
Courtesy of Adtranz (ABB Daimler-Benz Transportation)

Figure 12.24 X2000 in service.
Courtesy of Adtranz (ABB Daimler-Benz Transportation)

For any of the systems currently in service, the operation of the tilting mechanism is controlled by an on-board processor which uses data on train speed and lateral acceleration to establish the required tilt angle to provide the desired resultant force. A simple controller implementation would take the form of a pendulum constrained to swing in the plane of the carriage cross-section as in Fig. 12.25(a). As the carriage entered the bend, the pendulum would be displaced towards the outside of the bend as in Fig. 12.25(b) and the tilt angle would then be adjusted to bring the pendulum back within the defined limits as in Fig. 12.25(c).

In practice, as suggested by Fig. 12.26, information on speed and lateral acceleration is used in association with the defined control algorithms to control the lateral actuators and variable rate springs and dampers. As individual carriages can be entering, on or exiting the curve at any particular instant it is particularly important for the control of the individual carriages to be integrated in order to achieve an acceptable ride. Indeed, one of the problems associated with the design of tilting trains was a motion sickness induced by the tilting motion.

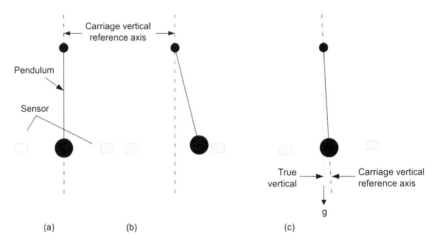

Figure 12.25 Tilt control using constrained pendulum.

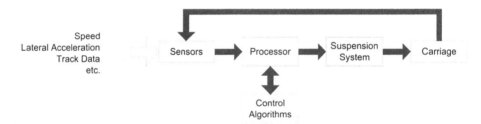

Figure 12.26 Tilt control configuration.

Some tilting trains

United Kingdom
The then British Rail began experimenting with tilting technology in the late 1960s with the *Advanced Passenger Train* (APT) programme. Following trials with the experimental APT-E in the 1970s, the APT-P, for prototype, version was introduced in December 1981 when its initial runs suffered from a range of problems, the majority of which were not associated with the tilting mechanism. Though the APT-P did re-enter service in 1982 and ran a regular schedule between Euston and Glasgow from 1982 to 1984, it was eventually withdrawn and the planned service version, the APT-S, was never built.

Italy

Following experiments in the early 1970s which showed that, using tilting technology, it was possible to negotiate curves at speeds up to 30% greater than with conventional trains, the ETR-401 was introduced into service between Rome and Naples in 1976. The ETR-401 was a 'one-off' prototype and consisted of a 4 car set carrying 120 passengers while the tilting mechanism was based on the use of a gyroscope and accelerometer at each bogie.

Following the cancellation of the APT programme in the UK, the technology was sold to Fiat Ferroviaria Spa where it was incorporated with the technology used in the ETR-401 to produce the Pendolini ETR-450 which entered service in 1988 to be followed in 1993 by the ETR-460 and in 1995 by the ETR-470 and ETR-480. In these later trains, the sensors are confined to the end cars and the tilting mechanism of the intervening carriages controlled by reference to the end cars.

Sweden

A tilting mechanism was developed by the then ASEA company in the early 1970s and this has been transferred to the X2000 train produced by Adtranz and shown in Fig. 12.24. A particular constraint in the design of the X2000 was the often extremely tight, in railway terms, curves of 600 m radius or less found on Swedish Railways which acted as a severe constraint on the speed of conventional trains.

A problem experienced with the X2000 following its entry into service was that bumps in the track caused the tilting mechanism to activate even though there was no curve, inducing motion sickness in some passengers. This was solved by relaying and regrading those parts of the track that caused a problem.

12.4 A PRACTICAL APPLICATION OF EXPERT SYSTEMS DESIGN

12.4.1 Background to a prototype development of an expert system

Expert systems augment, rather than replace, the experience and judgement of the human expert in that they can provide a reservoir of proven expertise and shorten the time spent in searching potential solution spaces. In some fields, the human expert is becoming rare, or the pace of development is such that novices are overwhelmed, established principles are overlooked and unnecessary 're-invention' occurs. While expert systems may have a part to play in capturing and making more accessible the accumulated experience of an industrial company, they are also a powerful educational tool able to range over a large knowledge field without commercial constraint.

This case study describes the process of building an expert system Schemebuilder® Expert– Fluid Power Module in a practical engineering field enabling hydraulic systems to be designed from a short statement of requirements in terms of loads attributes through to the automatic generation of feasible circuits. The expert system generates an output formatted for the internet using the hypertext markup language (HTML) and automatically creates for consideration by the design engineer a set of feasible hydraulic systems based on well proven principles of circuit design, allows for changes to the power supply circuit, one of the key aspects in the design of hydraulic systems, and automatically redefines the component lists. These operations are implemented according to the established practices used in hydraulic systems design.

Expert systems are a powerful aid in the process of concurrent engineering but their efficient development should take account of organisational as well as technological aspects, especially where different specialists must co-operate from the early phases of the design process. The ultimate objective of concurrent engineering is to provide an environment for the efficient management of interdisciplinary projects, ensuring total quality and reducing project duration whilst minimising cost. Knowledge systemisation across traditional interdisciplinary boundaries presents a great challenge to the principles of concurrent engineering and the development of the powerful AI tools which are essential for its effective realisation.

As described in Chapter 10, an AI tool, Schemebuilder® Mechatronics, focuses primarily on the domain of energetic systems covering the wide range of actuation configurations which are at the core of mechatronics. The commonly available energetic domains are electrical and fluid power, normally hydraulic and pneumatic. In order for a high level specification tool to be able to access domain knowledge, a common language is needed such as bond graph methodology, together with a modularised knowledge base. This case study describes the implementation of an incremental procedure in the expert system development process, focusing on the hydraulic systems domain. As such, it forms one of the Schemebuilder® Expert modules which can be accessed by the high-level system.

12.4.2 Expert system development process

A number of approaches for expert system development were analysed and an incremental procedure was chosen whose basic principle is to develop software in increments of functional capability and which has been used very successfully in large conventional software projects. In this approach, the progressive addition of rules increases the capabilities of the system, in terms of human analogy, from assistant to colleague and from colleague to expert. The primary advantage of the incremental model is that the increases in functional capability are easier to test, verify and validate than the products of individual stages in the waterfall model. This decreases the cost of incorporating corrections in the system. One of the primary matters for decision concerns the rules chaining process. To address this issue, some guidelines were considered from the literature, in particular:

- If the system will be asked to justify the decisions made, then it is important to opt for the chaining process which resembles the way the user thinks.
- What kind of event will fire the search for a solution? If it is a new fact, then a forward chaining process must be chosen.

With these guidelines and considering that the system aimed to be a design assistant tool, the forward chaining process was chosen in which the user needs, defined as functional requirements, were applied to start up the matching process between the required functions and the means of performing them. Also the forward chaining method is easily implemented in CLIPS, compared to the implementation of backward chaining which is not directly available.

12.4.3. Domain knowledge: why hydraulics?

The following points explain the basis by which the area of hydraulic systems was chosen to develop a working prototype tool:

- The hydraulic systems design area has a well established theoretical foundation. This aspect is important for a pilot-project intended to be used in a concurrent engineering environment, as well as for expert system development. Both purposes were intrinsic in the original objectives of the Schemebuilder® project.
- Hydraulic systems are composed of circuits, each having a specific function. This facilitates the building of functional blocks and also the application of object-oriented methodology; definition of *classes-and-objects*, *attributes* and *methods*.
- Hydraulics is a very broad area covering many fields, including mobile plant, machine tools, marine and avionics.

The application was chosen to show the potential of Schemebuilder® in one domain without restricting it, but rather facilitating its extension and enabling the integration of different tools, such as simulation, for supporting the design process. The aim was to build a system where a designer in industry would feel comfortable and confident in using it. Among the basic decisions to be considered in developing an expert system is the definition of the knowledge domain and its representation.

Quite often in the expert systems literature there are warnings against becoming one's own expert. However, in some cases depending on the timescale, the knowledge engineer's background and the system complexity this approach may prove worthwhile. This particular development project centred around a knowledge engineer of unusual background who had practical international research laboratory experience in hydraulic systems and also experience and strong interests in AI and expert systems development. Also accessible was in-house academic and research expertise in both fields which greatly helped the project in its early stages. This proved to be useful in presenting the project objectives to the industrial experts in a fairly well defined form and set out the basis for the incremental procedure. In parallel, contacts were made with specific industry sectors to define more specific market application domains, such as excavators or backhoe loaders.

12.4.4. Schemebuilder® concepts

The design methodology concepts derived from Schemebuilder® definitions, are mainly the concept of *scheme* and *working principle*. As defined in Chapter 3, a scheme is an outline of a solution to a design problem, carried to a point where the means of performing each major function has been fixed. A working principle consists of one or more required sub-functions, which may have certain attributes. Here, a scheme corresponds to a hydraulic system, i.e. a set of circuits, while each circuit relates to one working principle, for each circuit has its well defined function, such as flow control, pressure control, power supply, and so forth defined by the user requirements and accomplished by its components. This modular approach allows the application of object-oriented techniques for analysis and programming.

The procedure adopted applied the declarative, procedural, and strategic knowledge structures necessary for modelling the design process and the combination of their methods for eliciting knowledge. Declarative knowledge defines, for example, why one specific type of circuit meets some load attributes. Procedural knowledge explains what are the steps in the design process, while strategic knowledge might be used in defining for example what kind of flow control method (valve or pump) must be chosen depending on the power requirements of the application. Therefore, in hydraulic systems design, these structures are equally important.

12.4.5 System description

The system core was developed in CLIPS which runs under UNIX as well as Windows 95 and also includes the interfaces for other modules, such as the dynamic modeller, matching tool and user interface. In this development, the agent-based paradigm is applied, in which each agent (a expert system, a computer program or an human expert) interacts with the others to solve a complex task. As developed, the system has the following steps:

- *Loads definition*: primarily, qualitative attributes defined by the user. *Results*: generation of *loads objects*.
- *Circuits generation*: rules fired according to the different loads attributes. *Results*: generation of *circuits objects*, allowing that the same load generates different circuits to accomplish its function, depending on the attributes. This phase is shown below in Fig. 12.27.

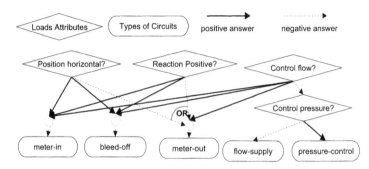

Figure 12.27 Knowledge diagram relating 'loads objects' and 'circuits

- *Alternatives generation*: checking what circuits match a specific load and have the same function. For example, bleed-off and meter-in control circuits are generated for the same combination of load position, reaction and control attributes.
- *Systems generation results*: generation of *systems objects*, according to the created alternatives as well as to single mapping (one load to one circuit) condition.
- *Systems presentation*: object presentation, diagrammatic form, explanation, component creation, dynamic model creation.

The knowledge base is platform independent. It is composed by a set of rules, facts bases, functions and messages for applications related to the hydraulic system design area, as shown in Fig. 12.28. The modules 'Trouble Shooting Fact Base' and 'Fluid Selection' deal with properties of the hydraulic system as a whole. Therefore, although they were implemented and are also important in a concurrent engineering approach, they will not be further described here. Additional modules such as 'Environmental Features', 'Cost Assistant' and 'Safety Module' can be included in an expanded system.

The dynamic-functional model agent consists of a set of messages, relating

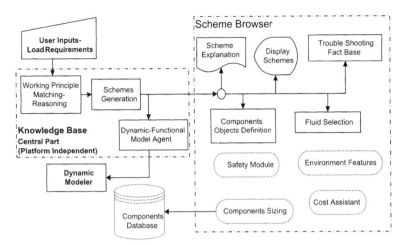

Figure 12.28 Expert system structure.

the specific circuits to create a file which defines the set of classes in the Dymola language. That is, the expert system generates files in the format compatible with the Dymola Modeller and the generated models allow the display of the hydraulic systems in a block diagram format. Models could readily be created in other simulation languages such as SIMULINK.

The expert system in development has four main entities:

Classes-&-Objects
Loads
Circuits
Systems-&-Components

which together define the structure of the user requirements, present the functionality and allow the user to obtain different schemes. Figure 12.27 shows the knowledge diagram, describing the *circuits definition rules*, based on some *loads attributes*. The knowledge base creates the systems objects based on the combination of the generated circuits in such form that every load, i.e. a set of functions, is accomplished by only one circuit, which provides the whole functionality for that specific load. For example, if the three load attributes in Fig. 12.27 define that a meter-out control circuit is needed and the load axis is vertical and therefore subject to gravitational loading, a counterbalance valve will be needed for this circuit. The same rule, which creates the specific circuit then modifies the value of the attribute *has_components* for this specific circuit, including the necessary valve.

This attribute is also used to create the set of *components objects* for each of the systems.

Among the objects, a special attention is given to the load and circuit objects, for they reflect the user needs and the working principle definition, respectively. Besides the specific attributes, each class has an *Id* attribute, that is an exclusive value given to each entity according to its class and the creation sequence (e.g. *load1, circuit1, system2*), which is used to define an object network as well as alternative design solutions for the same needs, i.e. the same set of loads attributes.

The class *circuit* has four subclasses:

pressure_control
flow_control
power_supply
flow_supply

Other subclasses can also be defined if required. Although these classes inherit properties from the class *circuit*, as a main feature of object-oriented modelling they also have their private properties and behaviours. For example, the default values of the attribute *has_components* is specific to each subclass and this attribute embraces the functionality attributable to each circuit as shown in Fig. 12.29.

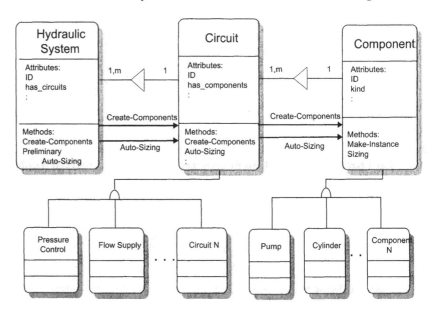

Figure 12.29 Inheritance and assembly relations among the classes.

The *explanation facility* is one of the key elements of every expert system. Here, this is achieved through the definition of the description attribute for the *load* and *circuit* classes. In the *load* class, the value of this attribute can be defined by the user. However, in the *circuit* class, this value is set by the expert system, inside each specific rule, in the creation of each circuit object. The value registers the *load_Id* of each circuit, together with the load attributes that can explain the existence of the specific circuit. The description attribute provides an explanation facility, as presented in a *results.dat* file generated by the system, of which part is shown in Fig. 12.30.

A most important part of any expert system is that it should have a user friendly user interface and this was a major aim of the prototype development. The use of the HTML format allows the viewing of the system output over the Internet and thus has the potential to be a powerful aid in a sales support environment.

12.4.6 Validation and testing

Major steps in the development process for this prototype hydraulic power expert system included:

********************************THESE WERE THE USER INPUTS********************************

load1, description: ("This is the main cylinder.")
load2, description: ("This is the winch.")
load3, description: ("The 3rd load is a vertical cylinder.")

LOADS ATTRIBUTES, SPECIFIED ACCORDING TO BOND GRAPHS TERMINOLOGY:

load_id	mode	reaction	position	domain
load1	flow	positive	horizontal	linear
load2	flow	negative	vertical	rotational
load3	effort	positive	vertical	linear

Based on these inputs and after some yes/no answers, the Expert System produces these results. For this task, there were created 2 alternative systems.

system1   has the following circuits: (circuit1 circuit2 circuit4 circuit5)
(circuit5 is a pressure-control circuit, matches load3 due to: ctr.=effort)
(circuit2 is a bleed_off circuit, matches load1 due to: ctr.=flow ,react=positive and pos.=horizontal)
(circuit4 is a rot_meter_out circuit, matches load2 due to: ctr.=flow ,react=negative ,pos.=vertical and domain=rotational)
(circuit1 has the POWER AND SAFETY FUNCTIONS, and it is ALWAYS created.)

system2 has the following circuits: (circuit1 circuit3 circuit4 circuit5)
(circuit5 is a pressure-control circuit, matches load3 due to: ctr.=effort)
(circuit3 is a meter-in circuit, matches load1 due to: ctr.=flow ,react=positive and pos.=horizontal)
(circuit4 is a rot_meter_out circuit, matches load2 due to: ctr.=flow ,react=negative ,pos.=vertical and domain=rotational)
(circuit1 has the POWER AND SAFETY FUNCTIONS, and it is ALWAYS created.

****

Figure 12.30 An example of the results the system provides, with the explanation facility.

- integration of the 'domain experts' from industry, to validate and expand the present knowledge base. This involved exposure of the system hands-on to a variety of industrial users and to postgraduate students, using benchmark examples and formal evaluation procedures;
- the progressive development of the user interface with input from the end users. An example of an output screen is given in Fig. 12.31;
- the definition of the *components data- base* structure. This can be expanded by industrial users to match their own product lines;
- the establishment of a common framework with the other energetic domains, i.e. pneumatics and electro-mechanics;
- the dynamic model library specification, creating object-oriented sub-models which are transportable into commonly-used simulation tools.

So far the incremental procedure and object-oriented modelling techniques have proved a satisfactory approach. However, in order to achieve a good quality system, the metrics set out in Table 12.2 have been considered throughout.

The main expected outcomes are the potential shortening of the design process for fluid power systems, the development of an instructional tool and

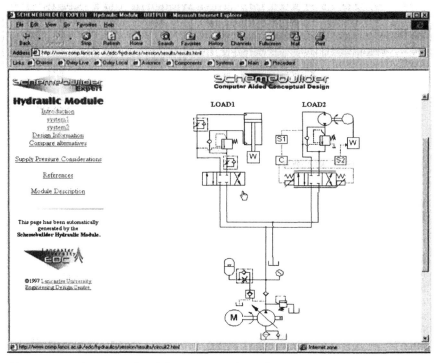

Figure 12.31 An example of the user interface: Output – Automatic Generation of Circuit Schematic.

Table 12.2 Quality metrics for an expert system

Some quality metrics for expert systems	How they have been taken into account
Correct outputs given correct inputs	Syntax test in the inputs strings
Complete output given correct input	Every load has its corresponding circuit
Consistent output given the same input again.	Repeatability tests have being carried out
Reliable so that it does not crash (often) due to bugs	
Usable for a range of expertise and preferably user-friendly	The GUI provides this facility
Maintainable and enhanceable	Object-oriented approach facilitates this
Reusable code for other hardware/ software environments	The knowledge base is independent of the Interface functions, PC and workstation platforms
Interfaceable with other software	Dynamic modeller interface available
Explanation facility	Use of the description attribute

the documentation of the design process in a concurrent engineering environment. This case study describes a prototype expert system which has proved useful as a teaching tool and is capable of further development and replication across the mechatronic domain.

The work summarised here comprised part of a doctorate research programme sponsored by the Brazilian Research Agency CAPES in conjunction with the Engineering Design Centre, Lancaster University, UK, through an exchange programme with the Federal University of Santa Catarina (UFSC), Brazil.

12.5 THE TOROTRAK INFINITELY VARIABLE TRANSMISSION: OPTIMISING POWER TRANSMISSION SYSTEMS, A GOAL WITHIN SIGHT?

Most rotary mechanical drive systems involve a substantial difference between the input and output speeds since in most cases the working range of the prime mover, such as an internal combustion engine, is ill-matched to the range of output speeds required. In the case of industrial drives driven predominantly by fixed speed 1450 r.p.m. or 2900 r.p.m. induction motors, a range of drive systems comprising single or multi-stage geared speed reducers, epicyclic, worm or harmonic gears may be employed as required by increasingly large reduction ratios of up to 200:1. Less commonly, step-up or speed increasing gear systems may be required.

A significant market is developing for these in the alternator drives for wind driven aerogenerators. Here the step-up ratio of 1:25 or more presents special challenges for the design and construction of highly efficient gear drives since the effect of frictional and windage losses in the high-speed output shaft and rotor is greatly magnified. Bradley *et al.* gives a short review of the many types of fixed and variable ratio change systems which are available.

For many applications, including the automotive, a variable ratio is essential so as to operate the engine within the contours of parameters which define its best efficiency over an infinitely variable range of load conditions and driver inputs. Figure 12.32 shows in outline the principle of a performance map for an automotive internal combustion engine. The driver's style or the requirements of the moment, for example while overtaking, may of course dictate maximising acceleration as the objective rather than optimising efficiency and any transmission control system will need to allow for such eventualities.

The traditional means of providing a degree of matching has been the use of stepped ratio gearboxes, now commonly with five ratios in manual versions and three or four in automatic. The common automatic gearbox has proven to be a highly reliable device, but in principle has remained unchanged for six decades. As a multistage epicyclic with brake band operation of the ratio changes, it is a step change gearbox in essence, even though the interposition of a torque converter between the engine input and the output to the final drive may make the changes less perceptible.

The incorporation of a torque converter however brings in the penalty of slip. Although in a notional 1:1 transmission nearly 100% of the torque is transmitted, there is a speed slip of 10–15% and a commensurate power loss which reflects directly in a higher fuel consumption. This is well recognised by users of conventional automatic transmission cars. Although the problem has been mitigated to some extent by the general adoption a mechanical clutch to lock up the torque converter in the highest gear ratio, this is chiefly of benefit in continuous high-speed autoroute driving, but does little for the predominantly urban motorist.

Automotive engineers have long recognised that an infinitely variable transmission (IVT) system would best enable the prime mover to be operated in its most favourable regime in all circumstances. The challenge has been how to obtain a robust and responsive system which would compete on cost with existing technology and provide a superior performance. Considering this from a 'mechatronic standpoint' it is apparent that a number of hydraulic or electrical systems could in principle meet the requirement for a wide ratio steplessly variable ratio system with a turn down of from 1:1 to 5:1. The additional costs, complexity and, in the case of electrical drives, power density make these options impracticable for normal automotive applications.

Many mechanical variators based on friction drives have been developed and are in use in industrial applications. Among these are systems based on wedge belts and cone pulley, ball and disc (Kopp Variator) or steel ring and conical disc principles (Heynau Drive). In these applications however, the power levels are limited to about 80 kW and the speed tends to be left constant once a process optimum has been found. The wedge belt and pulley system found an automotive application in van Doorne's range of DAF cars which enjoyed some popularity in the 1970s. However, the system was generally limited to about 45 kW in practice and the vehicle response in terms of driver perception was very different from conventional manual or automatic transmissions.

A transmission system which appeared to offer the potential for a wide range of variation and the ability to transmit high power was first proposed in an 1899 patent. This concept employs a set of rollers located between a pair of opposed toroidal discs. By varying the angle which the axes of rotation of the rollers make with the machine axis, it is possible to effect a steplessly variable ratio change between input and output as in Fig. 12.33.

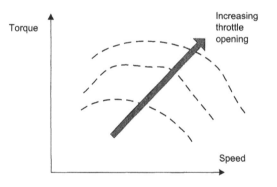

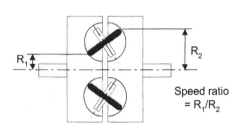

Figure 12.33 Stepless ratio change.

Figure 12.32 Generalised performance map for an automative internal combusion engine.

The principle was given practical application in the Hayes transmission (Austin, c1930) and the Perbury variator, 1960, and formed the basis of subsequent work in Rover Group. The last development attracted support from British Technology Group (BTG) in 1987 and a methodical process of development was undertaken to address the problems which had for decades inhibited the commercial breakthrough for what appeared to be an exceedingly promising concept. BTG has vested its IVT interests in two companies, Torotrak (Holdings) Ltd, to be responsible for licensing the technology, and Torotrak (Development) Ltd, to provide technical leadership and support.

It is often the case that an attractive concept has to await the development of supporting or enabling technologies before it can be convincingly demonstrated and introduced successfully to the user environment. The history of Torotrak amply illustrates this point. It could be said that three primary areas of technological development have brought the system to a stage where it is poised to transform the market for automotive transmissions.

12.5.1 Lubrication technology (tribology)

The torque is transmitted between the roller and disc interfaces by tractive forces. These are not, as may be thought, transmitted through conventional friction effects across metal-to-metal contacts. In order to generate and transmit the tractive forces, it is necessary to apply large forces acting across or normal to the surfaces. The contact patch between each roller and the surface of the toroid is thus subject to high levels of stress. This presents a challenge in terms of materials technology, but also has a beneficial effect on the lubrication regime. The mechanism involved is elasto-hydrodynamic lubrication in which the lubricant film experiences a virtually instantaneous increase in viscosity of several orders of magnitude as it enters and is subjected to the intense local pressure within the contact area. This maintains a fluid film, separating the two surfaces, but at the same time enabling the transmission of large shear forces.

The overall effect is to allow the tractive effort to be transmitted with no metal-to-metal contact and with negligible relative slip. In conjunction with Shell Research, special traction fluids have been developed which enhance this effect and have the stability to provide long-term reliability.

12.5.2 Surface engineering

The high interface pressures and thus the stress levels needed for the generation of sufficient tractive forces have presented a particular challenge in surface engineering. Some of the industrial drives mentioned earlier have a tendency to form wear tracks on the contact surfaces, if operated continually at one speed ratio, through the effect of the high Hertzian contact stresses causing fatigue pitting similar to that experienced in heavily loaded ball and roller bearings. In the Torotrak IVT, advanced surface treatments have been used to provide surfaces which have the necessary hardness and elasticity and substrate materials which have the toughness essential to their support.

12.5.3 Embedded microprocessor technology

A major contribution to the success of the Torotrak concept has been the principle of controlling the contact forces under all conditions to a level which is sufficient to control relative slip to within defined limits. The contact forces are generated by a hydraulically pressurised balance piston which applies an axial force to the toroid/roller assemblies. The pressure is modulated according to speed sensors which monitor the input/output speed conditions from which slip can be determined. The microprocessor-based control system is integrated with the vehicle engine management system to optimise ratio selection according to the driver's intent as determined by the use of the throttle.

12.5.4 Torotrak infinitely variable transmission

Torotrak in its present form is shown in Fig. 12.34. At the heart of the system is the variator based on two input toroidal discs driven via a torsional damper from the engine. The complete Torotrak mechanical package employs the principle of a split-power transmission in which the engine output can be combined with the variator output by means of an epicyclic gear. By the use of two clutches, two speed regimes can be selected. The output discs of the variator are connected via a fixed ratio chain to the central or 'sun' gear of the epicyclic gear set. The engine is connected to the variator input discs and also via an input gearset and parallel shaft to the mid or 'planet' carrier in the epicyclic gearset. The outer annulus of the epicyclic gearset is connected to the output transmission shaft and the road wheels.

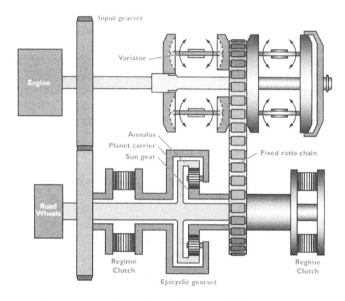

Figure 12.34 Torotrak.
(*Courtesy of Torotrak Ltd*)

The rotational speed of the annulus, and thereby of the output to the road wheels, is the sum of the relative speeds of the sun gear and the planet carrier. This arrangement enables the transmission to produce a full range of motion from reverse to zero speed to forward. In the zero output speed condition, a minimal amount of power is re-circulated by the variator via the parallel epicyclic shunt without having the need for mechanical disconnection as with a conventional clutch which is therefore dispensed with.

A key aspect of the Torotrak IVT is the way in which the ratio changes are effected by modulating the hydraulic pressure to the pistons actuating the roller carriages. By this means, the variator output torque can be controlled with a high rate of response, offering useful advantages when driving under difficult conditions as in snow and ice.

In automotive applications, Torotrak has two operating regimes. Low regime covers neutral, reverse and forward speeds up to the equivalent of second gear and high regime for all higher speeds up to overdrive. As the vehicle accelerates from rest, the variator ratio moves towards the limiting value set by the geometric limits of the roller carriage movement. When the highest forward ratio in low regime is reached in which the annulus, sun and planet carrier speeds are the same, the low regime clutch is automatically operated to disconnect the epicyclic gearbox while the high regime clutch simultaneously engages to connect the output shaft to the road wheels directly to the variator output. The roller's carriage angles then sweep in the opposite direction through the centre 1:1 ratio to provide a high overdrive.

The control mechanism for the IVT is based upon modulation of pressure in the actuation circuit to the roller carriages and it is the change of the radius at which the rollers run on the toroids that controls the system output torque. From dynamometer tests carried out on prospective engines to which Torotrak IVT may be applied, three-dimensional surfaces have been mapped for output speed and torque and throttle position. These are contained in the controller memory. Thus when an accelerator position demand is made the controller registers the engine speed and looks up the mapped torque output and from this an appropriate value of hydraulic pressure is selected to set the variator torque output. An electrically operated throttle valve is used to control the engine so as to operate at all times very close to its optimum condition on the operating map.

This 'drive-by-wire' system means that the driver controls the transmission to achieve the required vehicle response and the control algorithms then match the engine output accordingly. The embedded algorithms also allow the driver to select a range of options to maximise performance as opposed to economy, by offering a choice of 'automatic IVT' or 'gear selection' modes which provide a familiar feel comparable with a more conventional transmission.

The main advantages offered by the Torotrak transmission are:

- environmentally friendly with fuel savings of up to 17% and a corresponding reduction in carbon dioxide emissions;
- smooth responsive and quiet in operation;
- more compact than a conventional transmission, aided by the lack of a clutch or torque converter required in manual or conventional automatic transmissions for starting from rest;
- less expensive to manufacture having less geared components and otherwise components which are easily machined;

- more durable, as the variator components are protected from metal-to-metal contact by the film provided by the traction fluid.

The upper power limits of the Torotrak concept have not yet been fully explored and it is planned to address the market for medium and lightweight trucks. Many applications can also be foreseen in industrial drives including those requiring speed increasing ratios, where the high efficiency available from Torotrak should show a distinct advantage.

Further information can be obtained from Torotrak Development, 3 Titan Way, Leyland, Lancashire PR5 3QW, UK or, on the internet, at www.torotrak.com.

Bibliography

JOURNAL ARTICLES

Artificial intelligence and expert systems

Adelsberger, H.H. and Neumann, G. (1985) Goal oriented simulation and modelling using Prolog, in Proceedings of the SCS Conference on Modelling and Simulation on Microcomputers, pp. 42–47.

Albus, J.S. (1991) Outline for a theory of intelligence, *IEEE Trans*, SMC-21, 473–509.

Brooks, R.A. (1991) Intelligence without representation, *Artificial Intelligence*, Vol 47, Nos 1–3, January, 139–159.

Burton, A.M., Shadbolt, N.R., Gedgecock, A.P., and Rugg, G. (1986) A formal evaluation of knowledge elicitation techniques for expert systems, Alvey project IKBS 134, Department of Psychology, University of Nottingham, UK.

Burton, A.M., Shadbolt, N.R., Hedgcock, A.P., and Rugg, G. (1987) Evaluation of knowledge elicitation techniques, Alvey project IKBS 134, Department of Psychology, University of Nottingham, UK.

Eisenstadt, M. and Brayshaw, M. (1990) A knowledge engineering toolkit, *Byte*, October, 268–282.

Green, P., Seward, D.W. and Bradley, D.A. (1990) Knowledge acquisition for a robot excavator, in *Proceedings of the 7th International Symposium on Automation and Robotics in Construction*, Bristol, pp. 351–357.

Habib, R.S., Habib, H.S. and Majeed, B.S. (1993) Experimental Self-Location Vehicle Based on Ultrasonic-Waves Guided System, *Int. J. Electrical Engineering Education*, Vol. 30 No 3, July, 224–235.

Hillis, W.D. (1988) Intelligence as an emergent behaviour, in Graubard, S. (ed.), *Artificial Intelligence*, MIT Press.

Hong, S.Y. (1993) Knowledge-based diagnosis of drill conditions, *Journal of Intelligent Manufacturing*, Vol. 4, 233–241.

Isermann, R. and Raab, U. (1992) Intelligent actuators: ways to autonomous actuating systems, in *Proceedings of the IFAC Conf. on Intelligent Components and Instruments for Control Applications*, Malaga, Spain.

Kusko, B. (1992) Invest in higher machine IQ, *Reason*, October, 42–44.

Maes, P. (1989) How to do the right thing, *Connection Science*, Vol. 1, No. 3.

Markoff, J. (1991) The creature that lives in Pittsburgh, *The New York Times*, 21 April.

McCulloch, W.S. and Pitts, W. (1943) A logical calculus of the ideas immanent in nervous activity, *Bulletin of Mathematical Biophysics*, Vol. 5, 115–133.

Shannon, R.E., Mayer, R. and Adelsberger, H.H. (1985) Expert systems and simulation, *Simulation*, Vol. 44, No. 6, 275–284.

Tomiyama, T., Kiriyama, T. and Umeda, Y. (1994) Toward knowledge intensive engineering, in *Proceedings of the Lancaster International Workshop on Engineering Design*.

Wloka, D.W. (1990) How to build robot models for knowledge-based simulation, in *Proceedings of the 1990 European Simulation Multiconference*, pp. 362–367.

Automation and robotics

Holenstein, A.A. and Badreddin, E. (1991) Collision avoidance in a behaviour based mobile robot design, in *Proceedings of the IEEE International Conference on Robotics and Automation*, Sacramento, pp. 898–903.

Boddy, C.L. (1991) A real-time trajectory planner and end-effector collision avoidance system for a robotic arm, in *Proceedings of the IEE International Conference CONTROL'91*, Edinburgh, pp. 69–74.

Bradley, D.A., Seward, D.W., Heikkilä, T. and Vähä, P. (1994) Control architectures and operational strategies for intelligent, high-powered manipulators, in *Proceedings of the 11th International Symposium on Automation and Robotics in Construction*, Brighton, pp. 367–372.

Bradshaw, A., Roskilly, A.P. and Counsell, J.M. (1990) Nonlinear modelling of robust controllers for robotic manipulators, in *Proceedings of the IMechE Conference on Mechatronics: Designing Intelligent Machines*, Cambridge, September, pp. 223–230.

Brooks, R.A. and Flynn, A. (1989) *Fast, cheap and out of control: a robot invasion of the solar system*, MIT AI Laboratory Memo 1148, October.

Brooks, R.A. (1985) A robust layered control system for a mobile robot, AI Memo No 864, MIT AI Laboratory.

Brooks, R.A. (1986) A layered intelligent control system for a mobile robot, in *Proceedings of the 3rd International Symposium on Robotics Research*, MIT Press.

Brooks, R.A. (1991) New Approaches to robotics, *Science*, Vol 253, No 5025, September, 1227–1232.

Brooks, R.A., Maes, P., Mataric, M.J. and More, G. (1990) Lunar base construction robots, in *Proceedings of the IEEE International Workshop on Intelligent Robots and Systems, IROS'90*.

Caldwell, D.G. (1993) Natural and artificial muscle elements as robot actuators, *Mechatronics*, Vol. 3, No. 3, 269–283.

Choi, W. and Latombe, J-C. (199) A reactive architecture for planning and executing robot motions with incomplete knowledge, in *Proceeding of the IEEE Workshop on Intelligent Robots and Systems, IROS'91*, Osaka, pp. 24–29.

Chong, K.P., Dewey, B.R. and Pell, K.M. (1989) ROBOSIM: a simulation packag for robots, in *Proceedings of the 7th Annual Conference on Computer-Aided Engineering Design and Manufacturing*, pp. 239–246.

Heikkilä, T., Vähä, P. and Okkonen, J. (1993) A skilled and intelligent paper roll manipulator, in *Proceedings of the International Conference on Intelligent Robots and Systems, IROS'93*, Japan.

Heikkilä, T., Vähä, P. and Okkonen, J. (1994) A heavy duty manipulator with computational skills and intelligence, in *Proceedings of the International Conference on Machine Automation, ICMA'94*, Tampere, Finland, February.

Hollingum, J. (1990) Robot simulation comes to Britain, *Industrial Robot*, December, pp. 181–183.

Li, S. and Xu, Y. (1990) A software package for the graphic simulation of robots, *Acta Automatica Sinica*, Vol. 16, No. 4, 380–382.

NATO Advanced Research Workshop on Sensor Systems for Robotic Control, October/November 1989.

Seward, D.W. and Garman, A. (1996) The software development process for intelligent robots, *IEE Computing and Control Engineering Journal*, Vol.7 No. 2, 86–92.

Song, K.T. and Li, C.E. (1993) Tracking control of a free-ranging automatic guided vehicle, *Control Engineering Practice Journal*, Vol. 2, No. 1, 163–169.

Yan, J. and Chen, G.L. (1990) A robot system simulation analysis, *Systems Analysis, Modelling and Simulation*, Vol. 7, No. 10, 775–782.

Engineering design

Bracewell, R.H., Chaplin, R.V. and Bradley, D.A. (1992) *Schemebuilder and Layout – Tools for the design of mechatronic systems*, IMechE Conf. on Mechatronics; Designing Intelligent Systems, Cambridge, pp. 1–7.

Bradley, D.A. and Dawson, D. (1991) Information based strategies in the design of mechatronic systems, *Design Studies*, Vol. 12, No. 1, 12–18.

Burge, S.E. and Woodhead, M.A. (1997) Product Introduction Processes in the UK Aerospace Industry: Complex, Integrated Human Activity Systems, EPSRC IMI Grant Final Report.

Buur, J. and Andreasen, M.M. (1989) Design models in mechatronic product development, *Design Studies*, Vol. 10, No. 3, 155–162.

Buur, J. (1989) Positioning mechatronics design between mechanics, electronics and software, in *Proceedings of the International Conference on Advanced Mechatronics*, Tokyo, pp. 189–194.

Buur, J. (1992) Mechatronics design methodology, in *Proceedings of the NATO Workshop – Mechatronics Design in Textile Engineering*, Side, Turkey.

Cooper, R.G., The new product process. A decision guide for management, *Journal of Marketing Management*, Vol. 3, No. 3, 238–255.

Curwen, P. (1990) System Development Using the CORE Method, British Aerospace Report BAe/WIT/ML/GEN/SWE/1227.

Dorey, A.P. and Bradley, D.A. (1994) Sensors and measurement systems: essential elements of mechatronics, *Measurement Science and Technology*, Vol. 5, 1415–1428.

Dowlatshahi, S. (1994) Comparison of approaches to concurrent engineering, *International Journal of Advanced Manufacturing Technology*, Vol. 9, 106–113.

Finkelstein, L. and Finkelstein, A.C.W. (1983) Review of design methodology, *IEE Proceedings*, Vol. 130, Pt A, No. 4.

Gardam, A. and Burge, S.E. (1997) Changes in the engineering design process, in *Proceedings of the 13th National Conference on Manufacturing Research*, Glasgow, 9–11 Sept.

Gould, J.D., Boies, S.J., Levy, S., Richards, J.T. and Schoonard, J. (1987) The 1984 olympic message system: a test of behavioural principles of system design, *Communications of the ACM*, Vol. 30, No. 9, 758–769.

Gregory, L. (1995) Virtual prototyping on personal computers, *Mechanical Engineering*, July, Vol. 117, No. 7, 78–85.

Hauser, J.R. and Clausing, D. (1988) The house of quality, *Harvard Business Review*, No. 3, May–June, 63–73.

Lyman, D., Buesinger, F.R. and Keating, J.P. (1994) QFD in strategic planning, *Quality Digest*, May, 45–520.

Oh, V. and Sharpe, J. (1995) Conflict management in an interdisciplinary design environment, in *Proceedings of Lancaster International Workshop on Engineering Design*.

Oh, V., Langdon, P. and Sharpe, J. (1994) Schemebuilder: an integrated computer environment for product design, in *Proceedings of Lancaster International Workshop on Engineering Design*.

Ross, D.T. and Schoman, K.E. (1977) Structured analysis for requirements definition, *IEEE Transactions on Software Engineering*, SE-3 (1).

Styger, .L. (1995) Firming designs – rapid prototyping, *IEE Review*, January, 38–39.

Thornley, J.K. (1993) A high-speed piezoelectrically clutching device, *Mechatronics*, Vol. 3, No. 3, 295–304.

Walton, A.D. and Whittingham, N.A. (1991) Computer simulation: a user's view, *Industrial Robot*, Vol. 18, No. 4, 3–4.

Human–computer interface and virtual reality

Buur, J. and Windum, J. (1992) *Methods for Designing the Man/Machine Interface*, Institute for Engineering Design, Technical University of Denmark.

Buur, J. and Windum, J. (1992) *Perception of the User in Designing Man/Machine Interface*, Institute for Engineering Design, Technical University of Denmark.

Neelamkavil, F. and Beare, L. (1988) Techniques for animation on microcomputers, *Computer Graphics Forum*, Vol. 7, 21–27.

Rasmussen, J. (1983) Skills, rules and knowledge; signals, signs and symbols, and other distinctions in human performance models, *IEEE Transactions on Systems, Man and Cybernetics*, SMC 13, No. 3, May/June, 257–266.

Lucie

Bradley, D.A. and Seward, D.W. (1992) The Lancaster University computerised intelligent excavator programme, in *IMechE Conference on Mechatronics – Designing Intelligent Machines*, Dundee, pp. 163–169.

Bradley, D.A., Seward, D.W., Mann, J.E. and Goodwin, M.R. (1994) Artificial intelligence in the control and operation of construction plant – the autonomous robot excavator, *Automation in Construction*, No. 2, 217–218.

Bullock, D.M. (1988) Supervisory control for cognitive excavation, MSc dissertation, Department of Civil Engineering, Carnegie-Mellon University, Pittsburgh, Pennsylvania.

Manufacturing systems

Gindy, N.N.Z. (1992) *Feature-based Component Model for Computer Aided Process Planning*, Department of Manufacturing Engineering, Loughborough University.

Hauser, J.R and Clausing, D. (1988) The house of quality, *Harvard Business Review*, May–June, pp. 63–73.

Hu, H.S. & Brady, M. 1997 (Dynamic global path planning with uncertainty for mobile robots in manufacturing) *IEEE Trans. Robotics & Automation*, Vol 13, No 5, 760–767.

Kopacek, P. 1999, (Intelligent manufacturing: Present state and future trends) *Journal of Intelligent & Robotic Systems*, Vol 26, No 3–4, 217–229.

Pridham, M. and Thompson, G. (1995) Laser forming, *Manufacturing Engineer*, June, pp. 137–139.

Pugh, P. (1995) Rapid prototyping: a world wide perspective, *Journal of Rapid Prototyping*, Vol. 1, No. 3, 47–48.

Sprow, E.E. (1992) Rapid prototyping, *Manufacturing Engineering*, November, pp. 58–64.

Stokic, D.M., Vukobatovic, M.K. and Lekovic, D.B. (1991) Simulation of robots in flexible manufacturing cells, *Robotics and Computer-integrated Manufacturing*, Vol. 8, No. 1, 1–8.

Stone, M. (1984) Competing with Japan – the rules of the game, *Long Range Planning*, Vol. 17, No. 2.

Swanson, D. (1994) Managing rapid prototyping, in *Proceedings of the IEEE Conference, Wescon'94*, pp. 276–281.

Microsystems, microsensors and MEMS

Brignell, J.E. (1993) Quo vadis smart sensors?, *Sensors and Actuators A*, Vol. 37/38, 6–8.

Design Engineering (1992) A big step for micromotors, *Design Engineering*, November, 12.

Heuberger, A. (1993) Silicon microsystems, *Microelectronic Engineering*, Vol. 21, 445–448.

Huijsing, J.H. (1992) Integrated smart sensors, *Sensors and Actuators A*, Vol. 30, 167–174.

Knieling, M. (1991) Technologies and market developments for microsystem technologies, *Proceedings of the 2nd International Conference on Micro Electro, Opto, Mechanical Systems and Components, Micro System Technologies '91*, Berlin, pp. 544–551.

Olbrich, Th., Bradley, D.A. and Richardson, A.M.D. (1994) BIST and diagnosis in safety critical microsystems using reliability indicators, *Proceedings of the 5th European Symposium on Reliability of Electronic Devices, ESREF'94, Failure Physics and Analysis*, Glasgow.

Olbrich, Th., Bradley, D.A. and Richardson, A.M.D. (1994) Built in self test and diagnostics for microsystems, *Proceedings of the 2nd Intl. Symposium on Advanced Transport Applications, ISATA*, Aachen, pp. 511–518.

Prosser, S.J. (1991) *Advances in Automotive Sensors, Sensors: Technology, Systems and Applications*, IOP Publishing, pp. 493–504.

Sharma, R. (1993) Theory, analysis and implementation of an on-line BIST techniques, *VLSI Design*, Vol. 1, No. 1, 9–22..

Westbrook, M.H. (1988) Automotive sensors, and overview, *IEE Proceedings*, Vol. 135, Pt D, No. 5, 339–347.

Neural networks and fuzzy systems

Chuen, C.L. (1990) Fuzzy logic in control systems: Fuzzy logic controller – Parts I and II, *IEEE Transactions on Systems, Man and Cybernetics*, Vol. 20, No. 2,

405–435.

Gaines, B.R. (1976) Foundations of fuzzy reasoning, *International Journal of Man–Machine Studies*, No. 8, 623–688.

Handelman, D.A., Lane, S.H. and Gelfand, J.J. (1990) Integrating neural networks and knowledge-based systems for intelligent robot control, *IEEE Control Systems*, Vol. 10, No. 3, 49–55.

Jain, A.K. (1996) Artificial neural networks: a tutorial, *Computer*, Vol. 29, No. 3.

Kohonen, T. (1982) Self-organised formation of topologically correct feature maps, *Biological Cybernetics*, Vol. 43, 59–69.

Lippman, R.P. (1987) An introduction to computing with neural nets, *IEEE ASSP Magazine*, 4–22.

Nguyen, D.H. and Widrow, B. (1990) Neural networks for self-learning control systems, *IEEE Control Systems Magazine*, April, 18–22.

Rosenblatt, F. (1958) The perceptron: a probabilistic model for the information storage and organisation in the brain, *Psychological Review*, Vol 65, 386–408.

Sugeno, M. (1985) An introductory survey of fuzzy control, *Information Sciences*, Vol 36, Nos 1–2, July–August, 59–83.

Widrow, B. (1962) Generalisation and information storage in networks of adeline 'neurons', *Self-organising Systems,*, 435–461.

Zadeh, L.A. (1965) Fuzzy sets, *Information and Control*, Vol. 8, 338–353.

Software engineering

Boehm, B. (1988) A spiral model for software development and enhancement, *Computer* , Vol. 21, No. 5, 61–72.

Buckley, F.J. and Poston, R. (1984) Software quality assurance, *IEEE Tranactions on Software Engineering*, Vol. 10, No. 1, 36–41.

Elmqvist, H., Cellier, F. E. and Otter, M. (1993) Object-oriented modeling of hybrid systems, in *Proceedings of the European Simulation Symposium*, Delft, The Netherlands, October 25–28.

Nance, R.E. (1985) A specification language to assist in the analysis of discrete event simulation models, *CACM*, Vol. 28, No. 2, 190–201.

Seideawitz, E. (1987) Object-oriented programming in Smalltalk and ADA, in *Proceedings Object Oriented Programming Systems '87*.

Tonella, G. and Domingo, C. (1990) GLIDER: A new simulation language, in *Proceedings of UKSC Conference on Computer Simulation*, Brighton, pp. 67–74.

System safety

Proccaccia, H., Aufort, S. and Arsenis, S. (1997) The European industry reliability data bank EIReDA, in *Advances in Safety and Reliability, Vol.1, Proceedings of the International Conference on Safety and Reliability, ESREL'97*, Lisbon, Portugal, pp. 1819–1826.

Seward, D.W., Bradley, D.A. and Margrave, F.W. (1994) Hazard analysis techniques for mobile construction robots, in *Proceedings of the 11th International Symposium on Robotics in Construction*, Brighton, England pp. 35–42.

Seward, D.W., Quayle, S., Sommerville, I. and Morrey, R. (1997) Developing the safety case for large mobile robots, in *Advances in Safety and Reliability, Vol.1, Proceedings of the ESREL '97 International Conference on Safety and Reliability*, Lisbon, Portugal, June 1997, pp. 2293–2300.

Stavridou, V. (1997) *A Tutorial on Software Safety Standards*, The Safety Critical Systems Club Meeting, London.

Yong, Y.F. (1985) CAD – An aid to robot safety, in Bonney, M.C. and Yong, Y.F. (eds), *Robot Safety*, IFS Springer-Verlag.

Other

Anon (1986) Operating a backhoe efficiently, *Water Engineering and Management*, 28–31.

Malaguti, F. (1993) Soil cutting model developed by dynamic system identification methods, in *Proceedings of the 11th International ISTVS Confereence*, Lake Tahoe, September.

Rudnev, V.K. (1985) *Digging of Soil by Earthmoving with Powered Parts*, Balkema.

BOOKS

Artificial intelligence and expert systems

Aleksander, I. and Burnett, P. (1984) *Reinventing Man*, Penguin Books.

Boolos, G., Jeffrey, R. (1974) *Computability and Logic*, Cambridge University Press

Davis, L. (1991) *Handbook of Genetic Algorithms*, Van Nostrand Reinhold (Chapman and Hall).

Dean, T., Allen, J. and Aloimonos, Y. (1995) *Artificial Intelligence, Theory and Practice*, The Benjamin Cummings Publishing Company Inc.

Dolan, A.K. and Aldous, J. (1993) *Networks and Algorithms: An Introductory Approach*, Wiley.

Dreyfus, H.L. (1979) *What Computers Can't Do: The Limits of Artificial Intelligence*, Harper and Row.

Englemore, R. and Morgan, T, (eds) (1988) *Blackboard Systems*, Addison-Wesley.

Forsyth, R. and Naylor, C. (1985) *The Hitch-Hiker's Guide to Artificial Intelligence*, Chapman and Hall.

Gevarter, W.B. (1985) *Intelligent Machines: An Introductory Perspective of Artificial Intelligence*, Prentice Hall.

Giarrantano, J. and Riley, G. (1994) *Expert Systems – Principles and Programming*, 2nd Edition, PWS Publishing .

Ginsberg, M. (1993) *Essentials of Artificial Intelligence*, Morgan Kaufman.

Goldberg, D.E. (1989) *Genetic Algorithms in Search, Optimisation and Machine Learning*, Addison-Wesley.

Gonzalez, A.J. and Dankel, D.D. (1993) *The Engineering of Knowledge-Based Systems – Theory and Practice*, Prentice-Hall.

Hart, A. (1989) *Knowledge Acquisition for Expert Systems*, Kogan Page.

Haugeland, J. (1985) *Artificial Intelligence: The Very Idea*, MIT Press.

Hopgood, A.A. (1993) *Knowledge-based Systems for Engineers and Scientists*, CRC Press.

Huang, G.Q. and Brandon, J.A. (1993) *Co-operating Expert System in Mechanical Design*, Wiley.

Ignizio, J.P. (1991) *Introduction to Expert Systems*, McGraw-Hill.

Kolodner, J.L. (1993) *Case-based Reasoning*, Morgan Kaufman.

McCorduck, P. (1979) *Machines Who Think*, W.H. Freeman.

Michie, D. and Johnson, R. (1985) *The Creative Computer*, Pelican.

Nilsson, N. (1965) *Learning Machines*, McGraw-Hill.

Nilsson, N.J. (1998) *Artificial Intelligence: A New Synthesis*, Morgan Kaufmann.

Parsaye, K., Chignell, M., Khoshafian, S. and Wong, H. (1989) *Intelligent Databases: Object-Oriented, Deductive and Hypermedia Technologies*, Wiley.

Payne, E.C. and McArthur, R.C. (1990) *Developing Expert Systems: A Knowledge Engineer's Handbook for Rules and Objects*, Wiley.

Reisbeck, C.K. and Schank, R.C. (1989) *Inside Case-based Reasoning*, Erlbaum.

Rich, E. and Knight, K. (1991) *Artificial Intelligence*, McGraw-Hill.

Rzevski, G. (1990) *Engineering Design and Artificial Intelligence*, The Open University.

Simon, H. (1969) *The Science of the Artificial*, MIT Press.

Staugaard, A.C. (1987) *Robotics and AI*, Prentice Hall.

Taylor, W.A. (1988) *What Every Engineer Should Know about Artificial Intelligence*, MIT Press.

Watson, I.D. (1997) *Applying Case-Based Reasoning: Techniques for Enterprise Systems*, Morgan-Kaufman.

Werbose, P.J. (1974) *Beyond Regression: New Tools for Prediction and Analysis in Behavioural Systems*, PhD Thesis, Harvard University.

Winston, P.H. (1994) *Artificial Intelligence*, Addison-Wesley.

Wright, P.K. and Bourne, D.A. (1988) *Manufacturing Intelligence*, Addison-Wesley.

Automation and robotics

Advanced Robotics Research Limited (1992) *Safety and Standards for Advanced Robots – A First Exposition*, Report ARRL.92.009.

Aleksander, I. and Burnett, P. (1984) *Reinventing Man*, Penguin.

Asfahl, C.R. (1985) *Robots and Manufacturing Automation*, Wiley.

Badler, N.I., Barsky, B.A. and Zeltzer, D. (eds) (1991) *Making them Move: Mechanics, Control and Animation of Articulated Figures*, Morgan Kaufmann.

Craig, J.J. (1989) *Introduction to Robotics*, Addison-Wesley.

Cutkowsky, M.R. (1985) *Robotic Grasping and Fine Manipulation*, Kluwer.

Dyson, G. (1997) *Darwin Among the Machines*, Allen Lane.

Groover, M.P., Weiss M., Nagel, R.N. and Odrey, N. (1986) *Industrial Robotics*, McGraw-Hill.

Groover, M.P. (1987) *Automation, Production Systems and Computer-integrated Manufacturing*, Prentice Hall.

McKerrow, P.J. (1993) *Introduction to Robotics*, Addison-Wesley.

Minsky, M.L. (1985) *Robotics*, Anchor Press.

Ray, C. (1992) *Robots and Manufacturing Automation*, Wiley.

Rosheim, M.E. (1994) *Robot Evolution: The Development of Anthrobotics*, Wiley-Interscience.

Snyder, W.E. (1985) *Industrial Robots*, Prentice Hall.

Staugaard, A.C. (1987) *Robotics and AI*, Prentice Hall.

Thring, M.W. (1983) *Robots and Telechirs*, Ellis Horwood.

Engineering design

Andreasen, M.M. and Hein, L. (1985) *Integrated Product Development*, IFS Publications.

Andreasen, M.M., Kahler, S. and Lund, T. (1988) *Design for Assembly*, IFS Publications.

Baxter, M. (1996) *Product Design*, Chapman and Hall.

Baynes, K. and Pugh, F. (1981) *The Art of the Engineer*, Lutterworth Press.

Baynes, K. (1976) *About Design*, The Design Council.

Bedworth, D.D., Henderson, M.R. and Wolfe, P.M. (1991) *Computer Integrated Design and Manufacturing*, McGraw-Hill.

Belbin, R.M. (1981) *Management Teams: Why They Succeed or Fail*, Heinemann.

Bendell, A., Disney, J. and Pridmore, W.A. (eds) (1989) *Taguchi Methods: Applications in World Industry*, IFS Publications.

Boothroyd, G. and Dewhurst, P. (1987) *Product Design for Assembly*, Wakefield.

Bossert, J.L. (1991) *Quality Function Deployment – A practitioner's approach*, ASQC Quality Press.

Bradley, D. A., Dawson, D., Burd, N.C. and Loader, A.J. (1991) *Mechatronics: Electronics in Products and Processes*, Chapman and Hall.

Buur, J. (1989) *Mechatronics Design in Japan*, Institute for Engineering Design, Technical University of Denmark.

Buur, J. (1990) *A Theoretical Approach to Mechatronics Design*, Institute for Engineering Design, Technical University of Denmark.

Carter, D.E. and Baker, S. (1992) *Concurrent Engineering: The Product Development Environment for the 90s*, Addison-Wesley.

Checkland, P.B. (1981) *Systems Thinking, Systems Practice*, Wiley.

Cohen, L. and Cohen, L. (1995) *QFD – How to Make QFD Work for You*, Addison Wesley.

Cohen, B., Harwood, W.T. and Jackson, M.I. (1986) *The Specification of Complex Systems*, Addison-Wesley.

Constantine, L.L. and Yourdon, E. (1979) *Structured Design*, Prentice Hall.

Corbett, J., Dooner, M., Melka, J. and Pym, C. (1991) *Design for Manufacture: Strategies, Principles and Techniques*, Addison-Wesley.

Coulson, A.J. (1979) *A Bibliography of Design in Britain 1851–1970*, The Design Council.

Cross, N. (1994) *Engineering Design Methods*, Wiley.

DeMarco, T. (1979) *Structured Analysis and System Specification*, Prentice Hall.

French, M.J. (1985) *Conceptual Design for Engineers*, 2nd edn, Design Council .

French, M.J. (1988) *Invention and Evolution: Design in Nature and Engineering*, Cambridge University Press.

Gause, D.C. and Weinberg, G.M. (1989) *Exploring Requirements – Quality Before Design*, Dorset House Publishing.

Hales, C. (1993) *Managing Engineering Design*, Longman Scientific.

Hartley, J.R. and Mortimer, J. (1990) *Simultaneous Engineering*, Department of Trade and Industry.

Hartley, J.R. (1992) *Concurrent Engineering: Shortening Lead Times, Raising Quality and Lowering Costs*, Productivity Press.

Henry, J. and Walker, D. (eds) (1990) *Managing Innovation*, Sage.

Heskett, J. (1980) *Industrial Design*, Thames and Hudson.

Hollins, W. and Pugh, S. (1990) *Successful Product Design: What to Do and When*, Butterworth.

Huang, G.Q. and Brandon, J.A. (1993) *Co-operating Expert System in Mechanical Design*, Wiley.

Hubka, V. (1982) *Principles of Engineering Design*, Butterworth.

James G. (1992) *Modern Engineering Mathematics*, Addison-Wesley.

Jones, J.C. (1970) *Design Methods – Seeds of Human Futures*, Wiley.

Karnopp, D. and Rosenberg, R.C. (1964) *Analysis and Simulation of Multiport Systems – The Bond Graph Approach to Physical Systems*, McGraw-Hill.

Law, A.M. and Kelton, W.D. (1991) *Simulation, Modelling and Analysis*, McGraw-Hill.

Manners, D., (1990) *Hitchhikers' Guide to Electronics in the 90's*, Computer Weekly Publications.

Mayr, O. (1969) *The Origins of Feedback Control*, MIT Press.

McMahon, C. and Browne, J. (1993) *CADCAM From Principles to Practice*, Addison-Wesley.

Neelamkavil, F. (1987) *Computer Simulation and Modelling*, Wiley.

Nevins, J.L. and Whitney, D.E. (eds) (1989) *Current Design of Products and Processes: A Strategy for the Next Generation in Manufacturing*, McGraw-Hill.

Oakland, J. (1989) *Total Quality Management*, Pitman.

Pahl, G. and Beitz, W. (1996) *Engineering Design*, Springer-Verlag.

Petroski, H. (1985) *To Engineer is Human: The Role of Failure in Successful Design*, Vantage Books.

Petroski, H. (1996) *Invention by Design: How Engineers get from Thought to Thing*, Harvard University Press.

Pugh, S. (1991) *Total Design*, Addison-Wesley.

Revelle, J.B ., Moran, J.W. and Cox, C (1998) *The QFD Handbook*, Wiley.

Rich, B. and Janos, L. (1996) *Skunk Works: My Years at Lockheed*, Little Brown and Co.

Rosenberg, D. and Hutchinson, C. (1994) *Design Issues in Computer Supported Co-operative Work*, Springer-Verlag.

Rouse, W.B. and Boff, K.R. (eds) (1987) *System Design: Behavioural Perspectives on Designers, Tools and Organisations*, North-Holland.

Rzevski, G. (1990) *Engineering Design and Artificial Intelligence*, The Open University.

Rzevski, G. (1995) *Mechatronics: Designing Intelligent Machines*, Vol. 1, Butterworth-Heinmann.

Rzevski, G. (1995) *Mechatronics: Designing Intelligent Machines*, Vol. 2, Butterworth-Heinmann.

Schaefer, H. (1970) *The Roots of Modern Design*, Studio Vista.

Schenker, P.R. (1990) *The Team Handbook*, Joiner Associates.

Scrivener, S. A. R. (ed.) (1994) *Computer Supported Co-operative Work: The Multimedia and Networking Paradigm*, Ashgate.

Shina, S.G. (1991) *Concurrent Engineering and Design for Manufacture of Electronic Products*, Van Nostrand Reinhold.

Sommerville, I. and Sawyer, P. (1997) *Requirements Engineering – A Good Practice Guide*, Wiley.

Souder, W. and Sherman, J. (1994) *Managing New Technology Development*, McGraw-Hill.

Spurr, K., Layzell, P., Jennison, L. and Richards, N. (1994) *Computer Support for Co-operative Work*, Wiley.

Sydenham, P.H., Hancock, N.H. and Thorn, R. (1992) *Introduction to Measurement Science and Engineering*, Wiley.

Taguchi, G. (1985) *System of Experimental Design*, UNIPUB Kraus.

Taguchi, G. (1986) *Introduction to Quality Engineering: Designing Quality into Products and Processes*, Asian Productivity Association.

Tidd, J., Bessant, J. and Pavitt, K. (1997) *Managing Innovation – Integrating Technology, Market and Organisational Change*, Wiley.

Tjalve, E. (1979) *A Short Course in Industrial Design*, Butterworth.

Turino, J. (1992) *Managing Concurrent Engineering*, Van Nostrand Reinhold.

Twigg, D. and Voss, C.A. (1992) *Managing Integration in CAD/CAM and Simultaneous Engineering*, Chapman and Hall.

US Govt. Information Superintendent of Documents, *NASA Systems Engineering Handbook*, Stock Number 033–000–00156–3.

Walker, J.A. (1989) *Design History and the History of Design*, Pluto Press.

Wright, I. (1998) *Design Methods Engineering and Product Design*, McGraw-Hill.

History of technology

al-Hassan, A. and Hill, D.R. (1986) *Islamic Technology*, Cambridge University Press.

Ashton, T.S. (1924) *Iron and Steel in the Industrial Revolution*, Manchester, The University Press.

Bassalla, G. (1988) *The Evolution of Technology*, Cambridge University Press.

Bayley, S. (1982) *Art and Industry*, Boilerhouse Project.

Bennett, S. (1986) *A History of Control Engineering 1800–1930*, Peter Peregrinus.

Bennett, S. (1993) *A History of Control Engineering 1930–1955*, Peter Peregrinus.

Bernal, J.D. (1979) *Science in History*, Pelican (4 volumes).

Boas, M. (1970) *The Scientific Renaissance 1450–1630*, Fontana.

Bowers, B. (1982) *A History of Electric Light and Power*, Peter Peregrinus.

Braun, W, von. and Ordway, F.I. (1967) *History of Rocketry and Space Travel*, Nelson.

Bronowski, J. (1973) *Ascent of Man*, BBC Publications.

Brumbaugh, R.S., 1968, *Ancient Greek Gadgets and Machines*, Greenwood Press.

Buderi, R. (1996) *The Invention that Changed the World – The Story of Radar from War to Peace*, Simon and Schuster.

Burstall, A.F. (1963) *A History of Mechanical Engineering*, Faber and Faber.

Bussey, G. (1990) *Wireless: The Crucial Decade 1924–34*, Peter Peregrinus.

Callick, E.B. (1990) *Metres to Microwaves*, Peter Peregrinus.

Carson, R. (1962) *Silent Spring*, Houghton Mifflin.

Chaikin, A. (1994) *A Man on the Moon*, Michael Joseph.

Chapuis, A. (1958) *Automata: A Historical and Technological Study*, Batsford.

Cianchi, M. (1988) *Leonardo da Vinci's Machines*, Becocci Editore.

Clark, R.W. (1985) *Works of Man*, Century Publishing.

Clarke, A.C. (ed.) (1967) *The Coming of the Space Age*, Gollancz.

Clayton, H. (1968) *Atlantic Bridgehead: The Story of Trans-Atlantic Communications*, Gernstone Press.

Daumas, M. (1969) *A History of Technology and Invention*, John Murray.

de Camp, L.S. (1963) *The Ancient Engineers*, Doubleday.

Derry, T.K. and Williams, T.I. (1970) *A Short History of Technology*, Oxford University Press.

Drachmann, A.G. (1963) *The Mechanical Technology of Greek and Roman Antiquity*, Munksgaard.

Dunsheath, P. (ed.) (1951) *A Century of Technology 1851–1951*, Hutchinson.

Dunsheath, P. (1962) *A History of Electrical Engineering*, Faber and Faber.

Forbes, R.G. (1964/72) *Studies of Ancient Technology*, Brill (9 volumes).

Gille, B. (1966) *The Renaissance Engineers*, Lund Humphries.

Gleick, J. (1988) *Chaos: Making a New Science*, Heinemann.

Gregory, M.S., 1971, *History and Development of Engineering*, Longman.

Hadfield, C. (1968) *The Canal Age*, David and Charles.

Hastings, P. (1972) *Railroads: An International History*, Ernest Benn.

Hill, D.R. (1984) *A History of Engineering in Classical and Medieval Times*, Croom Helm.

Hinsley, F.H. and Stripp, A. (eds) (1993) *Codebreakers: The Inside Story of Bletchley Park*, Oxford University Press.

Hodges, A. (1985) *Alan Turing: The Enigma of Intelligence*, Unwin.

Hodges, H. (1971) *Technology in the Ancient World*, Pelican.

James, P. and Thorpe, N. (1995) *Ancient Inventions*, Michael O'Mara Books.

James, P., Thorpe, N., Kokkinos, N., Morkot, R. and Frankish, J. (1991) *Centuries of Darkness*, Jonathan Cape.

Johnson, B. (1978) *The Secret War*, BBC Publications.

Jones, R.V. (1978) *Most Secret War*, Hamish Hamilton.

Kahn, D. (1973) *The Codebreakers*, Weidenfeld and Nicholson.

Kaplan, M.H. (1978) *Space Shuttle*, Aero Publishers.

Landels, J.G. (1978) *Engineering in the Ancient World*, Chatto and Windus.

Lindsay, J. (1975) *Blast-power and Ballistics*, Frederick Muller.

Livy, *History of Rome*.

Lloyd, G.E.R. (1973) *Greek Science After Aristotle*, Chatto and Windus.

Marsden, E.W. (1969) *Greek and Roman Artillery, Historical Development*, Clarendon Press.

Mayr, O. (1989) *Authority, Liberty and Automatic Machinery in Early Modern Europe*, Johns Hopkins University Press.

McConnell, M. (1987) *Challenger: A Major Malfunction*, Guild Publishing.

Morris, P.R. (1990) *A History of the World Semiconductor Industry*, Peter Peregrinus.

Needham, J. (1954) *Science and Civilisation in China*, Cambridge University Press (6 volumes).

Pacey, A. (1990) *Technology and World Civilisation*, Basil Blackwell.

Pannell, J.P.M. (1964) *An Illustrated History of Civil Engineering*, Thames and Hudson.

Pannell, J.P.M. (1977) *Man the Builder: An Illustrated History of Engineering*, Thames and Hudson.

Plommer, W.H. (1973) *Vitruvius and Later Roman Building Manuals*, Cambridge University Press.

Price, A. (1970) *Aircraft versus Submarine*, Jane's.

Price, A. (1977) *Instruments of Darkness*, Macdonald and Jane's.

Rolt, L.T.C. (1963) *Thomas Newcomen*, Macdonald.

Rolt, L.T.C. (1965) *A Short History of Machine Tools*, MIT Press.

Rolt, L.T.C. (1970) *Isambard Kingdom Brunel*, Pelican.

Rolt, L.T.C. (1974) *Victorian Engineering*, Penguin.

Rolt, L.T.C. (1978) *George and Robert Stephenson*, Pelican.

Rolt, L.T.C. (1979) *Thomas Telford*, Pelican.

Ronan, C. (1983) *The Cambridge Illustrated History of the World's Science*, Cambridge University Press.

Sarton, G. (1970) *A History of Science*, The Norton Library (2 volumes).

Scott, J.D. (1958) *The Siemens Brothers 1858–1958: An Essay in the History of Industry*, Weidenfeld and Nicholson.

Singh, S. (1999) *The Code Book*, Fourth Estate.

Smiles, S. (ed.) (1874) *Lives of the Engineers*, John Murray.

Stewart, I. (1989) *Does God Play Dice?*, Basil Blackwell.

Swain, D. (2000) *The Cogwheel Brain: Charles Babbage and the Quest to Build the First Computer*, Little Brown.

Swords, S.S. (1986) *Technical History of the Beginnings of Radar*, Peter Peregrinus.

van Creveld, M. (1991) *Technology and War: From 2000 B.C. to the Present*, Brassey's.

Vitruvius Pollio, *The Ten Books on Architecture*, Translated by M.H. Morgan, Dover Publications.

White, K.D. (1984) *Greek and Roman Technology*, Thames and Hudson.

Wood, J. (1992) *History of International Broadcasting*, Peter Perigrinus.

Human–Computer interface and virtual reality

Aston, R. and Schwarz, J. (eds) (1994) *Multimedia: Gateway to the Next Millennium*, AP Professional.

Card, S. K., Moran, T. P. and Newell, A. (1983) *The Psychology of Human–Computer Interaction*, Erlbaum.

Carr, K. and England, R. (eds) (1995) *Simulated and Virtual Realities: Elements of Perception*, Taylor and Francis.

Dix, A. (1997) *Human–Computer Interaction*, Prentice Hall.

Downton, A. (ed.) (1991) *Engineering the Human–Computer Interface*, McGraw-Hill.

Goodwin, M. (1995) *Making Multimedia Work*, IDG Books.

Holsinger, E. (1994) *How Multimedia Works*, Ziff-Davis Press.

Johnson, P., (1992) *Human–Computer interaction: Psychology, Task Analysis and Software Engineering*, McGraw-Hill.

Kalawsky, R.S. (1993) *The Science of Virtual Reality*, Addison-Wesley.

Lansdale, M.W. and Ormerod, T.C. (1994) *Understanding Interfaces: A Handbook of Human–Computer Dialogue*, Academic Press.

Laurel, B. (ed.) (1990) *The Art of Human–Computer Interface Design*, Addison-Wesley.

MacDonald, L. and Vince, J. (eds) (1994) *Interacting with Virtual Environments*, Wiley.

Preece, J., Rogers, Y., Sharp, H., Benyon, D., Holland, S. and Carey, T. (1994) *Human Computer Interaction*, Addison-Wesley.

Shneiderman, B. (1992) *Designing the User Interface: Strategies for Effective Human–Computer Interaction*, Addison-Wesley.

Stampe, Rochl, B. and Eagan, J. (1993) *Virtual Reality Creations*, The Waite Group.

Tway, L. (1995) *Multimedia in Action*, Academic Press.

Vince, J. A. (1995) *Virtual Reality Systems*, Addison-Wesley.

Warwick, K., Gray, J. and Roberts, D. (eds) (1994) *Virtual Reality in Engineering*, Peter Perigrinus.

Wexeblat, A. (ed.) (1993) *Virtual Reality: Applications and Explorations*, Academic Publishers.

Manufacturing systems

Abegglen, J.C. and Stalk, G. (1987) *Kaisha – The Japanese Corporation*, Charles E. Tuttle.

Andreasen, M.M. and Hein, L. (1985) *Integrated Product Development*, IFS Publications.

Andreasen, M.M., Kahler, S. and Lund, T. (1988) *Design for Assembly*, IFS Publications.

Asfahl, C.R. (1985) *Robots and Manufacturing Automation*, Wiley.

Batchelor, B.G., Hill, D.A. and Hodgson, D.C. (1985) *Automated Visual Inspection*, IFS Publications.

Bedworth, D.D., Henderson, M.R. and Wolfe, P.M. (1991) *Computer Integrated Design and Manufacturing*, McGraw-Hill.

Bendell, A., Disney, J. and Pridmore, W.A. (eds) (1989) *Taguchi Methods: Applications in World Industry*, IFS Publications.

Bignell, D., Dooner, M., Hughes, J., Pym, S. and Stone, S. (1985) *Manufacturing Systems, Context, Applications and Techniques*, Blackwell..

Boothroyd, G. and Dewhurst, P. (1987) *Product Design for Assembly*, Wakefield.

Browne, J., Harhen, J. and Shivnan, J. (1988) *Production Management Systems*, Addison-Wesley.

Chang T-C., Wysk, R.A. and Wang, H-P. (1992) *Computer-aided Manufacturing*, Prentice Hall.

Corbett, J., Dooner, M., Melka, J. and Pym, C. (1991) *Design for Manufacture: Strategies, Principles and Techniques*, Addison-Wesley.

Coward, D.G. (1998) *Manufacturing Management*, Macmillan.

Department of Trade and Industry (1989) *Manufacturing into the Late 1990s*, HMSO.

Derr, K.W. (1995) *Applying OMT*, SIGS Books.

Edwards, P.R. (1991) *Manufacturing Technology in the Electronics Industry*, Chapman and Hall.

Gray, P.A. (1991) *Open Systems: A Business Strategy for the 1990s*, McGraw-Hill.

Groover, M.P. (1987) *Automation, Production Systems and Computer-integrated Manufacturing*, Prentice Hall.

Hannam, R. (1997) *Computer Integrated Manufacturing*, Addison-Wesley.

Hill, T. (1993) *Manufacturing Strategy*, Macmillan.

McMahon, C. and Browne, J. (1993) *CADCAM From Principles to Practice*, Addison-Wesley.

Ranky, P.G. (1990) *Flexible Manufacturing Cells and Systems in CIM*, CIMware.

Ray, C. (1992) *Robots and Manufacturing Automation*, Wiley.

Rembold, U., Nnaji, B.O. and Storr, A. (1993) *Computer Integrated Manufacturing and Engineering*, Addison-Wesley.

Schlesinger, G. (1988) *Testing Machine Tools*, 10th edn, Machinery Publishing Co.

Senker, P. (1985) *Towards the Automated Factory*, IFS Publications.

Straker, D. A. (1995) *Toolbook for Quality Improvement and Problem Solving*, Prentice-Hall.

Suzaki, K. (1987) *The New Manufacturing Challenge: Techniques for Continuous Improvement*, The Free Press.

US Air Force (1981) *Integrated Computer Aided Manufacturing (ICAM) Functional Modelling Manual (IDEF0)*, Air Force Materials Laboratory, Wright-Patterson AFB, AFWAL-TR-81-4023.

Williams, D.J. (1994) *Manufacturing Systems*, Chapman and Hall.

Wright, P.K. and Bourne, D.A. (1988) *Manufacturing Intelligence*, Addison-Wesley.

Wu, B. (1994) *Manufacturing Systems Design and Analysis*, Chapman and Hall.

Mechatronics

Auslander, D. and Kempf, C.J. (1995) *Mechatronics: Mechanical System Interfacing*, Prentice Hall.

Bradley, D. A., Dawson, D., Burd, N.C. and Loader, A.J. (1991) *Mechatronics: Electronics in Products and Processes*, Chapman and Hall.

Buur, J. (1989) *Mechatronics Design in Japan*, Institute for Engineering Design, Technical University of Denmark.

Buur, J. (1990) *A Theoretical Approach to Mechatronics Design*, Institute for Engineering Design, Technical University of Denmark.

Histand, M.B. and Alciatore, D. (1998) *Introduction to Mechatronics and Measurement Systems*, McGraw-Hill.

Mui, D.K. and Temesvary, V. (1992) *Mechatronics: Electromechanics and Contromechanics*, Springer-Verlag.

Popovic, D., Vlacic, L. and Popovic, D. (1998) *Mechatronics in Engineering Design and Product Development*, Marcel Dekker.

Rzevski, G. (1995) *Mechatronics: Designing Intelligent Machines*, Vol. 1, Butterworth–Heinmann.

Rzevski, G. (1995) *Mechatronics: Designing Intelligent Machines*, Vol. 2, Butterworth–Heinmann.

Shetty, D. and Volk, R. (1997) *Mechatronic System Design*, PWS Publishing Co.

Microsystems, microsensors and MEMS

Brendley, K.W. (1993) *Military Applications of Microelectromechanical Systems*, Rand Corp.

Chimsky, M. and Regis, E. (1996) *Nano: The Emerging Science of Nanotechnology*, Little Brown and Co.

Darling, D. (1995) *Micromachines and Nanotechnology: The Amazing World of the Ultrasmall*, Dillon.

Drexler, K.E. (1987) *Engines of Creation*, Anchor Press.

Drexler, K.E. (1992) *Nanosystems: Molecular Macinery, Manufacturing and Computation*, Wiley.

Forester, T. (ed.) (1980) *The Micro Electronics Revolution: The Complete Guide to the New Technology and its Impact on Society*, Basil Blackwell.

Gardner, J.W. (1994) *Microsensors: Principles and Applications*, Wiley.

Kovacs, G.T.A. (1998) *Micromachined Transducers Sourcebook*, McGraw-Hill.

Madou, M. (1997) *Fundamentals of Microfabrication*, CRC Press.

Muller, R.S., Howe, R.T., Senturia, S.D. and Smith, R. (1991) *Microsensors*, Institute of Electrical and Electronic Engineers.

Trimmer, W. (ed.) (1997) *Micromechanics and MEMS: Classic and Seminal Papers*, Institute of Electrical and Electronic Engineers.

Neural networks and fuzzy systems

Aleksander, I. and Morton, H. (1990) *An Introduction to Neural Computing*, Chapman and Hall.

Allman, W.F. (1989) *Apprentices of Wonder: Inside the Neural Network Revolution*, Bantam Books.

Altrock, C.V. (1995) *Fuzzy Logic and NeuroFuzzy Applications Explained*, Prentice Hall.

Beale, R. and Jackson, T. (1990) *Neural Computing: An Introduction*, Institute of Physics Press.

Bishop, C.M. (1995) *Neural Networks for Pattern Recognition*, Clarendon.

Brown, M. and Harris, C.J. (1994) *NeuroFuzzy Adaptive Modelling and Control*, Prentice Hall.

Cox, E. (1994) *The Fuzzy Systems Handbook*, Academic Press.

Dayhoff, J. (1990) *Neural Network Architectures: An Introduction*, Van Nostrand Reinhold (Chapman and Hall).

Freeman, J.A. and Skapura, D.M. (1991) *Neural Networks: Algorithms, Applica-*

tions and Programming Techniques, Addison-Wesley.

Harris, C.J., Moore, C.G. and Brown, M. (1993) *Intelligent Control: Aspects of Fuzzy Logic and Neural nets*, World Scientific.

Haykin, S. (1994) *Neural Networks: A Comprehensive Foundation*, Macmillan.

Klir, G.J. and Yuan, B. (eds) (1996) *Fuzzy Sets, Fuzzy Logic and Fuzzy Systems: Selected Papers by Lotfi A. Zadeh*, World Scientific.

Kohonen, T. (1989) *Self-organisation and Associative Memory*, Springer-Verlag.

Kosko, B. (1992) *Neural Networks and Fuzzy Systems: A Dynamical Systems Approach to Machine Intelligence*, Prentice Hall.

Kosko, B. (1994) *Fuzzy Thinking: The New Science of Fuzzy Logic*, Harper Collins.

Kruse, R., Gebhart, J. and Klawonn, F. (1994) *Foundations of Fuzzy Systems*, Wiley.

Mead, C. (1989) *Analogue VLSI and Neural Systems*, Addison-Wesley.

Minsky, M.L. and Papert, S. (1969) *Perceptrons: An Introduction to Computational Geometry*, MIT Press.

Nie, J. and Linkens, D. (1995) *Fuzzy-neural Control: Principles, Algorithms and Applications*, Prentice Hall.

Picton, P.D. (1994) *Introduction to Neural Networks*, Macmillan.

Rosenblatt, F. (1959) *Principles of Neurodynamics*, Spartan Books.

Terano, T., Asai, K. and Sugeno, M. (1994), *Applied Fuzzy Systems*, Academic Press.

Wasserman, P.D. (1989) *Neural Computing: Theory and Practice*, Van Nostrand Reinhold (Chapman and Hall).

Zadeh, L.A. (1987) *Fuzzy Sets and Applications: Selected Papers*, Wiley.

Software engineering

Blanchard, B.S. and Fabrycky, W.J. (1990) *Systems Engineering and Analysis*, Prentice Hall.

Braek, R. and Haugen, O. (1993) *Engineering Real Time Systems*, Prentice Hall.

Calvez, J. P. (1993) *Embedded Real-Time Systems*, Wiley.

Coad, P. and Yourdon, E. (1990) *Object-oriented Analysis*, Prentice Hall.

Cooling, J.E. (1990) *Software Design for Real-time Systems*, Chapman and Hall.

DeMarco, T. (1979) *Structured Analysis and System Specification*, Prentice Hall.

Hatley, D.J. and Pirbhai, I.A. (1987) *Strategies for Real-Time Systems Specification*, Dorset House.

IEE (1989) *Software in Safety Related Systems*, Institution of Electrical Engineers.

Ince, D.C. (1989) *Software Engineering*, Van Nostrand Reinhold.

Jackson, M.A. (1975) *Principles of Program Design*, Academic Press.

Jackson, M.A. (1983) *System Development*, Prentice Hall.

Jones, C.B. (1986) *Systematic Software Development Using VDM*, Prentice Hall.

Khoshafian, S and Abnous, R. (1990) *Object Orientation: Concepts, Languages, Databases, User Interfaces*, Wiley.

National Computing Centre, *SSADM Version 4 Manual*, Manchester.

Pressman, R.S. (1997) *Software Engineering. A Practitioner's Approach*, Fourth Edition McGraw-Hill.

Robinson, P.J. (1992) *Hierarchical Object-oriented Design*, Prentice Hall.

Rumbaugh, J., Blaha, M., Premerlani, W., Eddy, F., and Lorensen, W. (1991) *Object-Oriented Modelling and Design*, Prentice Hall.

Schulmeyer, G.G. (1990) *Zero-defect Software*, McGraw-Hill.

Simon, J.C. (1986) *Patterns and Operators: The Foundations of Data Representation*, Kogan Page.

Society for Computer Simulation (1985) Catalogue of Simulation Software, *Simulation*, Vol. 45, No. 4, pp. 196–209.

Sommerville, I. And Sawyer, P. (1997) *Requirements Engineering . A Good Practice Guide*, Wiley.

Sommerville, I. (1995) *Software Engineering*, 4th edn, Addison-Wesley.

Neumann, P.G. (1995) *Computer Related Risks*, Addison-Wesley.

Spriet, J.A. and Vansteenkiste, G.C. (1982) *Computer-aided Modelling and Simulation*, Academic Press.

Ward, P.T. and Mellor, S.J. (1985) *Structured Development for Real-time Systems*, Yourdon Press.

Yourdon, E. and Constantine, L. (1979) *Structured Design*, Prentice Hall.

Yourdon, E. (1989) *Modern Structured Analysis*, Prentice Hall.

System safety

Advanced Robotics Research Limited (1992) *Safety and Standards for Advanced Robots – A First Exposition*, Report ARRL.92.009.

Bishop, P.G. (ed.) (1990) *Dependability of Critical Computer Systems 3: Techniques Directory*, Elsevier Applied Science.

IEE (1989) *Software in Safety Related Systems*, Institution of Electrical Engineers.

Redmill, F.J. (ed.) (1988) *Dependability of Critical Computer Systems 1*, Elsevier Applied Science.

Redmill, F.J. (ed.) (1989) *Dependability of Critical Computer Systems 2*, Elsevier Applied Science.

Schlager, N. (ed.) (1994) *When Technology Fails: Significant Technological Disasters, Accidents and Failures of the Twentieth Century*, Gate Research.

Storey, N. (1996) *Safety-critical Computer Systems*, Addison-Wesley.

Other

Caterpillar Inc. (1988) *Caterpillar Performance Handbook*, Caterpillar.

Construction Industry Training Board (1990) *Digger Loader Training Notes*, CITB.

Nichols, H.L., Day, D.A. & Day, D.H., 1998, *Moving the Earth*, McGraw-Hill.

Pace, C. (1997) *A Safety Manager for a Robot Excavator*, MSc Dissertation, Lancaster University, Department of Engineering, June 1997.

Zelenin, A.N., Balovnev, V.I. and Kerov, L. (1986) *Machines for Moving the Earth*, Balkema.

STANDARDS

IEC draft standard IEC 1508, Parts 1–7 (1995) *Functional Safety of Programmable Electronic Systems*, British Standards Institute.

ISO 90041 (1994) *Quality Management and Quality Systems Elements*, ISO.

Ministry of Defence (UK) (1997) *Requirements for Safety Related Software in Defence Systems*, Defence Standard 00-55 Parts 1 and 2.

Ministry of Defence (UK) (1994) *Hazop Studies on Systems Containing Programmable Electronics*, Defence Standard 00-58 Parts 1 and 2.

Ministry of Defence (UK) (1996) *Safety Management Requirements for Defence Systems*, Defence Standard 00-56 Parts 1 and 2.

The Chemical Industries Association (1977) *HAZARD and Operability Studies*, CIA.

Index